Universitext

Series Editors

Sheldon Axler
San Francisco State University, San Francisco, CA, USA

Carles Casacuberta
Universitat de Barcelona, Barcelona, Spain

Angus MacIntyre
Queen Mary University of London, London, UK

Kenneth Ribet
University of California, Berkeley, CA, USA

Claude Sabbah
CNRS, École polytechnique Institut Polytechnique de Paris, Palaiseau, France

Endre Süli
University of Oxford, Oxford, UK

Wojbor A. Woyczyński
Case Western Reserve University, Cleveland, OH, USA

Universitext is a series of textbooks that presents material from a wide variety of mathematical disciplines at master's level and beyond. The books, often well class-tested by their author, may have an informal, personal even experimental approach to their subject matter. Some of the most successful and established books in the series have evolved through several editions, always following the evolution of teaching curricula, to very polished texts.

Thus as research topics trickle down into graduate-level teaching, first textbooks written for new, cutting-edge courses may make their way into Universitext.

For further volumes in this series go to www.springer.com/series/223

Universitext

Universitext is a series of textbooks that presents material from a wide variety of mathematical disciplines at master's level and beyond. The books, often well class-tested by their author, may have an informal, personal even experimental approach to their subject matter. Some of the most successful and established books in the series have evolved through several editions, always following the evolution of teaching curricula, into very polished texts.

Thus as research topics trickle down into graduate-level teaching, first textbooks written for new, cutting-edge courses may make their way into *Universitext.*

More information about this series at http://www.springer.com/series/223

Eduardo Casas-Alvero

Algebraic Curves, the Brill and Noether Way

 Springer

Eduardo Casas-Alvero
Matemàtiques i Informàtica
Universitat de Barcelona
Barcelona, Spain

ISSN 0172-5939   ISSN 2191-6675   (electronic)
Universitext
ISBN 978-3-030-29015-3   ISBN 978-3-030-29016-0   (eBook)
https://doi.org/10.1007/978-3-030-29016-0

Mathematics Subject Classification (2010): 14-01, 51-01, 14H50, 14H05, 14C17, 14C40, 14H20

This Springer imprint is published by the registered company Springer Nature Switzerland AG
The registered company address is: Gewerbestrasse 11, 6330 Cham, Switzerland

Again, to Antón and Adrià

Preface

This book is based on the notes of one-semester introductory courses in algebraic geometry I gave to students of the University of Barcelona in their first postgraduate year, during the academic years 2014/15 to 2017/18. Two main facts conditioned the choice of the contents and its presentation: first, the students had uneven – often rather limited – algebro-geometric backgrounds; second, it was assumed that most of the students would not follow further courses on algebraic geometry. This meant that, on one hand, just a few basic facts could be assumed as prerequisites, while, on the other, the course had to contain as many results interesting by themselves as possible. My choice was to take the basic facts of the intrinsic (i.e. birationally invariant) geometry on an irreducible complex curve as the main goal of the course, and reach it through the approach given by A. Brill and M. Noether in their fundamental memoir [2], followed, with small variations, in many classical treatises such as [10], [17], [19] and [21]. Brill and Noether's approach develops the theory on plane curves and makes intensive use of their local and projective properties. Therefore, following Brill and Noether requires a previous presentation of the essentials of the local and projective theories of plane algebraic curves, which is fortunate, because these can be developed from very basic prerequisites and stand by themselves as very relevant parts of the study of algebraic curves, with applications to computer algebra and computer graphics. An added advantage of Brill and Noether's approach is that it provides a very clear reading of the intrinsic geometry on plane curves.

Accordingly, besides a preliminary chapter (Chapter 1) presenting very elementary facts about hypersurfaces, the book is divided into three main chapters, devoted in this order to the local, projective and intrinsic geometry of algebraic curves. A description of the contents is as follows:

Chapter 2 starts with Newton polygons and the theorem of Puiseux about local parameterization of plane curves using convergent fractionary power series. It leads to the notion of branch (of a plane curve at a point), which is fundamental in Chapter 4 because rational functions on a curve are evaluated along its branches, not at its points. The characteristic exponents of a branch are introduced, but the analysis of the singularities of a plane curve is pursued no further. The local parameterizations allow us to define the intersection multiplicity – a numerical local invariant of two curves at a point – and prove its main properties. Intersection multiplicity will be of constant use from this point onwards.

Chapter 3 contains two main theorems: Bézout's theorem, which counts the number of intersection points of two curves, and Noether's Fundamental Theorem, which relates the equations of curves sharing certain points. Bézout's theorem needs intersection multiplicity for a correct statement; it is proved using Sylvester's resultant, whose definition and main properties are given. Polar curves are introduced: they will be related to adjoints and differentials in Chapter 4 and are used here, together with Bézout's theorem, to count the tangents to a curve through a point (Plücker's first formula). This is a representative example of the computations of numbers which, in the second half of the nineteenth century, gave rise to a branch of algebraic geometry named *enumerative geometry*. Noether's Fundamental Theorem gives local sufficient conditions for the equation of a curve to belong to the ideal generated by the equations of two other curves. It is the main tool used to prove Brill and Noether's Restsatz in Chapter 4 and has many other applications, some of which are included as exercises.

Chapter 4 presents the essentials of the intrinsic geometry on an algebraic curve, taking care of the birational invariance of all notions and results. Rational and birational maps are introduced first, and particular examples of the latter (*ordinary quadratic transformations*) are used to reduce the singularities of the plane curves, allowing us to deal in the sequel with curves having quite simple singularities (*ordinary singularities*). The main objects of the intrinsic geometry – Riemann surface, divisors, linear series – are then introduced, and used to reformulate, in Brill and Noether's terms, Riemann's main problem about the dimension of the spaces of rational functions with allowed poles. Adjoint curves are curves projectively related to a given curve; they lead to a construction of complete linear series in what is called the *Brill–Noether Restsatz* (the Brill–Noether remainder theorem). The Restsatz follows from Noether's fundamental theorem; it leads in turn to the definition of the genus of a curve (its main invariant) and to a partial yet very powerful solution of Riemann's problem: the Riemann inequality. Riemann's inequality is turned into an equality – the Riemann–Roch theorem – after introducing Jacobian groups, differentials and the canonical series, and proving one further result bearing Noether's name: the Noether Reduction Lemma. The last three sections of the book are devoted to the study of rational maps between curves, including the Riemann–Hurwitz formula, Lüroth's theorem and results on the rational images of curves associated to linear series on them.

There are exercises at the end of each chapter. Most are just so, exercises; other are extensions or applications of the already presented matter which have their own interest: the latter include continuity of algebraic functions (2.6), Plücker formulas (3.18, 4.35), tied points (4.15, 4.17), addition on elliptic curves (4.26), Weierstrass gaps (4.28), the classifications of elliptic curves and plane cubics (4.31, 4.32), and duality for plane curves (4.33, 4.34). Results from the exercises are not used in the text, but only in other exercises.

A few dates and very short historical comments are included at some points. For more on the rich and enlightening history of the matter presented here, the reader is referred to [9] and [13].

As said, only a very limited background is required to read this book: the contents of a basic course on algebra, including rings, fields, ideals and polynomials, some on complex analytic functions, mainly in a single variable, and their representation by power series, and just the most basic facts about projective spaces and homogeneous coordinates; the latter are, however, quickly recalled at the beginning of Chapter 1. The few results from commutative algebra that are needed will be presented and proved in the text.

I am very grateful to Profs. J.C. Naranjo, J. Roé and G. Welters for fruitful discussions on the subject and for their careful readings of parts of the manuscript, which resulted in valuable suggestions. In addition, G. Welters prepared a nice collection of exercises for the courses and kindly allowed me to include them here. Thanks are also given to E. Griniari and the staff of Springer-Verlag for their efficient and cooperative editorial work.

General Conventions

As usual, $\mathbb{R}$ and $\mathbb{C}$ will denote the fields of real and complex numbers, respectively; $\mathbb{Z}$ will denote the ring of integers and $\mathbb{N}$ the set of non-negative integers. The base field will be the field of complex numbers throughout. Complex numbers will be also referred to as *scalars* or *constants*. The imaginary unit will be denoted by $\boldsymbol{i}$, as the ordinary i will be often used for other purposes. The symbol ∞ will be taken to be different from any complex number and strictly higher than any real number.

The identity map on a set X is written Id_X, or just Id if X is clear.

Angle brackets $\langle \ \rangle$ will mean 'subspace generated by'. Usually, the entry in row i and column j of a matrix will be written in the form a_j^i, and also $a_{i,j}$ if the matrix is symmetric. The unit n-dimensional matrix will be denoted by $\mathbf{1}_n$, or just by $\mathbf{1}$ if no reference to the dimension is needed.

$k[X_1, \ldots, X_n]$ will denote the ring of polynomials in the variables $X_1, \ldots, X_n$ with coefficients in the ring k and, when k is a field, $k(X_1, \ldots, X_n)$ will denote the corresponding field of rational fractions, namely the field of quotients of $k[X_1, \ldots, X_n]$. We will refer to the highest degree monomial (resp. coefficient) of a non-zero polynomial in one variable as its *leading monomial* (resp. *leading coefficient*). The polynomials which are non-zero and have leading coefficient equal to one are called *monic*. Greatest common divisors, minimal common multiples and irreducible factors of polynomials in a single variable are always assumed to be monic. The polynomial 0 will be taken as a homogeneous polynomial of degree m for any non-negative integer m.

Unless otherwise stated, the roots of polynomials will be counted according to their multiplicities, that is, a root a of a polynomial $P(X)$ will be counted as many times as the number of factors $X - a$ appearing in the decomposition of $P(X)$ in irreducible factors.

The ring of formal power series in variables $x_1, \ldots, x_n$ with coefficients in a ring A will be denoted $A[[x_1, \ldots, x_n]]$, while $\mathbb{C}\{x_1, \ldots, x_n\}$ will denote the ring of convergent power series in the variables $x_1, \ldots, x_n$ with complex coefficients. The *initial form* of a non-zero series s is the sum of its non-zero monomials of minimal degree, denoted in the sequel by $\mathrm{In}(s)$. If s has a single variable x we will write $o_x s$ for the degree of its initial form, and take $o_x(0) = \infty$. In both cases $o_x s$ will be called the *order of s* (relative to x).

All rings will be assumed to be commutative and to have unit. Ideals other than the ring itself will be called *proper*. The principal ideal generated by an element f in a ring A will be denoted (f) if no confusion may arise; otherwise we will use the notation fA. If A is any entire ring and K its field of quotients, we will, as usual, not distinguish between an element $a \in A$ and the quotient $a/1 \in K$, the ring A being thus identified with a subset of K. The rings with a unique non-zero maximal ideal are called *local rings*.

We will very often deal with rings that contain the field $\mathbb{C}$ of complex numbers (or an isomorphic copy of it) as a subring: they are called $\mathbb{C}$-*algebras* because they inherit a complex vector space structure by restricting the first factor of the product to be a complex number. The ring homomorphisms between $\mathbb{C}$-algebras restricting to the identity of $\mathbb{C}$ will be called $\mathbb{C}$-*algebra homomorphisms*: they are homomorphisms for both the ring and the vector space structures.

Contents

Chapter 1

Hypersurfaces, Elementary Facts

1.1 Projective Hypersurfaces

The reader is referred to any book on projective geometry (for instance [4]) for the most basic facts about projective spaces and homogeneous coordinates. $\mathbb{P}_n$ will denote a complex n-dimensional projective space (line if $n = 1$, plane if $n = 2$); once a reference has been fixed in it, each point $p \in \mathbb{P}_n$ is determined by $n+1$ coordinates $x_0, \ldots, x_n$, $x_i \in \mathbb{C}$, one at least non-zero; they are called *homogeneous* or *projective coordinates* of p, and are in turn determined by p up to a non-zero common factor. We will use the notation $[x_0, \ldots, x_n]$ to denote the point with homogeneous coordinates $x_0, \ldots, x_n$. If a second reference is taken, the old coordinates are linearly related to the new ones $y_0 \ldots, y_n$ in the form

$$\begin{pmatrix} x_0 \\ \vdots \\ x_n \end{pmatrix} = M \begin{pmatrix} y_0 \\ \vdots \\ y_n \end{pmatrix} \tag{1.1}$$

where M is an $(n+1) \times (n+1)$ regular matrix depending on the relative position of the references and determined up to a non-zero constant factor.

A *projectivity* $f : \mathbb{P}_n \to \mathbb{P}'_n$, between n-dimensional projective spaces $\mathbb{P}_n$ and $\mathbb{P}'_n$, each with coordinates fixed, maps any point $[x_0, \ldots, x_n] \in \mathbb{P}_n$ to the point of $\mathbb{P}'_n$ with coordinates

$$\begin{pmatrix} \bar{x}_0 \\ \vdots \\ \bar{x}_n \end{pmatrix} = A \begin{pmatrix} x_0 \\ \vdots \\ x_n \end{pmatrix} \tag{1.2}$$

where A is an $(n+1) \times (n+1)$ regular matrix called *a matrix* of f (it is determined up to a non-zero constant factor) relative to the fixed coordinates. The formal identity between the equations (1.1) and (1.2) ensures

© Springer Nature Switzerland AG 2019
E. Casas-Alvero, *Algebraic Curves, the Brill and Noether Way*, Universitext,
https://doi.org/10.1007/978-3-030-29016-0_1

that notions independent of the choice of the coordinates are preserved by projectivities and conversely; it allows us to interpret any homogeneous linear and invertible substitution of $n+1$ variables as either a change of coordinates in an n-dimensional projective space, or a projectivity between n-dimensional projective spaces.

Assume we have fixed coordinates $x_0, \ldots, x_n$ on a projective space $\mathbb{P}_n$. As usual, an s-dimensional *linear variety* – or *projective subspace* – of $\mathbb{P}_n$ is the set of points of $\mathbb{P}_n$ whose coordinates satisfy a certain system of $n-s$ independent, linear and homogeneous equations in $n+1$ variables: when $s = 1, 2, n-1$, they are called *lines*, *planes* and *hyperplanes*, respectively.

An *algebraic projective hypersurface* V of $\mathbb{P}_n$ (we will say *hypersurface of* $\mathbb{P}_n$ for short) is the class modulo $\mathbb{C}$-proportionality of a non-constant homogeneous polynomial $F \in \mathbb{C}[x_0, \ldots, x_n]$. The polynomial F is said to be an *equation* of V: by the definition, it determines V and is determined by V up to a non-zero scalar factor. We will use the notation $V : F = 0$ to indicate that V is the hypersurface of equation F and will sometimes refer to V as *the hypersurface* $F = 0$. Often the equality $F = 0$ is also called an equation of V, which causes no confusion. The hypersurfaces of $\mathbb{P}_2$ are called *algebraic projective plane curves*. They will be often called curves of $\mathbb{P}_2$, or just *curves*, in the sequel. For other types of curves, such as affine curves, or analytic curves, or space curves, we will use their complete names unless no confusion may occur.

Assume that other coordinates $y_0, \ldots, y_n$ are taken in $\mathbb{P}_n$, related to $x_0, \ldots, x_n$ by (1.1) as above. Let G the polynomial obtained from F by performing the substitution of variables (1.1). We agree in taking the same hypersurface V as being the one defined by the equation G when coordinates $y_0, \ldots, y_n$ are used, this being consistent with performing successive changes of coordinates. In the sequel, hypersurfaces will always be handled using already fixed coordinates. The notions relative to them introduced below do not depend on the choice of the coordinates; checking this fact is in most cases straightforward from the fact that a substitution of coordinates such as (1.1) above induces a degree-preserving $\mathbb{C}$-algebra isomorphism $\mathbb{C}[x_0, \ldots, x_n] \simeq \mathbb{C}[y_0, \ldots, y_n]$, and is left to the reader.

The degree d of the equation F is called the *degree* of $V : F = 0$, denoted $\deg V$ in the sequel. The hypersurfaces of $\mathbb{P}_n$ of degree one are the hyperplanes (lines if $n = 2$, planes if $n = 3$) of $\mathbb{P}_n$. The hypersurfaces of degrees $2, 3, 4, 5 \ldots$ are called, in this order, *quadrics* (*conics* if $n = 2$), *cubics*, *quartics*, *quintics*, and so on.

A point $p = [x_0, \ldots, x_n]$ is said *to belong to* (or *to lie on*, or *to be a point of*) a hypersurface $V : F = 0$ if and only if $F(x_0, \ldots, x_n) = 0$, this condition is obviously independent of the choice of F, and is also independent of the arbitrary factor involved in the coordinates of the point just due to the fact that F is homogeneous. The set of points of a hypersurface V will be denoted $|V|$. We will usually write $p \in V$ for $p \in |V|$. Saying that V *contains* (or *goes through*) p is equivalent to $p \in V$. If p does not belong to V, written $p \notin V$, it is often said that V *misses* p. We will usually write $\mathbb{P}_n - |V| = \mathbb{P}_n - V$

and call it the *complement of V*.

Since the basis of a principal ideal (F) of $\mathbb{C}[x_0, \ldots, x_n]$ is determined up to multiplication by a non-zero complex number, the hypersurface $V : F = 0$ and the ideal (F) determine each other: (F) is then called *the ideal of* (or *associated to*) V. The ideals associated to hypersurfaces are thus the principal ideals of $\mathbb{C}[x_0, \ldots, x_n]$ generated by a non-zero homogeneous polynomial.

Obviously a hypersurface determines its set of points, while the converse is not true, as shown by the easy example of $x_0 = 0$ and $x_0^2 = 0$, two hypersurfaces with the same set of points and clearly non-proportional equations. The fact is that a hypersurface, as defined above – an equation taken up to proportionality – is an object richer than just its set of points, and therefore a naive definition taking just the set of points as being the hypersurface causes a loss of essential information.

Let $V_1 : F_1 = 0$ and $V_2 : F_2 = 0$ be hypersurfaces of $\mathbb{P}_n$. It is said that V_1 is *contained* – or *included* – in V_2, denoted $V_1 \subset V_2$, if and only if F_1 divides F_2, this condition being clearly independent of the choices of the equations. In such a case, it is equivalently said that V_2 *contains* V_1. Obviously, the inclusion is an ordering on the set of hypersurfaces of $\mathbb{P}_n$ and $V_1 \subset V_2$ implies $|V_1| \subset |V_2|$.

Assume that $V_1 : F_1 = 0$ and $V_2 : F_2 = 0$ are hypersurfaces of $\mathbb{P}_n$. The hypersurface $V : F_1 F_2 = 0$ does not depend on the choices of the equations F_1, F_2; it is called the hypersurface *composed of* V_1 and V_2, denoted $V = V_1 + V_2$. As is clear, $\deg(V_1 + V_2) = \deg V_1 + \deg V_2$, $V_1 \subset V_1 + V_2$, $V_2 \subset V_1 + V_2$ and $|V_1 + V_2| = |V_1| \cup |V_2|$. Accordingly, if, for $i = 1, \ldots, m$, $V_i : F_i = 0$ are hypersurfaces of $\mathbb{P}_n$ and μ_i positive integers, then the hypersurface $V : F_1^{\mu_1} \ldots F_m^{\mu_m} = 0$ will be written $V = \mu_1 V_1 + \cdots + \mu_m V_m$ and called the hypersurface *composed of* $V_i : F_i = 0$, $i = 1, \ldots, m$, each V_i counted μ_i times, or *taken with multiplicity* μ_i.

The polynomial ring $\mathbb{C}[x_0, \ldots, x_n]$ being factorial, any polynomial is a product of irreducible factors which are uniquely determined by the polynomial up to a non-zero constant factor (see [15], V.6, for instance). We will make use of the following fact:

Lemma 1.1.1 *If F is a homogeneous polynomial, then all its irreducible factors are homogeneous too.*

PROOF: By induction on $d = \deg F$. The case $d = 1$ is obvious, as then F is its only irreducible factor. Assume $d > 1$; if F is irreducible, then still F is its only irreducible factor and the claim is satisfied. Otherwise $F = GP$ with G and P polynomials and $r = \deg G < d$, $s = \deg P < d$. Write both G and P as sums of homogeneous polynomials, say

$$G = G_r + \cdots + G_0, \quad P = P_s + \cdots + P_0,$$

the subindices indicating the degree. From $F = GP$ follows the equality of the homogeneous parts of degree $d = r + s$ of both sides: $F = G_r P_s$. Then the claim follows by using the induction hypothesis on both G_r and P_s. ◇

A hypersurface $V : F = 0$ is called *irreducible* if and only if its equation F is irreducible as a polynomial, or, equivalently, if and only if its associated ideal is a prime ideal, these conditions being obviously independent of the choice of the equation.

Assume that $V : F = 0$ is an arbitrary hypersurface. If $F = F_1^{\mu_1} \ldots F_m^{\mu_m}$ is the decomposition of F into irreducible factors. By 1.1.1 above, all F_i, $i = 1, \ldots, m$, are homogeneous. For $i = 1, \ldots, m$, the irreducible hypersurfaces $V_i : F_i = 0$ are called the *irreducible components* of V, and each μ_i the *multiplicity* of V_i in (or *as an irreducible component of*) V; the irreducible components V_i with $\mu_i > 1$ are called *multiple irreducible components* of V. By the uniqueness of the decomposition of F into irreducible factors, both the irreducible components of V and their multiplicities are determined by V. The equality $V = \mu_1 V_1 + \cdots + \mu_m V_m$ is referred to as the decomposition of V in its irreducible components. Clearly, $|V| = |V_1| \cup \cdots \cup |V_m|$. Elementary properties of divisibility of polynomials easily give:

Proposition 1.1.2 *If V_1 and V_2 are hypersurfaces of $\mathbb{P}_n$ and $V_1 \subset V_2$, then*

(a) *All irreducible components of V_1 are irreducible components of V_2.*

(b) *If V_2 is irreducible, then $V_1 = V_2$.*

A hypersurface is said to be *reduced* if and only if all the multiplicities of its irreducible components are 1. If, as above, $V = \mu_1 V_1 + \cdots + \mu_m V_m$ is the decomposition of V into irreducible components, it is usual to write $V_{red} = V_1 + \cdots + V_m$. It is clear that $|V| = |V_{red}|$ and therefore, for any hypersurface W of $\mathbb{P}_n$, $V_{red} = W_{red}$ implies $|V| = |W|$. The converse is true and allows us to identify the reduced hypersurfaces with their sets of points. However it is not so easy to prove: it is a particular case of a deep theorem due to Hilbert (Hilbert's Nullstellensatz, see for instance [23], Vol. II, VII.3). We will prove a particular case of the Nullstellensatz, for plane curves, in forthcoming Corollary 3.2.7.

Let us have a look on the case $n = 1$. It is a well-known fact that any homogeneous polynomial $F \in \mathbb{C}[x_0, x_1]$ is a product of linear factors, these factors being thus its irreducible factors (for a proof, just write $F = x_0^d F(1, x_1/x_0)$, $d = \deg F$, decompose the second factor, as a polynomial in the single variable x_1/x_0, into linear factors and then cancel the denominators).

As a consequence the irreducible homogeneous polynomials in x_0, x_1 are those and only those which have degree one, and therefore the form $ax_0 + bx_1$, $(a, b) \neq (0, 0)$. The irreducible hypersurfaces of $\mathbb{P}_1$ are thus those and only those of the form $V : ax_0 + bx_1 = 0$. Since the hypersurface $V : ax_0 + bx_1 = 0$ has $p = [b, -a]$ as its only point and in turn that point p determines V, in the sequel no distinction will be made between the irreducible hypersurfaces and the points of $\mathbb{P}_1$, the irreducible hypersurfaces of $\mathbb{P}_1$ themselves being called just points.

The irreducible components of an arbitrary hypersurface $V : F = 0$ of $\mathbb{P}_1$ are thus points. More precisely, if $F = (a_1 x_0 + b_1 x_1)^{\mu_1} \ldots (a_m x_0 + b_m x_1)^{\mu_m}$, then taking $p_i = [b_i, -a_i]$, it is $V = \mu_1 p_1 + \ldots \mu_m p_m$: any hypersurface of $\mathbb{P}_1$ is composed of finitely many points counted with multiplicities, the points and multiplicities being determined by, and determining in turn, the hypersurface. Because of this the hypersurfaces of $\mathbb{P}_1$ are usually called *groups of points* or *effective divisors* of $\mathbb{P}_1$, the name hypersurface being seldom used. (A divisor is a formal linear combination of points with integral coefficients, we will deal with divisors on curves in Chapter 4.) Note that $\deg V = \mu_1 + \cdots + \mu_m$, so the degree of a group of points equals the number of points, provided each point is counted as many times as indicated by its multiplicity, which is called *counting the points with* (or *according to*) *multiplicities*.

Back to the n-dimensional case, let as before $V : F = 0$ be a hypersurface of $\mathbb{P}_n$ and assume to have fixed a line ℓ of $\mathbb{P}_n$. The line ℓ is in particular a one-dimensional projective space; if it is given by parametric equations

$$x_i = c_i^0 t_0 + c_i^1 t_1, \quad i = 0, \ldots, n, \tag{1.3}$$

then the parameters t_0, t_1 may be taken as projective coordinates of the point $[x_0, \ldots, x_n]$ in ℓ. By substituting the above equations (1.3) into the equation F of V one gets a polynomial in t_0, t_1

$$\bar{F} = F(c_0^0 t_0 + c_0^1 t_1, \ldots, c_n^0 t_0 + c_n^1 t_1)$$

which either equals zero or is homogeneous of degree $d = \deg V$. In the first case all the points of ℓ belong to V and it is said that ℓ is *contained* in V, written $\ell \subset V$. Otherwise, $\bar{F}$ defines a group of points of ℓ which is called the *intersection* of ℓ and V, and also the *section* of ℓ by V, denoted by $V \cap \ell$. Note that the number of points of the section, counted with multiplicities, is $d = \deg V$.

Next we check that $\ell \cap V$ does not depend on the coordinates. Assume we have new projective coordinates $y_0, \ldots, y_n$ and write x and y the column matrices with entries the coordinates $x_0, \ldots, x_n$ and $y_0, \ldots, y_n$ of a point. As said before, they are related by an equality $x = My$, M a regular matrix, in such a way that making in an equation F of V the substitution $x = My$ gives an equation G of V relative to the new coordinates. Put t for the column matrix with entries t_0, t_1 and write $x = Ct$ for the parameterization (1.3). Then $y = M^{-1}Ct$ is a parameterization of ℓ using the new coordinates, and it is clear that the results of substituting $x = Ct$ into F and substituting $y = M^{-1}Ct$ into G are the same.

Still take V and ℓ as above and assume $\ell \not\subset V$. For each $p \in V \cap \ell$, the *intersection multiplicity* of V and ℓ at p is defined as being the multiplicity of p in $V \cap \ell$, denoted $[V \cdot \ell]_p$. If $\ell \subset V$, we take $V \cap \ell = \ell$ and $[V \cdot \ell]_p = \infty$ for any $p \in \ell$. The intersection multiplicity of V and ℓ at any $p \notin V \cap \ell$ is taken to be zero. It follows:

Proposition 1.1.3 *Given a hypersuperface V of a projective space $\mathbb{P}_n$, any line ℓ of $\mathbb{P}_n$ either is contained in V or intersects V in $\deg V$ points, the points counted according to the intersection multiplicities of ℓ and V at them.*

Remark 1.1.4 As directly follows from the definitions,

$$(V_1 + V_2) \cap \ell = V_1 \cap \ell + V_2 \cap \ell$$

and

$$[(V_1 + V_2) \cdot \ell]_p = [V_1 \cdot \ell]_p + [V_2 \cdot \ell]_p$$

for any two hypersurfaces V_1, V_2, any line ℓ not contained in either of them, and any point p, all of $\mathbb{P}_n$.

A direct consequence of 1.1.3 is:

Corollary 1.1.5 $|V| \neq \emptyset$ *for any hypersurface V of $\mathbb{P}_n$.*

The field of complex numbers $\mathbb{C}$ being infinite, if a polynomial $P \in \mathbb{C}[x_0, \ldots, x_n]$ satisfies $P(a_0, \ldots, a_n) = 0$ for all $(a_0, \ldots, a_n) \in \mathbb{C}^{n+1}$, then $P = 0$ (see for instance [15],V.4). This in particular implies that $\mathbb{P}_n - V \neq \emptyset$ for any hypersurface V of $\mathbb{P}_n$, and also the following lemma that will be useful later on:

Lemma 1.1.6 *If $V : F = 0$ is a hypersurface of $\mathbb{P}_n$ and G a homogeneous polynomial that vanishes on all the points of $\mathbb{P}_n - V$, then $G = 0$.*

PROOF: Otherwise the set of points of the hypersurface $FG = 0$ would be $\mathbb{P}_n$, against the above.   ◇

1.2 Affine Hypersurfaces

Let $\mathbb{A}_n$ be a (complex) n-dimensional affine space with fixed (affine) coordinates $X_1, \ldots, X_n$. An (algebraic) *hypersurface of $\mathbb{A}_n$* – or *affine hypersurface* – is defined as the class modulo $\mathbb{C}$-proportionality of a non-constant polynomial $f \in \mathbb{C}[X_1, \ldots, X_n]$. As in the projective case, the polynomial f is said to be an *equation* of the hypersurface. If $n = 2$ the affine hypersurfaces are called *affine curves* or curves of $\mathbb{A}_2$. We leave to the reader the straightforward translation of all the definitions and considerations of Section 1.1 to the affine case, by dropping throughout the homogeneity condition on the equations. In particular the ideal (f) of $\mathbb{C}[X_1, \ldots, X_n]$ and the affine hypersurface $V : f = 0$ determine each other and (f) is still called *the ideal of* or *associated to V*: the ideals associated to affine hypersurfaces are thus the non-zero principal ideals of $\mathbb{C}[X_1, \ldots, X_n]$ and those associated to irreducible affine hypersurfaces are the non-zero prime and principal ideals.

The only essential difference with the projective case comes when considering the intersection of a line ℓ and a hypersurface V of $\mathbb{A}_n$. Then the polynomial in a single variable $\bar{f}$, obtained by substituting affine parametric equations of ℓ into an equation f of V, has degree at most $\deg V = \deg f$. The case $\deg \bar{f} = 0$ may occur: then ℓ and V share no points and their intersection is retained as empty. If, otherwise, $0 < \deg \bar{f} \leq \deg f$, one may proceed as in the projective case, the intersection is an affine hypersurface

$V \cap \ell$ of ℓ and the statement similar to 1.1.3 gives just an upper bound for the number of its points, still counted with multiplicities.

Our interest in affine hypersurfaces mainly comes from the fact that they appear as parts of projective hypersurfaces and, for certain purposes, are easier to handle. Next we describe how to cover a projective hypersurface using affine hypersurfaces.

Assume we have fixed projective coordinates $x_0, \ldots, x_n$ in a projective space $\mathbb{P}_n$. The complementaries of the hyperplanes $x_i = 0$, namely

$$\mathbb{A}_n^i = \{p = [x_0, \ldots, x_n] \mid x_i \neq 0\}, \quad i = 0, \ldots, n,$$

are sets covering the whole of $\mathbb{P}_n$. They are called the *affine charts* of $\mathbb{P}_n$ relative to the coordinates $x_0, \ldots, x_n$, the set $\mathbb{A}_n^i$ being referred to as the i-th affine chart, or the affine chart $x_i \neq 0$. Note that a different choice of coordinates will result in a different set of affine charts.

An *affine chart* of a projective space $\mathbb{P}_n$ is any of the affine charts of $\mathbb{P}_n$ relative to some given projective coordinates. Up to reordering the coordinates, the affine chart may of course be assumed to be the 0-th affine chart.

For the remainder of this section we assume that coordinates $x_0, \ldots, x_n$ on $\mathbb{P}_n$ have been chosen and we will fix our attention on the 0-th affine chart $\mathbb{A}_n^0$, denoted by $\mathbb{A}_n$ in the sequel, and its complement $\Pi : x_0 = 0$. All our considerations may be readily extended to the other charts. $\mathbb{A}_n$ appears as an n-dimensional affine space by taking

$$X_i = \frac{x_i}{x_0}, \quad i = 1, \ldots, n \tag{1.4}$$

as the affine coordinates of $p = [x_0, \ldots, x_n] \in \mathbb{A}_n$. Note that then $p = [1, X_1, \ldots, X_n]$.

Remark 1.2.1 Any other affine coordinates $Y_1, \ldots, Y_n$ on the affine chart $\mathbb{A}_n$ are similarly related to projective coordinates on $\mathbb{P}_n$. Indeed, if

$$Y_i = b_i + \sum_{j=1}^{n} a_j^i X_j, \quad i = 1, \ldots, n, \quad \det(a_j^i) \neq 0,$$

it is enough to take as new projective coordinates on $\mathbb{P}_n$

$$y_0 = x_0 \quad \text{and} \quad y_i = b_i x_0 + \sum_{j=1}^{n} a_j^i x_j, \quad i = 1, \ldots, n,$$

to have

$$Y_i = \frac{y_i}{y_0}, \quad i = 1, \ldots, n,$$

and still

$$\mathbb{A}_n = \{p = [y_0, \ldots, y_n] \mid y_0 \neq 0\}.$$

1.2.2 Absolute coordinate Once homogeneous coordinates x_0, x_1 are fixed on a one-dimensional projective space $\mathbb{P}_1$, it is usual to use a single coordinate, called an *absolute coordinate*, associated to the homogeneous coordinates x_0, x_1: one takes x_1/x_0 as the absolute coordinate of the point $[x_0, x_1]$ if $x_0 \neq 0$, and the symbol ∞ as the absolute coordinate of $[0, 1]$. The absolute coordinate extends the usual affine coordinate $X_1 = x_1/x_0$, on the affine chart $x_0 \neq 0$, to the whole of $\mathbb{P}_1$ and, clearly, mapping each point to its absolute coordinate is a bijection between $\mathbb{P}_1$ and $\mathbb{C} \cup \{\infty\}$. In the sequel, choosing an absolute coordinate will mean choosing homogeneous coordinates and taking the absolute coordinate associated to them. The reader may easily check that if z is an absolute coordinate on $\mathbb{P}_1$, then so is $z' = (az + b)/(cz + d)$ for any $a, b, c, d \in \mathbb{C}$ with $ad - bc \neq 0$, and any absolute coordinate on $\mathbb{P}_1$ arises in this way.

Let $V : F = 0$ be a hypersurface of $\mathbb{P}_n$ and take

$$f(X_1, \ldots, X_n) = F(1, X_1, \ldots, X_n).$$

The polynomial f is constant if and only if, up to a constant factor, $F = x_0^d$. In such a case no point of V belongs to $\mathbb{A}_n$ and we retain the *0-th affine part* V_0 of V as empty. Otherwise f defines a hypersurface V_0 of $\mathbb{A}_n$ which we will call the *0-th affine part of* V. In the sequel we will drop the reference to the 0-th chart by saying just *affine part* unless some confusion may result.

Remark 1.2.3 If V and W are hypersurfaces of $\mathbb{P}_n$, then it is straightforward to check that $(V + W)_0 = V_0 + W_0$, the equality still being true if some of the affine parts are empty, provided the convention $T + \emptyset = \emptyset + T = T$, T an affine linear variety or $T = \emptyset$, is adopted. Also, in all cases $|V_0| = |V| \cap \mathbb{A}_n$.

If $d = \deg V$, the reader may note that

$$F(x_0, \ldots, x_n) = x_0^d f(x_1/x_0, \ldots, x_n/x_0) \tag{1.5}$$

and therefore V can be recovered from V_0 and $\deg V$, even if $V_0 = \emptyset$. Using twice the equality (1.5) proves that projective hypersurfaces $V : F = 0$ and $W : G = 0$ have the same affine part, $V_0 = W_0$, if and only if, up to a constant factor, $F = G x_0^r$, $r \in \mathbb{Z}$. If Π is the hyperplane $\Pi : x_0 = 0$, we have:

Proposition 1.2.4 *Given a hypersurface $B : f = 0$ of $\mathbb{A}_n^0$, take $\bar{B}$ to be the hypersurface of $\mathbb{P}_n$ defined by*

$$F(x_0, \ldots, x_n) = x_0^d f(x_1/x_0, \ldots, x_n/x_0) \tag{1.6}$$

with $d = \deg f$. Then $\bar{B}$ is the only hypersurface of $\mathbb{P}_n$ which has $\deg \bar{B} = \deg B$ and $\bar{B}_0 = B$. A hypersurface V of $\mathbb{P}_n$ has $V_0 = B$ if and only if $V = \bar{B} + r\Pi$ for some $r \geq 0$. In particular, if V does not have Π as an irreducible component and $V_0 = B$, then $V = \bar{B}$.

PROOF: After its definition, it is straightforward to check that $\deg \bar{B} = d$ and $\bar{B}_0 = B$. It is also clear that $\bar{B}$ does not have Π as an irreducible component. The uniqueness of $\bar{B}$ and the remaining part of the claim follow from the preceding considerations.   ◇

The hypersurface $\bar{B}$ defined in 1.2.4 is called the *projective closure* of B.

Remark 1.2.5 If B and C are hypersurfaces of $\mathbb{A}_n$, then it follows from the definition of projective closure that $\overline{B + C} = \bar{B} + \bar{C}$ and $(\bar{B})_0 = B$.

Corollary 1.2.6 *Taking projective closures and taking affine parts are reciprocal bijections between the set of all irreducible hypersurfaces of $\mathbb{P}_n$ other than Π, and the set of all irreducible hypersurfaces of $\mathbb{A}_n$.*

PROOF: If the projective hypersurface V is irreducible and $V \neq \Pi$, then $V_0 \neq \emptyset$ and $V = \overline{(V_0)}$ by 1.2.4. After this, a decomposition $V_0 = C + C'$ would give, by 1.2.5, $V = \bar{C} + \bar{C}'$ against the irreducibility of V: this proves that V_0 is irreducible.

Assume now that an affine hypersurface B is irreducible. If $\bar{B} = V + W$, then, by 1.2.3, $B = (\bar{B})_0 = V_0 + W_0$ against the irreducibility of B, which proves the irreducibility of $\bar{B}$.

We have seen that the two maps of the claim are well defined. That they are reciprocal has been seen in 1.2.4 and 1.2.3, as already noted in the course of the proof.   ◇

The proof of the next corollary is straightforward and left to the reader:

Corollary 1.2.7 *The irreducible components of the affine part V_0 of a projective hypersurface V are the affine parts of the irreducible components of V other than $\Pi : x_0 = 0$. The multiplicity of each irreducible component of V other than $\Pi : x_0 = 0$ equals the multiplicity of its affine part as an irreducible component of V_0.*

Remark 1.2.8 In the situation of 1.2.7, each irreducible component of V other than Π may be recovered as the closure of its affine part, by 1.2.4. Taking affine parts and closures are thus a pair of reciprocal bijections between the irreducible components of V other than Π and the irreducible components of its affine part V_0.

Let ℓ be a line of $\mathbb{P}_n$ not contained in $\Pi : x_0 = 0$. The intersection $\ell \cap \Pi$ is a point; taking it and another point to span ℓ gives rise to parametric equations of ℓ of the form

$$x_0 = t_0, \quad x_i = c_i^0 t_0 + c_i^1 t_1, \quad i = 1, \dots, n. \tag{1.7}$$

Then $\ell_0 = \ell \cap \mathbb{A}_n$ is an affine line and is also the 0-th chart of ℓ relative to the coordinates t_0, t_1, because the points of ℓ_0 are those with $t_0 \neq 0$. Parametric equations of ℓ_0 are

$$X_i = c_i^0 + c_i^1 t, \quad i = 1, \dots, n, \tag{1.8}$$

where $t = t_1/t_0$ is the affine coordinate in the affine chart ℓ_0.

Take a hypersurface $V : F = 0$ and, as above, $f(X_1, \ldots, X_n) = F(1, X_1, \ldots, X_n)$ as an equation of its affine part. The obvious equality

$$F(t_0, c_1^0 t_0 + c_1^1 t_1, \ldots, c_n^0 t_0 + c_n^1 t_1)_{|t_0=1,t_1=t} = f(c_1^0 + c_1^1 t, \ldots, c_n^0 + c_n^1 t) \quad (1.9)$$

shows that the affine part of $V \cap \ell$ is $V_0 \cap \ell_0$. Therefore, the hypothesis and notations being as above, 1.2.7 gives:

Lemma 1.2.9 *The points of $V_0 \cap \ell_0$ are the points of $V \cap \ell$ other than $\ell \cap \Pi$; their multiplicities in $V_0 \cap \ell_0$ and $V \cap \ell$ are equal.*

Since we are mainly interested in projective hypersurfaces (in plane projective curves in fact), in the sequel we will often refer to affine parts of projective hypersurfaces rather than to affine hypersurfaces of an arbitrary affine space $\mathbb{A}_n$. However, such an affine space $\mathbb{A}_n$ may always be identified with the 0-th affine chart of a projective space $\mathbb{P}_n$ by mapping $(X_1, \ldots, X_n)$ to $[1, X_1, \ldots, X_n]$, after which any affine hypersurface of $\mathbb{A}_n$ appears as the 0-th affine part of its projective closure. Thus dealing with just the affine parts of projective hypersurfaces, instead of dealing with arbitrary affine hypersurfaces, is not a more restricted situation.

1.3   Singular Points

Let V be an affine or projective hypersurface and p a point of its ambient space. The *multiplicity of p on V*, sometimes also called the *multiplicity of V at p*, is defined as being the minimal value of the intersection multiplicities at p of V and the lines through p. It will be denoted $e_p(V)$. As follows from the definition of intersection multiplicity, $e_p(V) = 0$ if and only if $p \notin V$; otherwise $e_p(V) \geq 1$. The point p is said to be a *smooth* – or *simple*, or *non-singular* – point of V if and only if $e_p(V) = 1$; V is then said to be *smooth* or *non-singular* at p. The point p is called a *multiple* or *singular* point of V if $e_p(V) > 1$. In such a case V is said to be *singular*, or to have a singularity, at p. A hypersurface is called *singular* if and only if it is singular at some point; otherwise it is said to be *non-singular* or *smooth*. Points p with $e_p(V) = e$ are called *e-fold points of V* (*double points* if $e_p(V) = 2$, *triple points* if $e_p(V) = 3$, and so on).

A line ℓ through p is said to be *tangent to V at p* if and only if the intersection multiplicity of ℓ and V at p is strictly higher than $e_p(V)$. Then the point p is said to be a *contact point* of ℓ. If V is a projective hypersurface, the above notions depend only on an affine part of V containing p:

Proposition 1.3.1 *If p is a point of the affine part V_0 of a hypersurface V of $\mathbb{P}_n$, then $e_p(V_0) = e_p(V)$. In particular, p is a singular (resp. non-singular) point of V_0 if and only it is a singular (resp. non-singular) point of V. A line ℓ of $\mathbb{P}_n$ is tangent to V at p if and only if its affine part is tangent to V_0 at p.*

PROOF: Due to 1.2.9, the intersection multiplicity at p of any line ℓ of $\mathbb{P}_n$ through p and V equals the intersection multiplicity at p of their affine parts. Since any affine line through p is the affine part of a line of $\mathbb{P}_n$ through p, all the claims follow.   ⋄

Let p be a point of an affine space $\mathbb{A}_n$ which has been taken as the origin of the affine coordinates $X_1, \ldots, X_n$. Let $V : f = 0$ be a hypersurface of $\mathbb{A}_n$ and write its equation as a sum of homogeneous polynomials of increasing degrees, $f = f_e + \cdots + f_d$, with each f_i homogeneous of degree i, $f_e \neq 0$ and $e \geq 0$. In particular, f_e is the initial form of f. Any line ℓ through the origin p has parametric equations of the form $X_i = a_i t$, $i = 1, \ldots, n$, and the parameter of p is $t = 0$. The intersection multiplicity of ℓ and V at p is thus the multiplicity of the factor t in

$$f(a_1 t, \ldots, a_n t)) = t^e f_e(a_1, \ldots, a_n) + \cdots + t^d f_d(a_1, \ldots, a_n),$$

and therefore it equals e if $f_e(a_1, \ldots, a_n) \neq 0$, and is strictly higher otherwise. As a consequence, the multiplicity of p on V is e. Furthermore, ℓ is tangent to V at p if and only if $f_e(a_1, \ldots, a_n) = 0$. It follows that the points of the lines tangent to V at p are just the points of the hypersurface $f_e = 0$. Summarizing:

Proposition 1.3.2 *If $V : f = 0$ is a hypersurface of an affine space $\mathbb{A}_n$, on which affine coordinates with origin p have been fixed, then:*

(a) *the multiplicity of p on V is the degree e of the initial form f_e of f, and*

(b) *a line through p is tangent to V at p if and only if it is contained in the hypersurface $f_e = 0$.*

Since the initial form of a product is the product of the initial forms of the factors, the next proposition is a direct consequence of 1.3.2 (using 1.3.1 for the projective case).

Proposition 1.3.3 *For any two hypersurfaces V_1, V_2 of the same affine or projective space and any point p of the latter,*

(a) $e_p(V_1 + V_2) = e_p(V_1) + e_p(V_2)$.

(b) *A point is a singular point of $V_1 + V_2$ if and only if either it belongs to both V_1 and V_2, or it is a singular point of one of the V_i, $i = 1, 2$. In particular, any point belonging either to a multiple irreducible component of a hypersurface V, or to two different irreducible components of V, is a singular point of V.*

(c) *A line ℓ is tangent to $V_1 + V_2$ at p if and only if it is tangent to either V_1 or V_2 at p.*

It is clear from the computation preceding 1.3.2 (and also from 1.1.3 and the definition of multiplicity) that for any affine or projective hypersurface V and any point p, $e_p(V) \leq \deg V$. When equality holds, V is said to be a (projective or affine, according to the case) *cone* with *vertex* p. Cones may be easily recognized if suitable coordinates are used:

Proposition 1.3.4 *Assume that a point $p \in \mathbb{A}_n$ has been taken as the origin of the coordinates. Then a hypersurface $V : f = 0$ of $\mathbb{A}_n$ is a cone with vertex p if and only if f is a homogeneous polynomial. If a point $p \in \mathbb{P}_n$ has been taken as the 0-th vertex of the projective reference (i.e., $p = [1, 0, \ldots, 0]$), and the coordinates are $x_0, \ldots, x_n$, then a hypersurface $V : F = 0$ of $\mathbb{P}_n$ is a cone with vertex p if and only if F does not depend on x_0.*

PROOF: In the affine case, again write $f = \sum_{i=e}^{d} f_i$, with each f_i homogeneous of degree i, $f_e, f_d \neq 0$. By 1.3.2 $e_p(V) = e$ and, clearly, $\deg V = d$. V is thus a cone with vertex p if and only if $e = d$ or, equivalently, $f = f_e$, hence the first claim.

For the second claim, write F as a polynomial in x_0,

$$F = x_0^{d-e} F_e(x_1, \ldots, x_n) + \cdots + F_d(x_1, \ldots, x_n), \quad F_e \neq 0.$$

Then, using the affine coordinates $X_i = x_i/x_0$, $i = 1, \ldots, n$, the point p is the origin of coordinates and an equation of the 0-th affine part V_0 of V is $f = F_e(X_1, \ldots, X_n) + \cdots + F_d(X_1, \ldots, X_n)$. Each F_j being homogeneous of degree j, still by 1.3.2, $e_p(V) = e_p(V_0) = e$. Hence, V is a cone with vertex p if and only if $d = e$, which in turn is equivalent to being $F = F_e$.   ◇

Remark 1.3.5 The reader may easily check that the set of points of an affine or projective cone with vertex p is composed of lines through p.

Remark 1.3.6 The projective closure of an affine cone is clearly a projective cone with the same vertex.

Remark 1.3.7 If $n = 2$, according to 1.3.4, both the affine and the projective cones have equations that are homogeneous polynomials in two variables. These equations, as already recalled in Section 1.1, factorize into homogeneous linear factors; it follows that all the irreducible components of a plane cone with vertex p are lines through p. The converse is true and may easily be checked by the reader. Once the irreducible components of a plane cone C are known to be lines, it is clear that no other line may be contained in C, because such a line shares at most one point with each irreducible component of C and there are finitely many of the latter.

The hypersurface $f_e = 0$ mentioned in 1.3.2 is a cone with vertex p, by 1.3.4. It is worth paying some attention to it, and also to a projective version of it:

– If p is a point of an affine hypersurface V, and affine coordinates with origin p have been taken, then the hypersurface $f_e = 0$, defined by the initial form f_e of an equation f of V, is called the *tangent cone* to V at p, denoted $TC_p(V)$.

– If p is a point of a projective hypersurface V and V_0 an affine part of V containing p, then the projective closure of the tangent cone to V_0 at p is called the *tangent cone* to V at p, still denoted $TC_p(V)$.

Clearly, in both cases, $TC_p(V)$ is a cone with vertex p. We need of course to show that the above definitions are independent of the choices involved in them:

Lemma 1.3.8 *The tangent cone $TC_p(V)$ remains the same if other affine coordinates and – in the projective case – another affine chart are used in its definition.*

PROOF: We will deal with the projective case; a similar – easier – argument applies to the affine case. By 1.2.1, assume that $x_0, \ldots, x_n$ are projective coordinates such that the affine chart and the affine coordinates with origin p, used to define the tangent cone, are $x_0 \neq 0$ and $X_i = x_i/x_0$, $i = 1, \ldots, n$. If an equation of V is written as a polynomial in x_0,

$$F = x_0^{d-e} F_e(x_1, \ldots, x_n) + \cdots + F_d(x_1, \ldots, x_n), \quad F_e \neq 0,$$

then the corresponding equation f of the affine part of V appears as a sum of homogeneous polynomials, each of degree equal to the index,

$$f = F_e(X_1, \ldots, X_n) + \cdots + F_d(X_1, \ldots, X_n).$$

Hence, $e = e_p(V)$, the tangent cone to the affine part of V at p is $F_e(X_1, \ldots, X_n) = 0$, and so the tangent cone to V at p is $F_e(x_1, \ldots, x_n) = 0$.

If another affine chart and other affine coordinates with origin p are used, then still there will be projective coordinates $y_0, \ldots, y_n$ for which the affine chart is $y_0 \neq 0$ and the affine coordinates $Y_i = y_i/y_0$, $i = 1, \ldots, n$. If an equation of V is now

$$G = y_0^{d-e'} G_{e'}(y_1, \ldots, y_n) + \cdots + G_d(y_1, \ldots, y_n), \quad G_{e'} \neq 0,$$

then, as before, $e' = e_p(V) = e$ and the tangent cone is $G_e(y_1, \ldots, y_n) = 0$.

New and old projective coordinates are related by equalities

$$x_i = \sum_{i=0}^{n} a_i^j y_j, \quad i = 0, \ldots, n, \quad \det(a_i^j) \neq 0. \tag{1.10}$$

In them, $a_i^0 = 0$ for $i = 1, \ldots n$ because, both systems of affine coordinates having origin p, we have $p = [1, 0, \ldots, 0]$ in either system of projective coordinates. We may thus assume in addition $a_0^0 = 1$. Up to multiplying it by a non-zero constant factor, we may assume that G comes from F by performing the substitution of variables (1.10), and then it is straightforward to check that G_e results from F_e by the same substitution of variables. Therefore both F_e and G_e define the same hypersurface of $\mathbb{P}_n$, as wanted.   ◇

Remark 1.3.9 The proof of 1.3.8 shows how to get an equation of $TC_p(V)$ in the projective case, if coordinates with $p = [1, 0, \ldots, 0]$ are used.

Remark 1.3.10 Let C be an affine or projective plane curve and $p \in C$. Remark 1.3.4 applies to $TC_p(C)$ showing that its irreducible components are lines, and, using 1.3.2, also that the irreducible components of $TC_p(C)$ are the lines tangent to C at p. Thus, in particular, there are finitely many of the latter.

The next two theorems provide a characterization of the e-fold points of a hypersurface in both the affine and the projective cases. Any polynomial is taken as being its only 0-th derivative:

Theorem 1.3.11 *A point p is singular for an affine hypersurface $V : f = 0$ if and only if f and all its first-order derivatives vanish at p. The same p is an e-fold point of V if and only if all the derivatives of f of orders $0, \ldots, e-1$ have value 0 at p, and one of the e-th derivatives has not.*

PROOF: Clearly, only the second claim needs to be proved. Assume that p has affine coordinates $\alpha_1, \ldots, \alpha_n$. Then the lines through p are those with parametric equations

$$X_i = \alpha_i + \beta_i t, \quad i = 1, \ldots, n,$$

with $\beta = (\beta_1, \ldots, \beta_n)$ an arbitrary non-zero vector. On any of them p has parameter $t = 0$ and therefore we need to ensure that the polynomial $\bar{f} = f(\alpha_1 + \beta_1 t, \ldots, \alpha_n + \beta_n t)$ has the factor t (or, equivalently, the root 0) with multiplicity at least e for all β, and equal to e for some β. Since this is equivalent to having

$$\left(\frac{d^i \bar{f}}{dt^i} \right)_{t=0} = \sum_{j_1, \ldots, j_i} \left(\frac{\partial^i f}{\partial X_{j_1} \ldots \partial X_{j_i}} \right)_p \beta_{j_1} \ldots \beta_{j_i} = 0, \qquad (1.11)$$

for all non-zero β and $i = 0, \ldots, e-1$, and furthermore

$$\left(\frac{d^e \bar{f}}{dt^e} \right)_{t=0} = \sum_{j_1, \ldots, j_e} \left(\frac{\partial^e f}{\partial X_{j_1} \ldots \partial X_{j_e}} \right)_p \beta_{j_1} \ldots \beta_{j_e} \neq 0, \qquad (1.12)$$

for some non-zero β, the claim follows.   ◇

Theorem 1.3.12 *A point p is singular for a projective hypersurface $V : F = 0$ if and only if all the first-order derivatives of F have value 0 at p. The same p is an e-fold point of V if and only if all the derivatives of F of order $e-1$ have value 0 at p, while at least one of the e-th derivatives has not.*

PROOF: Recall first that for any homogeneous $G \in \mathbb{C}[x_0, \ldots, x_n]$ of degree d, $dG = x_0(\partial G/\partial x_0) + \cdots + x_n(\partial G/\partial x_n)$ (*Euler's formula*). This in particular implies that, if all the derivatives of a certain order i of a homogeneous polynomial F are zero at a point, then so are all the derivatives of F of order at most i. This fact and arguments similar to those used in the proof of 1.3.11, this time using parametric equations

$$x_i = \alpha_i t_0 + \beta_i t_1, \quad i = 0, \ldots, n,$$

with $p = [\alpha_0, \ldots, \alpha_n]$, and t_1 in the place of t, prove the claim.   ◇

1.4   Rational Functions on Projective Spaces

Assume, as before, that we have fixed coordinates $x_0, \ldots, x_n$ in $\mathbb{P}_n$. For any two homogeneous polynomials $F, G \in \mathbb{C}[x_0, \ldots, x_n]$, of the same degree d and with $G \neq 0$, consider the map

$$(F : G) : \mathbb{P}_n - |G = 0| \longrightarrow \mathbb{C}$$

$$[a_0, \ldots, a_n] \longmapsto \frac{F(a_0, \ldots, a_n)}{G(a_0, \ldots, a_n)}.$$

The polynomials F and G being both homogeneous of degree d, the arbitrary factor up to which the coordinates of the point are defined is irrelevant to the definition, because for any non-zero $\lambda \in \mathbb{C}$,

$$\frac{F(\lambda a_0, \ldots, \lambda a_n)}{G(\lambda a_0, \ldots, \lambda a_n)} = \frac{\lambda^d F(a_0, \ldots, a_n)}{\lambda^d G(a_0, \ldots, a_n)} = \frac{F(a_0, \ldots, a_n)}{G(a_0, \ldots, a_n)}.$$

Remark 1.4.1 The set of all maps $(F : G)$, for $F, G \in \mathbb{C}[x_0, \ldots, x_n]$ homogeneous of the same degree, $G \neq 0$, is the same if different coordinates on $\mathbb{P}_n$ are used, as a linear and invertible substitution of the variables does not affect the conditions on F and G.

Usually these maps are taken up to the extensions or restrictions resulting from cancelling or adding common non-zero homogeneous factors to F and G. To this end, take as equivalent two of the above maps if and only if there is a hypersurface V of $\mathbb{P}_n$ such that the maps are both defined and agree on all points of $\mathbb{P}_n - V$. Symmetric and reflexive properties are obviously satisfied; for the transitivity, if $(F : G)$ and $(F' : G')$ are defined and agree on $\mathbb{P}_n - V$ and the same holds for $(F' : G')$ and $(F'' : G'')$ on $\mathbb{P}_n - V'$, just note that $(F : G)$ and $(F'' : G'')$ are defined and agree on $\mathbb{P}_n - (V + V')$. Two important facts are:

Proposition 1.4.2

(a) *Maps $(F : G)$ and $(F' : G')$ are equivalent if and only if $F/G = F'/G'$ as elements of $\mathbb{C}(x_0, \ldots, x_n)$.*

(b) *If maps $(F : G)$ and $(F' : G')$ are equivalent, then they take the same value at any point at which both are defined.*

PROOF: If $(F : G)$ and $(F' : G')$ are equivalent, then the polynomial $FG' - F'G$ takes value zero on the complement of a hypersurface, and therefore, due to 1.1.6, $FG' - F'G = 0$. From this, claim (b) and the *only if* part of claim (a) follow. The *if* part of claim (a) is clear.   ⋄

Let χ be one of the above equivalence classes. Claim (b) of 1.4.2 allows us to patch together all the representatives of χ to obtain a single map φ defined on a subset of $\mathbb{P}_n$. Note that any of the representatives of χ determines χ, and therefore also φ. The map φ is called *the rational function on $\mathbb{P}_n$*

represented or *defined* by any of the representatives of χ, which in turn are called representatives of φ. Clearly, the rational function φ determines its representatives – as these are the maps $(F : G)$ that are restrictions of φ – and therefore determines χ. Thus φ and χ determine each other and the difference between them is purely formal: φ is a map locally defined by maps $(F : G)$, while χ is the class of all the maps $(F : G)$ that patch together to define φ. In the sequel we will refer to the rational function φ, its corresponding class χ being seldom mentioned.

A rational function φ is thus *defined* – as a map – at a point p if and only if so is one of its representatives. When this is the case, φ is equivalently said to be *regular* at p, p is called a *regular point* of φ, and the value of φ at p, $\varphi(p)$, is the common value at p of all the representatives of φ that are defined at p. In the sequel, when writing $\varphi(p)$ we will often implicitly assume that φ is defined at p.

Remark 1.4.3 The fact that the ring $\mathbb{C}[x_0, \ldots, x_n]$ is factorial allows a simpler description of the rational functions on $\mathbb{P}_n$ that fails in other cases (for the rational functions on a curve, for instance). By claim (a) of 1.4.2 and 1.1.1, any rational function φ has a representative of the form $(F : G)$ with F and G coprime (*irreducible representative*). This representative is then maximal in the sense that it is defined at any point at which a representative of φ – and hence φ itself – is defined: indeed, again by 1.4.2(a) and 1.1.1, any representative of φ has then the form $(FA : GA)$ with A homogeneous.

If φ and ψ are rational functions on $\mathbb{P}_n$ defined, respectively, by $(F : G)$ and $(F' : G')$, then $\varphi + \psi$ is defined as being the rational function represented by $(FG' + F'G : GG')$, and $\varphi\psi$ as being the one represented by $(FF' : GG')$.

The reader may take care of checking that these definitions do not depend on the representatives, and that $(\varphi + \psi)(p) = \varphi(p) + \psi(p)$ and $(\varphi\psi)(p) = \varphi(p)\psi(p)$ for all $p \in \mathbb{P}_n$ at which both φ and ψ are defined. It is also direct to check that the set of all the rational functions on $\mathbb{P}_n$ equipped with these operations is a field: it is called the *field of rational functions of* $\mathbb{P}_n$ and denoted $\mathbb{C}(\mathbb{P}_n)$. We will identify each complex number a with the rational function represented by $(a : 1)$, which is the constant function with value a. This identification being obviously compatible with the operations, it turns $\mathbb{C}(\mathbb{P}_n)$ into a $\mathbb{C}$-algebra.

Among the regular points of a rational function φ, those at which the value of φ is 0 are called *zeros* of φ. If $\varphi \neq 0$, then the zeros of φ^{-1} are called *poles* of φ. Clearly φ is not defined at its poles, as no complex number a satisfies $a0 = 1$. It is usual to take the symbol ∞ as the value of φ at each of its poles: this extends the rational function to a map with values in $\mathbb{C} \cup \{\infty\}$ which still may fail to be defined in the whole of $\mathbb{P}_n$ (see below). The reader may see $\mathbb{C} \cup \{\infty\}$ either as a Riemann sphere, or as a projective line $\mathbb{P}_1$ on which an absolute coordinate has been fixed.

The points which are neither regular points nor poles of a rational function φ are called *indetermination points* of φ. For an easy example, the rational function on $\mathbb{P}_2$ represented by $(x_1 : x_2)$ has $p = [1, 0, 0]$ as its only indetermination point, the points of $x_1 = 0$ other than p as zeros, and the points of $x_2 = 0$ other than p as poles.

The notion of rational function and the related notions introduced above are all independent of the choice of the homogeneous coordinates due to 1.4.1. Once coordinates $x_0, \ldots, x_n$ have been fixed on $\mathbb{P}_n$, by 1.4.2(a), it is clear that mapping the rational function represented by $(F : G)$ to the element $F/G \in \mathbb{C}(x_0, \ldots x_n)$ is an isomorphism of $\mathbb{C}$-algebras between $\mathbb{C}(\mathbb{P}_n)$ and the subfield of $\mathbb{C}(x_0, \ldots x_n)$ of all the quotients with numerator and denominator homogeneous and of the same degree. Being strictly formal, one should say that F/G is *the representation in the coordinates* $x_0, \ldots x_n$ – or a *projective representation* – of the function represented by the map $(F : G)$; the representation in other coordinates is of course obtained by the corresponding linear substitution of variables. Once the coordinates have been fixed, we will identify the functions with their representations, thus making no distinction between the rational function defined by $(F : G)$ and the quotient $F/G \in \mathbb{C}(x_0, \ldots, x_n)$. Accordingly, $\mathbb{C}(\mathbb{P}_n)$ will be identified with the subfield of $\mathbb{C}(x_0, \ldots x_n)$ of all the quotients F/G with F, G homogeneous of the same degree and $G \neq 0$.

Fix a point $p \in \mathbb{P}_n$ and consider the subset of $\mathbb{C}(\mathbb{P}_n)$ of all rational functions which are defined at p: they obviously compose a subring of $\mathbb{C}(\mathbb{P}_n)$ containing $\mathbb{C}$ (hence a $\mathbb{C}$-algebra), which is called the *local ring* of $\mathbb{P}_n$ at p, and sometimes also the local ring of p in $\mathbb{P}_n$. It will be denoted $\mathcal{O}_{\mathbb{P}_n,p}$. Still $\mathcal{O}_{\mathbb{P}_n,p}$ does not depend on the coordinates. Once coordinates $x_0, \ldots x_n$ are fixed, the elements of $\mathcal{O}_{\mathbb{P}_n,p}$ are (identified with) the quotients F/G, with $F, G \in \mathbb{C}[x_0, \ldots x_n]$ homogeneous of the same degree and $G(p) \neq 0$.

The subset

$$\mathfrak{M}_{\mathbb{P}_n,p} = \{\varphi \in \mathcal{O}_{\mathbb{P}_n,p} \mid \varphi(p) = 0\}$$

clearly is an ideal of $\mathcal{O}_{\mathbb{P}_n,p}$, and any element of $\mathcal{O}_{\mathbb{P}_n,p} - \mathfrak{M}_{\mathbb{P}_n,p}$ is invertible: therefore $\mathfrak{M}_{\mathbb{P}_n,p}$ is the only maximal ideal of $\mathcal{O}_{\mathbb{P}_n,p}$ and the latter is, by definition, a local ring, hence its name.

Assume we have fixed projective coordinates $x_0, \ldots, x_n$ on $\mathbb{P}_n$. The other affine charts giving rise to similar considerations, let us fix our attention on the 0-th affine chart $x_0 \neq 0$, denoted $\mathbb{A}_n$ and equipped as usual with the affine coordinates $X_i = x_i/x_0$, $i = 1, \ldots, n$. It is straightforward to check that the maps

$$F/G = \frac{F(x_0, \ldots, x_n)}{G(x_0, \ldots, x_n)} \longmapsto \widehat{F/G} = \frac{F(1, X_1, \ldots, X_n)}{G(1, X_1, \ldots, X_n)} \tag{1.13}$$

and

$$\frac{f(X_1, \ldots, X_n)}{g(X_1, \ldots, X_n)} \longmapsto \frac{f(x_1/x_0, \ldots, x_n/x_0)}{g(x_1/x_0, \ldots, x_n/x_0)}, \tag{1.14}$$

are reciprocal isomorphisms of $\mathbb{C}$-algebras between $\mathbb{C}(\mathbb{P}_n)$ and $\mathbb{C}(X_1, \ldots, X_n)$. The rational fraction $\widehat{F/G}$ allows us to compute in the obvious way the values of F/G on the points of $\mathbb{A}_n$, using the affine coordinates of the points; we will call it the *affine representation* of the rational function F/G in coordinates

$X_1, \ldots, X_n$. In the sequel, once the projective coordinates and the affine chart have been fixed, we will identify each rational function F/G with its affine representation $\widehat{F/G}$, and therefore the $\mathbb{C}$-algebras $\mathbb{C}(\mathbb{P}_n)$ and $\mathbb{C}(X_1, \ldots, X_n)$ through the above isomorphisms.

Example 1.4.4 If $n = 1$, with the above conventions, any absolute coordinate z is a rational function and $\mathbb{C}(\mathbb{P}_1) = \mathbb{C}(z)$.

Example 1.4.5 Using affine representations, the local ring of a point $p \in \mathbb{A}_n$ appears as being

$$\mathcal{O}_{\mathbb{P}_n, p} = \{ f/g \in \mathbb{C}(X_1, \ldots, X_n) \mid g(p) \neq 0 \}.$$

It is sometimes called the local ring of $\mathbb{A}_n$ at p and written $\mathcal{O}_{\mathbb{A}_n, p}$.

The rational functions of $\mathbb{P}_n$ that are regular at all points of $\mathbb{A}_n$ clearly describe a subring of $\mathbb{C}(\mathbb{P}_n)$ that contains $\mathbb{C}$: it is called the *affine ring* of $\mathbb{A}_n$, denoted in the sequel by $\mathcal{A}(\mathbb{A}_n)$. After its definition, the affine ring $\mathcal{A}(\mathbb{A}_n)$ does not depend on the choice of the coordinates $x_0, \ldots, x_n$, but only on the choice of the affine chart $\mathbb{A}_n$. The next proposition describes the elements of $\mathcal{A}(\mathbb{A}_n)$ by means of their representations as elements of $\mathbb{C}(x_0, \ldots, x_n)$:

Proposition 1.4.6 *A rational function $f \in \mathbb{C}(\mathbb{P}_n)$ belongs to $\mathcal{A}(\mathbb{A}_n)$ if and only if it is represented by an irreducible fraction of the form F/x_0^d with $F \in \mathbb{C}[x_0, \ldots, x_n]$ homogeneous of degree d.*

PROOF: The *if* part is obvious. For the converse we leave the reader to check that any homogeneous polynomial in $\mathbb{C}[x_0, \ldots, x_n]$ effectively depending on one of the variables x_i, $i > 0$, is zero at at least one point of $\mathbb{A}_n$. If $f = P/Q$, P and Q are coprime and Q effectively depends on one variable other than x_0, then, by the above, $Q(p) = 0$ for a point $p \in \mathbb{A}_n$ and, the representative $(P : Q)$ being irreducible (1.4.3), f is not regular at p.   ◇

Using affine representations with affine coordinates $X_i = x_i/x_0$, $i = 1, \ldots, n$ gives an easier description of $\mathcal{A}(\mathbb{A}_n)$ that follows directly from 1.4.6:

Corollary 1.4.7 *After identifying $\mathbb{C}(\mathbb{P}_n) = \mathbb{C}(X_1, \ldots, X_n)$, we have $\mathcal{A}(\mathbb{A}_n) = \mathbb{C}[X_1, \ldots, X_n]$.*

1.5   Linear Families and Linear Conditions

By its own definition, the projective hypersurfaces of degree d of $\mathbb{P}_n$ are the points of the projective space associated to the vector space of the homogeneous polynomials in $x_0, \ldots, x_n$ of degree d, the equations of a hypersurface being its representatives. We will denote it by $\mathcal{H}_d(\mathbb{P}_n)$ or just $\mathcal{H}_d^n$ if no mention of $\mathbb{P}_n$ is needed.

An easy count shows that the number of monomials of degree d in $x_0, \ldots, x_n$ is $\binom{n+d}{n}$. Since these monomials compose a basis of the vector space of all homogeneous polynomials of degree d, the dimension of the projective space of the hypersurfaces of degree d of $\mathbb{P}_n$ is $\dim \mathcal{H}_d(\mathbb{P}_n) = \binom{n+d}{n} - 1$. In particular, the curves of degree d of $\mathbb{P}_2$ describe a projective space of dimension $\binom{d+2}{2} - 1 = (d^2 + 3d)/2$.

The coefficients of a homogeneous polynomial F being its components relative to the above basis of monomials, for $F \neq 0$, the coefficients of F may be taken as homogeneous coordinates of the hypersurface $F = 0$, which we will often do in the sequel without further mention.

Having a projective structure on the set of all hypersurfaces of $\mathbb{P}_n$ of fixed degree is a very relevant fact regarding their geometry. This structure is handled, as said above, by taking either any equation F as a representative of the hypersurface $F = 0$, or the coefficients of F as homogeneous coordinates of $F = 0$. From now on, notions that apply to the points of a projective space will be freely used for hypersurfaces, by taking them as elements of the corresponding $\mathcal{H}_d^n$. For instance, hypersurfaces $F_0 = 0, \ldots, F_r = 0$, of the same degree d, are said to be *linearly independent* when they are so as elements of $\mathcal{H}_d^n$, which is in turn equivalent to the linear independence of their equations $F_0, \ldots, F_r$.

Remark 1.5.1 Fix the affine chart $\mathbb{A}_n^0 : x_0 \neq 0$ of $\mathbb{P}_n$ and conventionally take the empty set $\emptyset$, with equation any non-zero complex number, as the affine part of the hypersurface $x_0^d = 0$. Then the rule

$$F(x_0, \ldots, x_n) \mapsto F(1, X_1, \ldots, X_n)$$

defines an isomorphism between the space of homogeneous polynomials of degree d in $x_0, \ldots, x_n$ and the space of polynomials of degree at most d in $X_1, \ldots, X_n$, its inverse being

$$f(X_1, \ldots, X_n) \mapsto x_0^d f(x_1/x_0, \ldots, x_n/x_0).$$

Since, as seen in Section 1.2, this isomorphism maps the equations of any projective hypersurface of degree d to the equations of its affine part, the projective structure of $\mathcal{H}_d^n$ may be handled by using as representatives the equations of the affine parts of the hypersurfaces $V \in \mathcal{H}_d^n$ instead of the equations of the hypersurfaces themselves. Note in particular that the coefficients of any equation F of a projective hypersurface V – coordinates of V – are the coefficients of the corresponding equation $F(1, X_1, \ldots, X_n)$ of the affine part of V.

The linear varieties of $\mathcal{H}_d(\mathbb{P}_n)$ are called *linear systems* (or *linear families*) of hypersurfaces of $\mathbb{P}_n$. Note that the empty set appears as a linear system, the only one with dimension -1. An r-dimensional linear system Λ of hypersurfaces of $\mathbb{P}_n$ may thus be presented in the form

$$\Lambda = \{V : F = 0 \mid F \in S - \{0\}\},$$

for S an $(r+1)$-dimensional subspace of the vector space of all homogeneous polynomials of a certain given degree, or, more explicitly, if $r \geq 0$,

$$\Lambda = \{V_{\lambda_0, \ldots, \lambda_r} : \lambda_0 F_0 + \cdots + \lambda_r F_r = 0 \mid (\lambda_0, \ldots, \lambda_r) \in \mathbb{C}^{r+1} - \{0\}\},$$

where $F_0, \ldots, F_r$ are linearly independent homogeneous polynomials of the same degree. In the latter case the linear system Λ is usually denoted

$$\Lambda : \lambda_0 F_0 + \cdots + \lambda_r F_r = 0,$$

$\lambda_0, \ldots, \lambda_r$ then being retained as free variables. According to the usual convention in linear projective geometry, the linearly independent hypersurfaces $F_0 = 0, \ldots, F_r = 0$ are said *to span* – or to be *generators* of – Λ. In the case when $F_0, \ldots, F_r$ are coprime, no hypersurface is contained in all members of Λ and Λ is said *to have no fixed part*. Otherwise, the hypersurface $G = \gcd(F_0, \ldots, F_r) = 0$ is called the *fixed part* of Λ: it is contained in all members of Λ and the linear system spanned by the hypersurfaces $F_0/G = 0, \ldots, F_r/G = 0$, obtained by removing the fixed part from all members of Λ and called the *variable part* of Λ, has no fixed part.

The one-dimensional linear systems of hypersurfaces – the lines of $\mathcal{H}_d(\mathbb{P}_n)$ – are called *pencils*. A pencil of hypersurfaces of degree d thus has the form

$$\mathcal{P} = \{V_{\lambda_0, \lambda_1} : \lambda_0 F_0 + \lambda_1 F_1 = 0 \mid (\lambda_0, \lambda_1) \in \mathbb{C}^2 - \{0\}\}.$$

The two-dimensional linear systems of hypersurfaces are called *nets*, while those of dimension three are called *webs*.

A condition imposed on the hypersurfaces of degree d of $\mathbb{P}_n$ is called *linear of order s* if and only if the hypersurfaces satisfying it describe a linear variety of codimension s of $\mathcal{H}_d(\mathbb{P}_n)$. The linear conditions of order 1 are called *simple*. The hypersurfaces satisfying a simple linear condition thus describe a hyperplane of $\mathcal{H}_d(\mathbb{P}_n)$.

In other words, a condition is linear of order s if and only if it translates into s independent linear equations to be satisfied by the coefficients of the equations of the hypersurfaces. Simple linear conditions translate into a single non-trivial linear equation.

The easiest example of a simple linear condition is *going through a given point p*: indeed, after fixing $p = [\alpha_0, \ldots, \alpha_n]$, the hypersurface $V : F = 0$, of degree d, goes through p if and only if $F(\alpha_0, \ldots, \alpha_n) = 0$, which is a linear equation in the coefficients of F. The equation is non-trivial because at least one of the α_i is non-zero. Such a condition is often referred to as the *condition imposed by p on the hypersurfaces of degree d*.

Example 1.5.2 Let e be a non-negative integer and take $n = 2$ for simplicity. *Having multiplicity at least e at a given point* is a linear condition of order $e(e+1)/2$ on the plane curves of a fixed degree $d \geq e - 1$. For, take the coordinates such that the point is the origin of coordinates of an affine chart of $\mathbb{P}_2$. By 1.3.1 and 1.3.12 we may use the equations of the affine parts of the curves and then, by 1.3.2 the condition is equivalent to equating to zero all the coefficients of degree less than e, a total of $e(e+1)/2$ obviously independent equations. The reader may deal similarly with the n-dimensional case.

Elementary facts about incidence of linear varieties of projective spaces give non-obvious results on hypersurfaces once applied to $\mathcal{H}_d(\mathbb{P}_n)$. Here are some examples:

Lemma 1.5.3 *A simple linear condition on the hypersurfaces of $\mathbb{P}_n$ is satisfied by either all or exactly one of the hypersurfaces of a given pencil.*

PROOF: Just use that, in a projective space, for any line ℓ and any hyperplane H, the intersection $\ell \cap H$ is either ℓ or a single point. ◇

Proposition 1.5.4 *There is always a plane curve of degree d through m given points of $\mathbb{P}_2$ provided $m \leq (d^2 + 3d)/2$. The curves through these points describe a linear system of dimension at least $(d^2 + 3d)/2 - m$; in particular there are infinitely many such curves if $m < (d^2 + 3d)/2$.*

PROOF: This follows from the fact that the intersection of m hyperplanes of a projective space is a linear variety of codimension at most m. ◇

Simple conditions imposed on hypersurfaces of degree d are said to be *linearly independent* – or just *independent* – when so are the linear equations they translate into, or, equivalently, when the intersection of the hyperplanes of $\mathcal{H}_d(\mathbb{P}_n)$ they determine has codimension equal to the number of hyperplanes. The next statement is then almost tautological:

Proposition 1.5.5 *$(d^2 + 3d)/2$ points of $\mathbb{P}_2$ impose independent conditions on the curves of degree d if and only if there is exactly one curve of degree d going through them.*

Remark 1.5.6 More generally, in an arbitrary projective space the intersection of linear varieties of codimensions $c_1, \ldots, c_r$ is a linear variety of codimension at most $c_1 + \cdots + c_r$: in our case this means that imposing simultaneously linear conditions of orders $c_1, \ldots, c_r$ makes a linear condition of order at most $c_1 + \cdots + c_r$. Also in this case, when the equality holds the conditions are said to be *independent*.

Remark 1.5.7 After 1.5.6 and 1.5.2, imposing curves of a given degree d to have given points $q_1, \ldots, q_r$ as points of multiplicities at least $e_1, \ldots, e_r$, respectively, is a linear condition of order at most $\sum_{i=1}^{r} e_i(e_i + 1)/2$.

Next we improve Remark 1.5.7 above by showing that the conditions of having each q_i as a e_i-fold point are independent if d is high enough. The argument of the proof has been taken from [1, XI.4]. For a different one, see [12, Chap. 5, Th. 1].

Proposition 1.5.8 *Imposing the curves of a fixed degree d to have given distinct points $q_1, \ldots, q_r$ as points of multiplicities at least $e_1, \ldots, e_r$, respectively, is a linear condition of order $\sum_{i=1}^{r} e_i(e_i + 1)/2$, provided $d \geq e_1 + \cdots + e_r - 1$.*

PROOF: We know from 1.5.7 that the curves of degree d that have multiplicities at least $e_1, \ldots, e_r$ at $q_1, \ldots, q_r$, respectively, describe a linear system L whose codimension in $\mathcal{H}_d(\mathbb{P}_2)$ is

$$\dim \mathcal{H}_d(\mathbb{P}_2) - \dim L \leq \sum_{i=1}^{r} \frac{e_i(e_i + 1)}{2}.$$

Assume $d \geq e_1 + \cdots + e_r - 1$. To prove that equality holds, we will use induction on r. The case $r = 1$ has been seen in 1.5.2; we thus assume $r > 1$. By the induction hypothesis, the curves of degree d having $q_1, \ldots, q_{r-1}$ as points of multiplicities at least $e_1, \ldots, e_{r-1}$ describe a linear system L' that contains L and whose codimension in $\mathcal{H}_d(\mathbb{P}_2)$ equals $\sum_{i=1}^{r-1} e_i(e_i + 1)/2$: we thus have

$$\dim L' - \dim L \leq \frac{e_r(e_r + 1)}{2} \qquad (1.15)$$

and it will suffice to prove that equality holds.

For each $i = 1, \ldots, r - 1$ choose a cone $K_i : G_i = 0$ of degree e_i with vertex q_i and missing q_r (use 1.3.4). The curves of the form

$$C = K_1 + \cdots + K_{r-1} + C',$$

where C' is an arbitrary curve of degree $d' = d - \sum_{i=1}^{r-1} e_i$, describe a linear system S' whose corresponding vector space is $G_1 \ldots G_{r-1}\mathbb{C}[x_0, x_1, x_2]_{d'}$, where $\mathbb{C}[x_0, x_1, x_2]_{d'}$ denotes the vector space of the homogeneous polynomials of degree d'. Since $e_{q_i}(K_i) = e_i$ for $i = 1, \ldots, r - 1$, it is clear that $S' \subset L'$.

The map

$$\sigma : \mathcal{H}_{d'}(\mathbb{P}_2) \longrightarrow S'$$

$$C' \longmapsto C = \sum_{i=1}^{r-1} K_i + C'$$

is the projectivity induced by the multiplication of equations by $\prod_{i=1}^{r-1} G_i$. Since $d' \geq e_r - 1$, again by 1.5.2, the curves of degree d' that have multiplicity at least e_r at q_r compose a linear system T of codimension $e_r(e_r + 1)/2$ in $\mathcal{H}_{d'}(\mathbb{P}_2)$. On the other hand, since none of the cones K_i goes through q_r, a curve $C \in S'$ has $e_{q_r}(C) \geq e_r$ if and only if its variable part $C' = \sigma^{-1}(C)$ has $e_{q_r}(C') \geq e_r$. This shows that the curves $C \in S'$ with $e_{q_r}(C) \geq e_r$ describe the linear system $S = \sigma(T)$ and that

$$\dim S' - \dim S = \frac{e_r(e_r + 1)}{2}. \qquad (1.16)$$

From the definitions, we have $L \cap S' = S$. Then, (1.16), elementary arguments of incidence of linear varieties and (1.15) give

$$\frac{e_r(e_r + 1)}{2} = \dim S' - \dim L \cap S' \leq \dim L' - \dim L \leq \frac{e_r(e_r + 1)}{2},$$

showing that equality holds in (1.15).   ◇

1.6   Analytic Curves and Analytic Maps

If $\mathbb{A}_2$ is an affine chart of a projective plane $\mathbb{P}_2$ (or just an affine plane), fix $p \in \mathbb{A}_2$ and let U be an open neighbourhood of p in $\mathbb{A}_2$. Once affine coordinates x, y for which $p = (0, 0)$ have been taken, we will consider the sets of the form

$$\gamma = \{(x, y) \in U | s(x, y) = 0\},$$

where s is a complex power series in x, y convergent in U; we will call them *analytic curves*, the series s then being called an *equation* of γ. Our considerations being local at p, these curves will be considered up to reduction of the neighbourhood in which they are defined. Being more formal, this means taking as equivalent two analytic curves γ and γ' when there is a small enough open neighbourhood V of p with $\gamma \cap V = \gamma' \cap V$, and dealing then with the resulting equivalence classes; these classes are called *germs of analytic curve at p* or just germs of curve if no confusion may arise.

We will also consider maps $\psi : W \to \mathbb{A}_n$ where W is an open neighbourhood of 0 in $\mathbb{C}$ and $\mathbb{A}_n$ is an affine chart of a projective space $\mathbb{P}_n$ – or just an affine space. When the affine coordinates of $\psi(t)$ are analytic functions on W we will say that ψ is an *analytic map*. If ψ is analytic, then there is a smaller neighbourhood $W' \subset W$ of 0 and power series $S_1, \ldots, S_n$, convergent in W', such that

$$\psi(t) = (S_1(t), \ldots, S_n(t))$$

for $t \in W$.

If $\mathbb{A}_n$ and $\mathbb{A}'_n$ are affine charts of a projective space $\mathbb{P}_n$, say the 0-th charts corresponding to projective coordinates $x_0, \ldots, x_n$ and $y_0, \ldots, y_n$, the affine coordinates $X_i = x_i/x_0$ and $Y_i = y_i/y_0$, $i = 1, \ldots, n$, of the points of $\mathbb{A}_n \cap \mathbb{A}'_n$ are related by the equalities

$$X_i = \frac{a_0^i + \sum_{j=1}^{n} a_j^i Y_j}{a_0^0 + \sum_{j=1}^{n} a_j^0 Y_j},$$

$i = 1, \ldots, n$, the coefficients a_i^j being the entries of the matrix changing coordinates $y_0, \ldots, y_n$ into coordinates $x_0, \ldots, x_n$. Since $x_0 = \sum_{j=0}^{n} a_j^0 y_j \neq 0$ for all points in $\mathbb{A}_n \cap \mathbb{A}'_n$, the right-hand sides of these equalities are rational – and hence analytic – functions on $\mathbb{A}_n \cap \mathbb{A}'_n$ and therefore it is clear that both the above definitions of analytic curve and analytic map are independent of the choices of the affine chart and the affine coordinates on it.

1.7   Exercises

1.1 Find the singular points of each of the curves of $\mathbb{P}_2$ defined by the equations below. Give the corresponding multiplicities and tangent cones.

(a) $x_0^2 x_1^2 + x_0^2 x_2^2 + x_2^4 = 0$,

(b) $x_0^4 - x_1^3 x_2 + x_0^2 x_1 x_2 = 0$,

(c) $x_0^2 x_2 + x_0 x_1^2 - 2x_0 x_2^2 - 2x_1^2 x_2 + x_2^3 = 0$.

1.2 Prove that the cubic of $\mathbb{P}_2$ $x_1^3 - x_0 x_2^2 - x_0^2 x_1 = 0$ is smooth. Use 1.1.3 to prove that an irreducible cubic of $\mathbb{P}_2$ has at most one singular point, and that the multiplicity of such a point, if it exists, is two. Give two examples of irreducible cubics having a double point, one with two different tangent lines at it, and the other with a single one (*nodal cubic* and *cuspidal cubic* respectively, see 3.5.3 and 3.5.4 for the names). Give two examples of reducible cubics, one with two different double points and the other with a single triple point.

1.3 Prove that all points of a multiple irreducible component W of a (projective or affine) hypersurface V are singular points of V of multiplicity at least the multiplicity of W in V.

1.4 Prove that if W is a (projective or affine) hypersurface, then the following conditions are equivalent:

(i) W is a cone with vertex p.

(ii) Any line pq, $q \in W$, $q \neq p$, is contained in W.

1.5 Let C be a curve of $\mathbb{P}_2$ of degree $d \geq 3$. Show that the following conditions are equivalent:

(i) C is irreducible and has a point of multiplicity $d - 1$.

(ii) Using suitable coordinates, C has equation $F_d + x_0 F_{d-1} = 0$, where $F_d, F_{d-1} \in \mathbb{C}[x_1 x_2]$ are homogeneous, coprime and have degrees d and $d - 1$, respectively.

1.6 Let C be a curve of $\mathbb{P}_2$ with equation $F_d + F_{d-r} x_0^r = 0$, where $F_d, F_{d-r} \in \mathbb{C}[x_1, x_2]$ are homogeneous of degrees d and $d - r$ respectively, and $d - 2 \geq r \geq 1$.

(1) Prove that $[1, 0, 0]$ is a singular point of C, of multiplicity $d - r$.

(2) Prove that the points $[0, a_1, a_2]$ for which $a_2 x_1 - a_1 x_2$ is a multiple factor of F_d, are singular points of C in the case when $r \geq 2$.

(3) Prove that if C is irreducible, then it has no singular point other than the above ones.

1.7 Consider the curve C of $\mathbb{P}_2$ defined by $x_1^d + x_0^{d-1} x_2 + x_0 x_2^{d-1} = 0$, $d \geq 2$.

(1) Prove that C is smooth, thus proving the existence of smooth curves of degree d in $\mathbb{P}_2$ for any $d > 0$.

(2) Describe the intersection of C and the line $x_2 = 0$.

(3) Prove that for any $d \geq 2$ there exist reducible curves of degree d in $\mathbb{P}_2$ which have a double point and no other singular point.

1.8 Let $L_1, \ldots, L_r$, $r > 1$, be simple linear conditions on the hypersurfaces of degree d of $\mathbb{P}_n$.

(1) Prove that $L_1, \ldots, L_r$ are dependent if and only if there is an i, $1 \leq i \leq r$, such that all the hypersurfaces satisfying L_j for all $j \neq i$, also satisfy L_i.

(2) Prove that $L_1, \ldots, L_r$ are independent if and only if for any i, $1 < i \leq r$, there is a hypersurface satisfying $L_1, \ldots, L_{i-1}$ and not satisfying L_i.

(3) Prove that different points $p_1, \ldots, p_5 \in \mathbb{P}_2$ impose independent conditions on the conics of $\mathbb{P}_2$ – or, equivalently, lie on a single conic – if and only if no four of them belong to the same line.

1.9 Assume given positive integers $e_1, \ldots, e_r$. Determine an integer k such that, for any distinct points $p_1, \ldots, p_r \in \mathbb{P}_2$ and any $d \geq k$, there is a curve of degree d of $\mathbb{P}_2$ having multiplicity at least e_i at p_i, $i = 1, \ldots, k$.

Proving a converse is more difficult. The following statement is known as the Nagata conjecture; it was conjectured by Nagata in 1959 and remains unproved at the time of writing (see [6], [7]):

For $r > 9$, if an integer d satisfies that for any distinct points $p_1, \ldots, p_r \in \mathbb{P}_2$ there is a curve of degree d having multiplicity at least e_i at p_i, $i = 1, \ldots, k$, then $d \geq (e_1 + \cdots + e_r)/\sqrt{r}$.

1.10 Extend 1.5.4, 1.5.7 and 1.5.8 to the n-dimensional case.

1.11 In a projective plane $\mathbb{P}_2$, with coordinates x_0, x_1, x_2, let

$$\mathcal{N} : \lambda_0 F_0 + \lambda_1 F_1 + \lambda_2 F_2 = 0$$

be a net of curves of degree d. Take $\mathbf{J}(\mathcal{N})$ – the *Jacobian locus* of $\mathcal{N}$ – to be the set of points of $\mathbb{P}_2$ at which the polynomial

$$\mathfrak{J}(\mathcal{N}) = \det\left(\frac{\partial F_i}{\partial x_j}\right)_{i,j=0,1,2}$$

is zero.

(1) Prove that $p \in \mathbf{J}(\mathcal{N})$ if and only if p is a singular point of at least one curve of $\mathcal{N}$. This in particular proves that $\mathbf{J}(\mathcal{N})$ is independent of the choices of the coordinates on $\mathbb{P}_2$ and the generators of $\mathcal{N}$.

(2) Let $\mathbb{A}_2$ be the affine chart $x_0 \neq 0$, $x = x_1/x_0$, $y = x_2/x_0$ and $f_i = F_i(1, x, y)$, $i = 0, 1, 2$. Prove that $\mathbf{J}(\mathcal{N}) \cap \mathbb{A}_2$ is the locus of zeros of the polynomial

$$\begin{vmatrix} f_0 & f_1 & f_2 \\ \frac{\partial f_0}{\partial x} & \frac{\partial f_1}{\partial x} & \frac{\partial f_2}{\partial x} \\ \frac{\partial f_0}{\partial y} & \frac{\partial f_1}{\partial y} & \frac{\partial f_2}{\partial y} \end{vmatrix}.$$

(3) Take

$$\mathcal{N}' : \lambda_0 x_0^2 + \lambda_1 x_0 x_1 + \lambda_2 x_1^2 = 0$$

and prove that $\mathbf{J}(\mathcal{N}') = \mathbb{P}_2$.

(4) When $\mathbf{J}(\mathcal{N}) \neq \mathbb{P}_2$ (or, equivalently, $\mathfrak{J}(\mathcal{N}) \neq 0$) the Jacobian locus $\mathbf{J}(\mathcal{N})$ is the set of points of the curve $J(\mathcal{N}) : \mathfrak{J}(\mathcal{N}) = 0$, which is called the *Jacobian curve* of $\mathcal{N}$. Prove that the Jacobian curve is independent of the choices of the coordinates on $\mathbb{P}_2$ and the generators of $\mathcal{N}$. (More on the Jacobian curve is in Exercise 4.14.)

1.12 The dual plane. Assume we have fixed homogeneous coordinates x_0, x_1, x_2 on a projective plane $\mathbb{P}_2$. The set of all lines of $\mathbb{P}_2$ is turned into a projective plane $\mathbb{P}_2^{\vee}$ (*dual plane* or *plane of lines*) by taking as homogeneous coordinates of a line (*line coordinates*) the coefficients of any of its equations.

(1) Prove that if new coordinates y_0, y_1, y_2 are taken on $\mathbb{P}_2$, related to the old ones by $(y_0, y_1, y_2)^t = A(x_0, x_1, x_2)^t$, A a regular 3×3 matrix, then the new line coordinates v_0, v_1, v_2 are related to the old ones u_0, u_1, u_2 by the rule $(u_0, u_1, u_2)^t = A^t(v_0, v_1, v_2)^t$. This shows that the projective structure of $\mathbb{P}_2^{\vee}$ does not depend on the choice of the coordinates.

(2) Prove that the lines of $\mathbb{P}_2^{\vee}$ are the pencils of lines of $\mathbb{P}_2$, namely the families of lines $p^* = \{L \in \mathbb{P}_2^{\vee} \mid p \in L\}$ for $p \in \mathbb{P}_2$.

(3) Prove that the line coordinates of p^* (as a line of $\mathbb{P}_2^{\vee}$) are the coordinates of p. This allows us to identify the bidual plane $\mathbb{P}_2^{\vee\vee}$ with $\mathbb{P}_2$ by identifying p^* with p.

Chapter 2

Local Properties of Plane Curves

2.1 Power Series

Let $f \in \mathbb{C}[x, y]$. Our first aim in this chapter is to find some sort of power series $s = s(x)$, in the variable x, such that $f(x, s(x))$ is identically zero in x; in other words s is a root of f viewed as a polynomial in y with coefficients in $\mathbb{C}[x]$. Actually, we will look for series s subjected to the further condition $s(0) = 0$. These series, once proved to be convergent, will provide a parametric representation of the affine curve $C : f(x, y) = 0$ locally at $(0, 0)$, in the sense that the points of C in a neighbourhood of $(0, 0)$ are those and only those of the form $(x, s(x))$, for s one of the series above and $|x|$ small enough. As the reader may have noted, the implicit function theorem gives a direct solution of our problem in the case when $(\partial f / \partial y)_{(0,0)} \neq 0$; nevertheless, we will make no assumption on the y-derivative of f and therefore what we will get is, for polynomials in two variables, an extension of the implicit function theorem.

Since at the first stages we will not be concerned with convergence, we will consider *formal power series* in x, namely expressions of the form

$$\sum_{i \geq 0} a_i x^i$$

where the a_i are complex numbers and x is a free variable. These series, with the usual sum and product, compose an integral ring we will denote, as already said, by $\mathbb{C}[[x]]$: the *ring of formal complex power series* in the variable x. Among these series, those which are convergent (that is, have non-zero radius of convergence) compose a subring $\mathbb{C}\{x\} \subset \mathbb{C}[[x]]$, called the ring of *convergent power series* in x.

Actually, to solve our problem we need to consider series of a more general type, called *fractionary power series*. Next we sketch the main facts we need about them. For a more formal introduction to fractionary power series, the

© Springer Nature Switzerland AG 2019
E. Casas-Alvero, *Algebraic Curves, the Brill and Noether Way*, Universitext,
https://doi.org/10.1007/978-3-030-29016-0_2

reader may see [3, Section 1.2]. *Fractionary power series in the variable x are complex formal power series in a fixed root of x*, namely elements of a ring $\mathbb{C}[[x^{1/n}]]$, n a positive integer, written in the form

$$\sum_{i\geq 0} a_i (x^{1/n})^i = \sum_{i\geq 0} a_i x^{i/n},$$

where the a_i are complex numbers. They are subjected to the convention

$$\sum_{i\geq 0} a_i x^{i/n} = \sum_{i\geq 0} a_i x^{ri/rn} \tag{2.1}$$

for any positive integer r.

Note that, by the definition, all the exponents effectively appearing in a fractionary power series may be written with the same denominator. If cancellations are made, they may appear with different denominators but in any case there is a multiple common to all denominators. Therefore, a series such as $\sum_{i\geq 0} x^{1/2^i}$ is not a fractionary power series according to our definition, Note also that, for any positive integer n, through the identification (2.1), $\mathbb{C}[[x]]$ appears as a subring of $\mathbb{C}[[x^{1/n}]]$; the elements of the former are sometimes called *integral power series* to avoid confusions.

Using (2.1) any two (or finitely many) fractionary power series s, s' may be written with the same common denominator in their exponents and hence viewed as elements of the same $\mathbb{C}[[x^{1/n}]]$. After this the sum $s + s'$ and the product ss' are defined and do not depend on the choice of the common denominator n, because the elements identified by the rule (2.1) are those corresponding by injective ring-homomorphisms

$$\mathbb{C}[[x^{1/n}]] \to \mathbb{C}[[x^{1/rn}]].$$

As a consequence, the set of all fractionary power series compose a domain; its fraction field is usually denoted $\mathbb{C}\langle\langle x \rangle\rangle$.

The *order in x* of a fractionary power series $s \neq 0$, $o_x s$, is the least degree in x of the monomials effectively appearing in s; it is a non-negative rational number. If $s = 0$ we take $o_x s = \infty$.

Remark 2.1.1 As is clear, for any fractionary power series s, s',

$$o_x(ss') = o_x s + o_x s' \quad \text{and} \quad o_x(s + s') \geq \min(o_x s, o_x s').$$

Furthermore, the inequality is an equality in the case when $o_x s \neq o_x s'$.

The minimal common denominator of all the exponents effectively appearing in a fractionary power series s is called its *polydromy order*, denoted $\nu(s)$ in the sequel.

A fractionary power series $s = \sum_{i\geq 0} a_i x^{i/n}$ is called *convergent* when so is the integral power series $\sum_{i\geq 0} a_i t^i$. If the latter has radius of convergence ρ, then for any complex number z with $|z| < n\rho$ and any n-th root $z^{1/n}$ of z the numerical series $\sum_{i\geq 0} a_i z^{i/n}$ is convergent. If its sum is written $s(z)$,

s defines in general a multi-valued complex function $z \mapsto s(z)$, defined in a neighbourhood of 0 in $\mathbb{C}$. Convergent fractionary power series obviously form a subring of the ring of formal fractionary power series.

For a fixed positive integer n and an n-th root of unity ε, it is straightforward to check that the map

$$\sigma_\varepsilon : \mathbb{C}[[x^{1/n}]] \longrightarrow \mathbb{C}[[x^{1/n}]]$$
$$\sum_{i \geq 0} a_i x^{i/n} \longmapsto \sum_{i \geq 0} a_i \varepsilon^i x^{i/n}$$

is a ring-automorphism that leaves invariant all the integral power series. It is also clear that $\sigma_\varepsilon \sigma_{\varepsilon'} = \sigma_{\varepsilon\varepsilon'}$ for any two n-th roots of unity $\varepsilon, \varepsilon'$, $\sigma_{\varepsilon^{-1}} = \sigma_\varepsilon^{-1}$ and $\sigma_1 = \mathrm{Id}$. The automorphisms σ_ε are called *conjugation automorphisms*. If $s \in \mathbb{C}[[x^{1/n}]]$, its images under the conjugation automorphisms

$$\sigma_\varepsilon(s), \quad \varepsilon \in \mathbb{C}, \quad \varepsilon^n = 1,$$

are called the *conjugates* of s. A relevant fact is

Lemma 2.1.2 *The set of conjugates of a fractionary power series s does not depend on the ring $\mathbb{C}[[x^{1/n}]]$, with $s \in \mathbb{C}[[x^{1/n}]]$, used to define them.*

PROOF: Assume s to be $\sum_{i \geq 0} a_i x^{i/\nu}$, where $\nu = \nu(s)$ is the polydromy order of s. Using $\mathbb{C}[[x^{1/\nu}]]$, the conjugates of s are

$$\sum_{i \geq 0} a_i \varepsilon^i x^{i/\nu}, \quad \varepsilon \in \mathbb{C}, \quad \varepsilon^\nu = 1.$$

If $s \in \mathbb{C}[[x^{1/n}]]$, then $n = r\nu$ for a positive integer r and, using this time $\mathbb{C}[[x^{1/n}]]$, the conjugates of s are

$$\sum_{i \geq 0} a_i \eta^{ri} x^{ri/r\nu}, \quad \eta \in \mathbb{C}, \quad \eta^{r\nu} = 1.$$

Since while η describes the set of the $(r\nu)$-th roots of unity, η^r describes the set of the ν-th roots of unity, the claim follows. ◇

Proposition 2.1.3 *The number of different conjugates of a fractionary power series s is its polydromy order $\nu(s)$.*

PROOF: As above take $\nu = \nu(s)$ and write $s = \sum_{i \geq 0} a_i x^{i/\nu}$. By the definition of $\nu(s)$, there are indices $i_1, \ldots, i_k$ such that $a_{i_j} \neq 0$, $j = 1, \ldots, k$, and $\gcd(\nu, i_1, \ldots, i_k) = 1$. Then, for suitable integers $c_0, \ldots, c_k$, we have

$$c_0 \nu + c_1 i_1 + \cdots + c_k i_k = 1. \tag{2.2}$$

On the other hand, the number of ν-th roots of unity being ν, assume we have

$$\sum_{i \geq 0} a_i \varepsilon^i x^{i/\nu} = \sum_{i \geq 0} a_i \eta^i x^{i/\nu},$$

with $\varepsilon^\nu = \eta^\nu = 1$. Then

$$\sum_{i \geq 0} a_i(\varepsilon^i - \eta^i)x^{i/\nu} = 0$$

and so, for every i, $a_i(\varepsilon^i - \eta^i) = 0$; in particular $\varepsilon^{i_j} = \eta^{i_j}$ for $j = 1, \ldots, k$. Using (2.2) gives $\varepsilon = \eta$ and hence the claim.   ◇

A direct consequence of 2.1.2 and 2.1.3 is:

Corollary 2.1.4 *A series* $s \in \mathbb{C}[[x^{1/n}]]$ *belongs to* $\mathbb{C}[[x]]$ *if and only if* $\sigma_\varepsilon(s) = s$ *for every* n-*th root of unity* ε.

2.2 The Newton–Puiseux Algorithm

Assume given $f \in \mathbb{C}[x, y]$. In this section we present the *Newton–Puiseux algorithm*, which, under a suitable hypothesis on f, gives a constructive proof of the existence of a fractionary power series s in x such that $f(x, s) = 0$ and $s(0) = 0$; s will thus be a root of f if f is taken as a polynomial in y with coefficients in $\mathbb{C}[x]$, which we will call a y-*root* of f in the sequel. Actually, we will place ourselves in a slightly more general setting by assuming that f is a polynomial in y with coefficients formal power series in x, namely $f \in \mathbb{C}[[x]][y]$, as this will be useful in the next section. In fact the Newton–Puiseux algorithm works even in the more general situation in which f is a convergent power series in x, y (see [3, Chap. 1]). The algorithm was developed and intensively used by Newton in the second half of the seventeenth century; proofs were given by Puiseux in 1850.

Thus take

$$f = \sum_{\alpha \geq 0, \, \beta = 0, \ldots, r} A_{\alpha, \beta} x^\alpha y^\beta \in \mathbb{C}[[x]][y].$$

On the real plane $\mathbb{R}^2$, plot the points (α, β) for which $A_{\alpha, \beta} \neq 0$, thus getting the set

$$\Delta(f) = \{(\alpha, \beta) \in \mathbb{R}^2 \mid A_{\alpha, \beta} \neq 0\},$$

called the *Newton diagram* of f.

The boundary of the convex envelope of Δ is called the *Newton polygon* $\mathbf{N}(f)$ of f. It is a broken line, a polygon in the case when $f \in \mathbb{C}[x, y]$. Since we are interested in just a part of it, we will take

$$\Delta'(f) = \Delta(f) + (\mathbb{R}^+)^2,$$

$\mathbb{R}^+$ being the set of non-negative real numbers. Then the boundary of the convex envelope of $\Delta'(f)$ is composed of two half-lines, each parallel to one of the coordinate axes, and a polygonal line – maybe reduced to a single point – joining their ends. This polygonal line will be called the *local Newton polygon of* f, and denoted $N(f)$.

In the sequel we will conventionally assume that the second axis of $\mathbb{R}^2$ is vertical, oriented upwards. We will take $N(f)$ oriented so that its first vertex

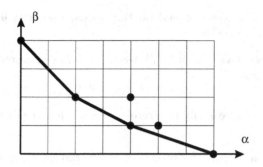

Figure 2.1: $\Delta(f)$ and $N(f)$ for $f = y^4 - x^2y^2 - 2x^4y^2 + x^4y + x^5y + x^7$.

is the one with highest second coordinate, while its last vertex is the one with highest first coordinate. When its last vertex belongs to the α-axis, we will say that $N(f)$ *ends on the axis.* As the reader may easily check, this occurs if and only if f has no factor y. Similarly, f has no factor x if and only if the first vertex of $N(f)$ lies on the β-axis.

The ordinate of the first vertex of $N(f)$ is called the *height* of $N(f)$, denoted $h(N(f))$. Also, for any side Γ of $N(f)$, the difference between the ordinates of the first and last vertices of Γ is called the *height* of Γ. The proof of the next lemma is direct:

Lemma 2.2.1 $h(N(fg)) = h(N(f)) + h(N(g))$.

Take z to be a free variable. If Γ is a side of $N(f)$ whose last end is (α_0, β_0), then

$$F_\Gamma = \sum_{(\alpha,\beta)\in\Gamma} A_{\alpha,\beta} z^{\beta-\beta_0}$$

will be called the *equation associated to* Γ. Clearly F_Γ is a non-zero polynomial whose degree equals the height of Γ. Note also that $F_\Gamma(0) = A_{\alpha_0,\beta_0} \neq 0$ and therefore F_Γ has no zero root. We will need:

Lemma 2.2.2 *If Γ has slope $-n/m$, $\gcd(n,m) = 1$, then $F_\Gamma \in \mathbb{C}[z^n]$.*

PROOF: If (α_0, β_0) is the last vertex of Γ, then any integral point on Γ has the form $(\alpha_0 - km, \beta_0 + kn)$, for a non-negative integer k, and therefore the corresponding monomial of F_Γ has degree kn, as claimed.   ◇

Remark 2.2.3 If $h(N(f)) = 0$, then $f = x^\ell f'$ with $f'(0,0) \neq 0$ and no fractional power series s with $s(0) = 0$ satisfies $f(x,s) = 0$; otherwise the last equality would give $f'(x,s) = 0$ and so $f'(0,0) = f'(0,s(0)) = 0$.

Thus assume that $f \in \mathbb{C}[x,y]$ has $h(N(f)) > 0$; then either $N(f)$ does not end on the α-axis or it has at least one side. Accordingly, the 0-th step of the *Newton–Puiseux algorithm* for a y-root s of f with $s(0) = 0$ is performed by either

step 0.a: if $N(f)$ does not end on the α-axis, take $s = 0$ and end the
   algorithm at this point, or

step 0.b: if there is a side Γ of $N(f)$, assuming it has equation $n\alpha + m\beta = k$,
   $(n, m) = 1$, take
   $$s = x^{m/n}(a + s^{(1)})$$
   where a is a – necessarily non-zero – root of F_Γ and $s^{(1)}$ is to be determined.

If step 0.b has been performed, take new variables x_1, y_1 related to x, y
by the rules
$$x = x_1^n,$$
$$y = x_1^m(a + y_1).$$

Then, since all $(\alpha, \beta) \in \Delta(f)$ are on or above the line spanned by Γ, all
have $n\alpha + m\beta \geq k$ and so

$$\bar{f} = f(x_1^n, x_1^m(a + y_1)) = \sum_{(\alpha,\beta)\in\Delta(f)} A_{\alpha,\beta} x_1^{n\alpha+m\beta}(a + y_1)^\beta$$

$$= x_1^k \sum_{(\alpha,\beta)\in\Delta(f)} A_{\alpha,\beta} x_1^{n\alpha+m\beta-k}(a + y_1)^\beta$$

$$= x_1^k f_1$$

with

$$f_1 = \sum_{(\alpha,\beta)\in\Delta(f)} A_{\alpha,\beta} x_1^{n\alpha+m\beta-k}(a + y_1)^\beta$$

$$= \sum_{(\alpha,\beta)\in\Gamma} A_{\alpha,\beta}(a + y_1)^\beta + \sum_{\substack{(\alpha,\beta)\in\Delta(f)\\n\alpha+m\beta>k}} A_{\alpha,\beta} x_1^{n\alpha+m\beta-k}(a + y_1)^\beta \in \mathbb{C}[[x_1]][y_1].$$

$$(2.3)$$

The above equalities directly give

$$f(x_1^n, s(x_1^n)) = \bar{f}(x_1, s^{(1)}) = x_1^k f_1(x_1, s^{(1)}),$$ (2.4)

which is in fact the reason behind the definition of x_1, y_1.

The following lemma will be useful next, and also later on:

Lemma 2.2.4 *The height of $N(f_1)$ is the multiplicity of a as a root of F_Γ.*

PROOF: By the definition F_Γ and (2.3) above,

$$f_1(0, y_1) = (a + y_1)^{\beta_0} F_\Gamma(a + y_1),$$

where β_0 is still the ordinate of the last end of Γ. In particular $f_1(0, y_1) \neq 0$,
which ensures that the first vertex of $N(f_1)$ lies on the second axis and
therefore $h(N(f_1))$ equals the multiplicity of 0 as a root of $f_1(0, y_1)$. Since

$a \neq 0$, by the equality above, the latter equals the multiplicity of a as a root of F_Γ, hence the claim. ◇

In particular, $h(N(f_1)) > 0$ and the procedure may be repeated from x_1, y_1 and f_1. Inductively, if no step j.a has already been performed, at step i, $i > 0$ either

step i.a: if $N(f_i)$ does not end on the α-axis, take $s_i = 0$ and end the algorithm at this point, or

step i.b: if there is a side Γ_i of $N(f_i)$, assuming it has equation $n_i\alpha + m_i\beta = k_i$, $(n_i, m_i) = 1$, take

$$s^{(i)} = x_i^{m_i/n_i}(a_i + s^{(i+1)})$$

where a_i is a – necessarily non-zero – root of F_{Γ_i} and s_{i+1} is to be determined.

If step i.b has been performed, then still

$$f_i(x_i, s^{(i)}) = x_{i+1}^{k_i} f_{i+1}(x_{i+1}, s^{(i+1)}) \tag{2.5}$$

and

$$\begin{aligned}
s &= x^{m/n}(a + x_1^{m_1/n_1}(a_1 + \cdots + x_i^{m_i/n_i}(a_i + s^{(i+1)})\ldots)) \\
&= x^{m/n}(a + x^{m_1/nn_1}(a_1 + \cdots + x^{m_i/n\ldots n_i}(a_i + s^{(i+1)})\ldots)).
\end{aligned} \tag{2.6}$$

If step i.a has been performed, then

$$\begin{aligned}
s &= x^{m/n}(a + x_1^{m_1/n_1}(a_1 + \cdots + x_{i-1}^{m_{i-1}/n_{i-1}}a_{i-1})\ldots) \\
&= x^{m/n}(a + x^{m_1/nn_1}(a_1 + \cdots + x^{m_{i-1}/n\ldots n_{i-1}}a_{i-1})\ldots).
\end{aligned} \tag{2.7}$$

Thus, if for some $i \geq 0$ the step i.a is performed, then s has i summands and is determined by the equality (2.7). Otherwise, the equalities (2.6), varying i, determine all summands composing s.

Now we are going to see that s, determined as above, satisfies our requirements. First of all, we need to show that s is a fractionary power series: this is clear from equality (2.7) if one step i.a is performed, but otherwise the denominators appearing in the equality (2.6) could be unbounded. The next lemma shows that this is never the case and therefore settles the problem:

Lemma 2.2.5 *If no step a is performed, then $n_i = 1$ for all but finitely many indices i.*

PROOF: By 2.2.2, $F_\Gamma = P(z^n)$. After factoring P into linear factors, it is clear that F_Γ has roots εa, $\varepsilon^n = 1$, all with the same multiplicity. This multiplicity has been seen to be $h(N(f_1))$ in Lemma 2.2.4, therefore $\deg F_\Gamma \geq nh(N(f_1))$. On the other hand it is clear that $\deg F_\Gamma$, which is the height of Γ, is at most $h(N(f))$. It follows that $h(N(f)) \geq nh(N(f_1))$. Repeatedly using this gives:

$$h(N(f)) \geq nh(N(f_1)) \geq nn_1 h(N(f_2)) \geq \cdots \geq nn_1\ldots n_i h(N(f_{i+1}))$$

for all i, and therefore the claim. ◇

Remark 2.2.6 In the proof of 2.2.5 we used the fact that the number of n-th roots of unity is n. Therefore the same argument would not apply if a field of positive characteristic were taken instead of $\mathbb{C}$; the fact is that Puiseux's theorem below does not hold in positive characteristic, see Exercise 2.7.

The next theorem completes our job in this section.

Theorem 2.2.7 (Puiseux, 1850) *If $f \in \mathbb{C}[[x]][y]$ has $h(N(f)) > 0$, then the Newton–Puiseux algorithm applied to f produces a fractionary power series s that satisfies $s(0) = 0$ and $f(x,s) = 0$.*

PROOF: We have seen above that any output s of the Newton–Puiseux theorem is a fractionary power series, and it is also clear that $s(0) = 0$. For the remaining claim, assume we have performed steps 0.b,..., i.b, $i \geq -1$. Then the equalities (2.4) and (2.5) give

$$f(x,s) = x_1^k x_2^{k_1} \ldots x_{i+1}^{k_i} f_{i+1}(x_{i+1}, s^{(i+1)}). \tag{2.8}$$

If step $(i+1)$.a is then performed, $y_{i+1}|f_{i+1}$ and $s^{(i+1)}$ is taken equal to 0, after which $f(x,s) = 0$, as wanted.

If, otherwise, no step a is ever performed, the equality (2.8) does hold for all i. By taking orders in x we get, for all $i > 0$,

$$o_x f(x,s) = \frac{k}{n} + \frac{k_1}{nn_1} + \cdots + \frac{k_i}{nn_1 \ldots n_i} + o_x f_{i+1}(x_{i+1}, s^{(i+1)}). \tag{2.9}$$

By 2.2.5, $n_i = 1$ for all but finitely many i. Then the right-hand sides of the equalities (2.9) are unbounded, which gives $o_x f(x,s) = \infty$ and therefore, also in this case, $f(x,s) = 0$ as claimed. ◇

Remark 2.2.8 There may be different choices at each step of the Newton–Puiseux algorithm, not only because steps a and b may be both possible, but also because when performing a step b, there may be many choices of both the side Γ of $N(f_i)$ and the root a of F_Γ. We have seen above that all these choices give fractionary power series s that are y-roots of f and have $s(0) = 0$. We will see in 2.3.7 below that there is no other such root.

2.3 Further y-roots. Convergence

Once we know how to get a y root of f, getting all y-roots of f – still with $s(0) = 0$ – is an easy task. Assume that s is a fractionary power series and has polydromy order $\nu(s) = \nu$. Still denote by σ_ε the conjugation automorphisms of $\mathbb{C}[[x^{1/\nu}]]$ (see Section 2.1). We will make use of the next lemma; its easy proof is left to the reader:

Lemma 2.3.1 *If s is a y-root of $P \in \mathbb{C}[[x]][y]$, then all the conjugates of s, $\sigma_\varepsilon(s)$, $\varepsilon^\nu = 1$, are y-roots of P too.*

Take

$$g_s = \prod_{\varepsilon^\nu = 1} (y - \sigma_\varepsilon(s)).$$

We have:

Lemma 2.3.2 *The polynomial g_s belongs to $\mathbb{C}[[x]][y]$ and is irreducible there.*

PROOF: Since the coefficients of g_s, as a polynomial in y, are symmetric functions of its roots $\sigma_\varepsilon(s)$, $\varepsilon^\nu = 1$, they are invariant under all the conjugation automorphisms and therefore, by 2.1.4, they belong to $\mathbb{C}[[x]]$. For the irreducibility, assume that $P \in \mathbb{C}[[x]][y]$ is a divisor of g_s that has the root s. Then, by 2.3.1, $P(\sigma_\varepsilon(s)) = 0$ for all ε. The ν conjugates $\sigma_\varepsilon(s)$ of s being pairwise different by 2.1.3, P has at least degree ν and therefore cannot be a proper divisor of g_s.   ◇

The next lemma gives an alternative expression of g_s that will be useful in the next Section 4.

Lemma 2.3.3 *If $s = ax^{m/n} + \ldots$, where $a \neq 0$, n, m are coprime and the dots mean terms of higher order, and $\nu = \nu(s)$, then*

$$g_s = (y^n - a^n x^m)^{\nu/n} + \sum_{ni + mj > \nu m} a_{i,j} x^i y^j$$

for some $a_{i,j} \in \mathbb{C}$.

PROOF: This follows by direct computation from the definition of g_s.   ◇

Here is a more complete version of Puiseux's theorem:

Theorem 2.3.4 (Puiseux's theorem, second version) *If $f \in \mathbb{C}[[x]][y]$ has $h(N(f)) > 0$, then there are fractionary power series in x, $s_1, \ldots, s_k$, $k > 0$, such that $s_i(0) = 0$, $i = 1, \ldots, k$ and*

$$f = x^\ell g_{s_1} \ldots g_{s_k} \tilde{f} \tag{2.10}$$

with $\ell \geq 0$, $\tilde{f} \in \mathbb{C}[[x]][y]$ and $\tilde{f}(0,0) \neq 0$.

PROOF: By 2.2.7, take a fractionary power series s_1 which is a y-root of f and $\nu_1 = \nu(s_1)$. The Euclidean division

$$f = g_{s_1} f' + R$$

may be performed in $\mathbb{C}[[x]][y]$ because g_{s_1} is monic. By 2.3.1, the conjugates $\sigma_\varepsilon(s_1)$, $\varepsilon^{\nu_1} = 1$, are y-roots of f and, clearly, also of g_{s_1}. The number of these conjugates being $\nu_1 = \deg g_{s_1} > \deg R$, we have $R = 0$ and so $f = g_{s_1} f'$. By 2.2.1, $h(N(f')) = h(N(f)) - h(N(g_{s_1})) = h(N(f)) - \nu_1$.

Now, if $h(N(f')) = 0$, f' has the form $f' = x^\ell \tilde{f}$, with $\tilde{f}(0,0) \neq 0$, and we are done. Otherwise f' has a y-root and we repeat the procedure from f'. Since the heights of the local Newton polygons are strictly decreasing, after finitely many factorizations we will eventually get a last factor $x^\ell \tilde{f}$ with $\tilde{f}(0,0) \neq 0$ and the proof is complete.   ◇

Remark 2.3.5 Since $x^\ell \tilde{f}$ has no y-roots s with $s(0) = 0$ (see 2.2.3), the y-roots s of f with $s(0) = 0$ are $s_1, \ldots, s_k$ and their conjugates. Each s_j and its conjugates have the same multiplicity, which equals the number of times that the factor g_{s_j} appears in (2.10).

Remark 2.3.6 Since $h(N(g_{s_j})) = \nu_j$, the number of y-roots s of f with $s(0) = 0$, counted with their multiplicities, equals $\sum_1^k h(N(g_{s_i})) = h(N(f))$, as obviously $h(N(x^\ell \tilde{f})) = 0$.

Actually, all the roots found above result from the Newton–Puiseux algorithm:

Proposition 2.3.7 *If $f \in \mathbb{C}[[x]][y]$ and a fractionary power series s, with $s(0) = 0$, is a y-root of f, then the Newton–Puiseux algorithm applied to f gives rise to s, provided suitable choices are made.*

PROOF: If $s = 0$ the claim is obvious. Otherwise assume $s = ax^{m/n} + \ldots$, where $a \neq 0$ and the dots indicate terms of higher degree. If $k = \min\{n\alpha + m\beta \mid (\alpha, \beta) \in \Delta(f)\}$, direct substitution into f gives

$$f(x, s) = \left(\sum_{n\alpha + m\beta = k} A_{\alpha, \beta} a^\beta \right) x^{k/n} + \ldots$$

where the dots still indicate terms of higher degree. Since the initial term exhibited above needs to be zero and $a \neq 0$, there should be at least two different non-zero $A_{\alpha, \beta}$ with $n\alpha + m\beta = k$, and so $N(f)$ has a side Γ on the line $n\alpha + m\beta = k$. This set, the annulation of the initial term above is equivalent to $F_\Gamma(a) = 0$. So the initial term of s arises from the 0.b step of the Newton–Puiseux algorithm with the choices of Γ and a. After this, take $s^{(1)}$, x_1, y_1 and f_1 as indicated by that 0.b step; then $f_1(x_1, s^{(1)}) = 0$, and the argument may be repeated.   ◇

Next we give a couple of results about the multiplicities of the y-roots. Still assume $f \in \mathbb{C}[[x]][y]$ and let $s \neq 0$, $s(0) = 0$, be a y-root of f. The notations being as above, let f_1 and $s^{(1)}$ arise from the 0.b step of the Newton–Puiseux algorithm for s. Then

Lemma 2.3.8 *The multiplicity of $s \neq 0$ as a root of f equals the multiplicity of $s^{(1)}$ as a root of f_1.*

PROOF: Assume $f = (y - s)^\mu f'$ with $f'(x, s) \neq 0$. Then

$$x_1^k f_1 = x_1^{\mu m}(y_1 - s^{(1)})^\mu f'(x_1^n, x_1^m(a + y_1))$$

and

$$f'(x_1^n, x_1^m(a + s^{(1)})) = f'(x, s) \neq 0,$$

hence the claim   ◇

Remark 2.3.9 Of course 2.3.8 may be applied at successive steps 1.b,...,$(i-1)$.b of the Newton–Puiseux algorithm, showing that the multiplicities of s and $s^{(i)}$, as roots of f and f_i respectively, are equal.

If in particular s is a simple y-root and the algorithm for it does not end at a step i.a, the $s^{(i)}$, $i \geq 0$, are simple y_i-roots of the corresponding f_i. Assume furthermore that i is taken such that no y-root of f different from s has the same first i monomials as s. Then, since different y_i-roots of f_i would give rise to different y-roots of f with the same first i monomials, $s^{(i)}$ is the only y_i-root of f_i, and further we already know it to be simple. By 2.3.6, $h(N(f_i)) = 1$, and so:

Lemma 2.3.10 *Assume that the Newton–Puiseux algorithm for a simple y-root s of f has no step j.a: then there is an i for which $h(N(f_i)) = 1$. Conversely if, for a certain i, the Newton–Puiseux algorithm gives $h(N(f_i)) = 1$, then the corresponding y-root of f is simple.*

PROOF: The first claim has been proved above. The second claim is straightforward from 2.3.6 and 2.3.9.   ◇

Often the computations for a y-root s of f are not continued once a local Newton polygon of height one appears, as then enough terms of s to distinguish it from all other y-roots of f have been computed. The y-root s is then said to have been *separated* from the other y-roots. This in particular implies that the terms computed so far are shared by no conjugate of s other than s itself, and hence that the minimal common denominator of their exponents is the polydromy order $\nu(s)$ of s. By 2.3.10, the above applies only to simple y-roots, and it is guaranteed to occur if the y-root is already known to be simple and no step i.a is performed.

Regarding convergence, we are now able to prove:

Theorem 2.3.11 *If $f \in \mathbb{C}\{x\}[y]$ – in particular if f is a polynomial – then all the y-roots s of f with $s(0) = 0$ are convergent.*

PROOF: Assume that s, $s(0) = 0$, is a y-root of f of multiplicity e. Then s is a y-root of $\partial^{e-1} f / \partial y^{e-1}$ of multiplicity 1. Since f is convergent, so is $\partial^{e-1} f / \partial y^{e-1}$. We may thus replace f with $\partial^{e-1} f / \partial y^{e-1}$ and assume without restriction that s is a simple root of f. The claim is obvious if the Newton–Puiseux algorithm for s involves a step i.a, as then s has finitely many monomials. Otherwise, by 2.3.11, there is an i for which $h(N(f_i)) = 1$. Then on one hand f_i has as only y_i-root $s^{(i)}$ while, on the other, $(\partial f_i / \partial y)_{(0,0)} \neq 0$. By the implicit function theorem, there is an integral convergent power series which is a y_i-root of f_i; since such a y_i-root needs to be $s^{(i)}$, the series $s^{(i)}$ is convergent. After this, s, related to $s^{(i)}$ by the equality (2.6), is convergent too.   ◇

2.4   Branches of a Plane Curve

In the first half of this section we will define the branches of an algebraic curve of $\mathbb{P}_2$ at a point and some objects associated to them, using an affine chart of $\mathbb{P}_2$ and coordinates on it. Later on, in 2.4.9, 2.5.4 and 2.6.3, we will see that neither the branches nor those associated objects depend on the choice of the affine chart and the coordinates. Our considerations being local, with obvious changes they apply to affine curves as well.

Let C be a curve of $\mathbb{P}_2$ and O a point of C. In the sequel we assume to have fixed an affine chart $\mathbb{A}_2$ of $\mathbb{P}_2$ containing O, with affine coordinates x, y, and that the corresponding affine part C_0 of C has equation $f = 0$. The coordinates are further assumed to be taken such that $O = (0,0)$ and the y-axis, $x = 0$, is not contained in C; then (2.2.3) $h(N(f)) > 0$ and x does not divide f. The y-roots s of f with $s(0) = 0$ are called the *Puiseux series of C* – or C_0 – at O. Since two conjugate power series carry the same information, it is usual to list just one Puiseux series per conjugacy class. If these are $s_1, \ldots, s_k$, $k > 0$, say with $s_j(x) = \sum_{i>0} a_i^j x^{i/\nu_j}$, $\nu_j = \nu(s_j)$, the factorization given by Puiseux's theorem 2.3.4 reads

$$f = g_{s_1} \cdots g_{s_k} \tilde{f}, \quad g_{s_i} = \prod_{\varepsilon^{\nu_i}=1} (y - \sigma_\varepsilon(s_i)). \tag{2.11}$$

The fact that $\tilde{f}(0,0) \neq 0$ causes the factor $\tilde{f}$ to have no zero in a neighbourhood of O, and hence to be irrelevant locally at O. Each factor g_{s_j}, $j = 1, \ldots, k$, is convergent by 2.3.11 and therefore defines an analytic – in general non-algebraic – curve $\gamma_j = \{(x,y) \in U \mid g_{s_j}(x,y)\} = 0$ in a suitable neighbourhood U of O. The germs of these analytic curves (i.e. their classes modulo reduction of the neighbourhood in which they are defined, see Section 1.6) are called the *branches of C* – or C_0 – at O. The point O is often referred to as being the *origin* of each of these branches.

Remark 2.4.1 In view of the definition of the g_{s_j}, each s_j gives a parameterization of the corresponding γ_j

$$\begin{cases} x & = x \\ y & = s_j(x) = \sum_{i>0} a_i^j x^{i/\nu_j} \end{cases}, \tag{2.12}$$

in the sense that the points of γ_j in a neighbourhood of O are those and only those of the form $(x, s_j(x))$ for $x \in \mathbb{C}$ and $|x|$ small enough.

To avoid fractionary powers, a ν-th root t of x is often taken as new parameter, after which (2.12) is turned into

$$\begin{cases} x & = t^{\nu_j} \\ y & = s_j(t^{\nu_j}) = \sum_{i>0} a_i^j t^i. \end{cases} \tag{2.13}$$

Both (2.12) and (2.13) are referred to as Puiseux parameterizations of γ_j. Obviously both the branches and their parameterizations are unaffected by a different choice of the s_j, $j = 1, \ldots, k$, within their conjugacy classes, but for a change of parameter $t \to \varepsilon t$, $\varepsilon^{\nu_j} = 1$, in (2.13). It is convenient to define the set of branches of a curve C at a point $O \notin C$ as empty.

The notations being as above, we have:

Lemma 2.4.2 *The polynomial $g_{s_j} \in \mathbb{C}\{x\}[y]$ is irreducible and monic, and takes value zero at all the points of γ_j in a certain neighbourhood of O. Furthermore g_{s_j} is the only element of $\mathbb{C}\{x\}[y]$ satisfying these conditions.*

PROOF: It has already been seen that g_{s_j} satisfies the conditions of the claim. If $h \in \mathbb{C}\{x\}[y]$ also does, then, by 2.4.1, we have $h(x, s_j(x)) = 0$ for all complex values of x with $|x|$ small enough; this shows that $h(x, s_j(x)) = 0$ as an element of $\mathbb{C}\{x\}$, and therefore s_j is a root of h. The conjugates of s_j are then roots of h too, after which g_{s_j} divides h. Since h is irreducible and both g_{s_j} and h are monic, we have $h = g_{s_j}$ and the claim is proved.   ◇

It is clear from the definition of γ_j that both the equation g_j and the conjugacy class of s_j determine it. Conversely, it follows from 2.4.2 that each branch γ_j determines its equation g_j. In turn g_j obviously determines its set of roots $\{\sigma_\varepsilon(s_j)\}_{\varepsilon^{\nu_j}=1}$, which is the conjugacy class of s_j. We thus see that the branches of C at O are in one to one correspondence with the conjugacy classes of the Puiseux series of C at O. Any of the conjugate Puiseux series corresponding to a branch is called a *Puiseux series of the branch*.

The *multiplicity of γ_j in*, or *as a branch of*, C – or C_0 – is defined as being the multiplicity of the corresponding factor g_{s_j} in the decomposition (2.11). It is, of course, the multiplicity of s_j as a y-root of f.

Remark 2.4.3 The equations g_{s_j} of the branches of C at O are the irreducible factors of f in $\mathbb{C}\{x, y\}$ that vanish at the origin. After this, clearly, γ is a branch at O of a curve $C = C_1 + C_2$ if and only if γ is a branch at O of either C_1 or C_2. Also, the multiplicity of γ as a branch of C is the sum of its multiplicities as a branch of C_1 and as a branch of C_2.

The definitions of multiplicity of a point and tangent cone at a point, already given for curves, are extended to the branches of a curve using their monic equations of 2.4.2:

The *multiplicity on γ_j of the origin O* of γ_j is $e_O(\gamma_j) = \deg \mathrm{In}(g_{s_j})$.

The *tangent cone* to γ_j at O is the algebraic curve $TC_O(\gamma_j) : \mathrm{In}(g_{s_j}) = 0$.

The multiplicity $e_O(\gamma_j)$ is often called the *multiplicity of γ_j*. It has no relation to – and therefore should not be confused with – the multiplicity of γ_j *as a branch of C*. It is also called the *order* of γ. Branches of multiplicity one are called *smooth branches*. As for algebraic curves, the tangent cone to a branch γ_j at O is an algebraic curve of degree $e_O(\gamma_j)$, composed of lines which are called the *tangent lines* to the branch. In fact, each branch of a curve has a single tangent line, as shown next:

Proposition 2.4.4

(a) *Each branch γ of a curve C at a point O has a single tangent line at O, which therefore has multiplicity $e_O(\gamma_j)$ in $TC_O(\gamma)$.*

(b) *The multiplicity of O on C is the sum of its multiplicities on the branches of C at O.*

(c) *The tangent cone to C at O is composed of the tangent cones to the branches of C at O. In particular, a line is tangent to C at O if and only if it is the tangent line to one of the branches of C at O.*

PROOF: For claim (a) use 2.3.3: using the conventions therein, if the Puiseux series of the branch is $s = ax^{m/n} + \ldots$, then its tangent cone is either

$$x^{\nu m/n} = 0 \text{ if } m/n < 1, \text{ or}$$

$$y^\nu = 0 \text{ if } m/n > 1, \text{ or}$$

$$(y - ax)^\nu = 0 \text{ in the case when } m/n = 1.$$

In all cases it is composed of a single line. For the remaining claims just use the obvious equality $\text{In}(f) = \prod_{i=1}^k \text{In}(g_{s_i})$.   ◇

Remark 2.4.5 It follows from 2.4.4(b) that if O is a smooth point of C, then C has a single branch at O. The converse is not true: for any $e \in \mathbb{Z}$, $e > 1$, the curve $C : x^{e+1} - y^e = 0$ has multiplicity e and a single branch at $(0,0)$, as the reader may easily check.

While proving 2.4.4 we have seen in particular:

Corollary 2.4.6 (of the proof of 2.4.4) *The multiplicity and tangent line at the origin of the branch with Puiseux series $s = ax^{m/n} + \ldots$, $a \neq 0$, $(m,n) = (1)$, $\nu(s) = \nu$, are, respectively:*

(a) *$\nu m/n$ and the y-axis if $m/n < 1$;*

(b) *ν and the x-axis if $m/n > 1$;*

(c) *ν and the line $y - ax = 0$ if $m/n = 1$.*

In the case in which x is an irreducible factor of multiplicity $\ell > 0$ of f, one may proceed as above; nevertheless, to have a complete representation of the points of the curve around the origin, the y-axis $x = 0$ should be taken as a further branch of C, of multiplicity ℓ as a branch of C. Such a branch, unlike the other ones, does not have an associated Puiseux series or a Puiseux parameterization with parameter x.

Assume that s is a Puiseux series of a branch γ: from the exponents of the non-zero monomials composing s we collect:

– the first fractionary exponent: m_1/n_1, where $\gcd(m_1, n_1) = 1$, $n_1 > 1$;

– the first exponent which cannot be reduced to denominator n_1: $m_2/n_1 n_2$, where $\gcd(m_2, n_2) = 1$, $n_2 > 1$;

– the first exponent which cannot be reduced to denominator $n_1 n_2$: $m_3/n_1 n_2 n_3$, where $\gcd(m_3, n_3) = 1$, $n_3 > 1$, and so on, until eventually getting

– the first exponent which cannot be reduced to denominator $n_1 \ldots n_{h-1}$: $m_h/n_1 \ldots n_{h-1} n_h$, where $\gcd(m_h, n_h) = 1$, $n_h > 1$ and $n_1 \ldots n_h = \nu(s)$.

The exponents $m_1/n_1, \ldots, m_h/n_1 \ldots n_h$ (none if $\nu(s) = 1$) are called the *characteristic exponents of* s; they were introduced by Smith in 1876. An important fact which will not be proved here (see [3, 5.5.3]) is that the characteristic exponents are the same for all Puiseux series of γ, no matter which coordinates are used, provided the coordinates are taken with the y axis non-tangent to γ. If this is the case, the characteristic exponents of the Puiseux series are called the *characteristic exponents of* γ. The characteristic exponents of a branch retain all its relevant algebro-geometric properties and, therefore, the branches of the plane curves are classified according to their characteristic exponents ([3, 5.5.4]).

As said, most of the above appears as dependent on the choice of the affine chart and the coordinates, and the Puiseux series as such certainly are: for an easy example consider the parabola $y - x^2 = 0$; it has x^2 as its only Puiseux series at the origin if coordinates x, y are used, while it has the two conjugate Puiseux series $\pm y^{1/2}$ if the coordinates are y, x. In the remainder of this section we will see that, even if the Puiseux series depend on the choice of the coordinates, their corresponding parameterizations (2.13) define analytic maps from neighbourhoods of 0 in $\mathbb{C}$ into the curve which are essentially intrinsic.

In order to avoid modifying domains of definition, we will use germs of maps instead of maps. Limiting ourselves to the case we will use – the definition may obviously be made much more general – and leaving the details to the reader, consider maps defined in a neighbourhood of $0 \in \mathbb{C}$ with image in a certain set S; take two of them as equivalent when they have the same restriction to a neighbourhood of 0 in $\mathbb{C}$ contained in the intersection of their definition domains. Each equivalence class is called a *germ of map* – or just a germ if no confusion may result – *at* 0, indicated

$$\varphi : (\mathbb{C}, 0) \longrightarrow (S, \varphi(0)),$$

where $\varphi(0)$ denotes the common image of 0 by all the representatives of φ. Germs of maps are usually understood as being maps defined locally at the point and taken up to shrinking their domains of definition.

We will say that φ has image in $S' \subset S$ when one of its representatives has image in S', and that it is *constant* when so is one of its representatives. The germ of an identical map will be denoted just by Id.

Our target S will be either $S = \mathbb{C}$, or $S = \mathbb{A}_2$, or $S = \mathbb{P}_2$; then we will say that a germ is analytic if and only if it has an analytic representative (see Section 1.6).

We will consider in particular germs of analytic maps $\varphi : (\mathbb{C}, 0) \to (\mathbb{C}, 0)$; their representatives are analytic functions $\bar{\varphi}$ defined in a neighbourhood of 0 and such that $\bar{\varphi}(0) = 0$. Any such φ may be represented by a map $t \mapsto \sum_{i>0} a_i t^i$ where the power series $u = \sum_{i>0} a_i x^i$ is convergent: then it is said that u represents φ. The series u clearly determines φ, and is in turn determined by φ because its coefficients are determined by the values at 0 of the derivatives of any representative of φ.

Germs of analytic maps $\varphi : (\mathbb{C}, 0) \to (\mathbb{C}, 0)$ and $\varphi' : (\mathbb{C}, 0) \to (S, \varphi'(0))$ may be composed by just composing suitable representatives and taking the corresponding germ; the composite germ $\varphi' \circ \varphi : (\mathbb{C}, 0) \to (S, \varphi'(0))$ is analytic too. In particular, a germ $\varphi : (\mathbb{C}, 0) \to (\mathbb{C}, 0)$ is called an analytic isomorphism if and only if it is analytic and there exists a $\varphi' : (\mathbb{C}, 0) \to (\mathbb{C}, 0)$, also analytic, such that both $\varphi' \circ \varphi$ and $\varphi \circ \varphi'$ are the germ of the identity.

The inverse map theorem proves that a germ of analytic map $\varphi : (\mathbb{C}, 0) \to (\mathbb{C}, 0)$ is an analytic isomorphism if and only if the value at 0 of the derivative of any representative of φ is not zero, or, equivalently, $a_1 \neq 0$ if φ is represented by $\sum_{i>0} a_i x^i$. This in particular shows that either of the conditions $\varphi' \circ \varphi = \text{Id}$ and $\varphi \circ \varphi' = \text{Id}$, posed in their definition above, suffices to characterize analytic isomorphisms.

We will say that a germ of map $\varphi : (\mathbb{C}, 0) \to (S, \varphi(0))$ is taken up to *isomorphism on the source* when it is taken up to replacement with $\varphi \circ \delta$, where $\delta : (\mathbb{C}, 0) \to (\mathbb{C}, 0)$ is an isomorphism. Taking a germ up to isomorphism on the source may be understood as taking it up to a change of the local parameter t at the source, $t = \sum_{i>0} b_i \bar{t}^i$, $b_1 \neq 0$.

In the sequel, if $s = \sum_{i \geq 0} a_i x^{i/\nu}$, $\nu = \nu(s)$ is a convergent fractionary power series we will write $s(t^\nu) = \sum_{i \geq 0} a_i t^i$ for $t \in \mathbb{C}$ and $|t|$ small enough. The next theorem being local, it holds for both affine and projective plane curves. Below is presented the projective version; the affine version results from it by obvious changes.

Theorem 2.4.7 *Let C be an algebraic curve of $\mathbb{P}_2$ and O a point of C.*

(a) *There are non-constant analytic germs*

$$\varphi_j : (\mathbb{C}, 0) \longrightarrow (\mathbb{P}_2, O),$$

$j = 1, \ldots, k$, with image in $|C|$ such that for any non-constant analytic germ

$$\psi : (\mathbb{C}, 0) \longrightarrow (\mathbb{P}_2, O)$$

with image in $|C|$, there is a unique $j = 1, \ldots, k$ and a unique analytic germ $\bar{\psi} : (\mathbb{C}, 0) \to (\mathbb{C}, 0)$ such that $\psi = \varphi_j \circ \bar{\psi}$.

(b) *Germs φ_j, $j = 1, \ldots, k$, satisfying the conditions of (a) above, are uniquely determined by C and O up to isomorphisms on the sources.*

(c) *Take an affine chart $\mathbb{A}_2$ of $\mathbb{P}_2$ containing O and affine coordinates on $\mathbb{A}_2$ such that $O = (0,0)$ and the y-axis is not contained in C. If s_j, $j = 1, \ldots, k$ are Puiseux series of C at O, one per conjugacy class, and $\nu_j = \nu(s_j)$, then the maps*

$$U_j \longrightarrow \mathbb{P}_2$$
$$t \longmapsto (t^{\nu_j}, s_j(t^{\nu_j})),$$

where the U_j are suitable neighbourhoods of 0 in $\mathbb{C}$, represent analytic germs $\varphi_1, \ldots, \varphi_k$ satisfying the conditions of part (a).

PROOF: We will begin by proving part (b). Assume we have two sets of germs $\{\varphi_j\}_{j=1,\ldots,k}$ and $\{\varphi'_j\}_{j=1,\ldots,k'}$, each satisfying the conditions of (a); up to swapping them over, we assume $k \geq k'$.

Remark 2.4.8 If there exists a $\delta : (\mathbb{C}, 0) \to (\mathbb{C}, 0)$ such that $\varphi_\ell = \varphi_j \circ \delta$, then, since also $\varphi_\ell = \varphi_\ell \circ \mathrm{Id}$, by the uniqueness claimed in (a), taking φ_ℓ in the role of ψ yields $\ell = j$ and $\delta = \mathrm{Id}$.

By the hypothesis, φ_1 factorizes through a certain φ'_{j_1},

$$\varphi_1 = \varphi'_{j_1} \circ \bar{\varphi}_1,$$

and φ'_{j_1} factorizes in turn through a φ_ℓ,

$$\varphi'_{j_1} = \varphi_\ell \circ \bar{\varphi}'_{j_1}.$$

It follows that

$$\varphi_1 = \varphi_\ell \circ \bar{\varphi}'_{j_1} \circ \bar{\varphi}_1$$

and therefore, by 2.4.8, $\ell = 1$ and $\bar{\varphi}'_{j_1} \circ \bar{\varphi}_1 = \mathrm{Id}$. It follows that both $\bar{\varphi}'_{j_1}$ and $\bar{\varphi}_1$ are isomorphisms, after which φ'_j and φ_1 differ by an isomorphism between their sources.

Now the argument is repeated from φ_2; just note that it needs to factor through a φ'_{j_2} with $j_2 \neq j_1$: otherwise, since φ'_{j_1} and φ_1 differ by an isomorphism between their sources, φ_2 would factor through φ_1 against 2.4.8. Continuing in this way, eventually each φ_ℓ is paired to a φ'_{j_ℓ} from which it differs by an isomorphism between the sources, and all these φ'_{j_ℓ} are different. In particular $k \leq k'$, which gives $k = k'$ and no φ'_j remains unpaired.

The existence of part (a) will result from proving (c), which we will do next. The situation being as described in (c), take, as above,

$$s_j(x) = \sum_{i>0} a_i^j x^{i/\nu_j}, \quad \nu_j = \nu(s_j).$$

Assume that the affine part of C has equation $f \in \mathbb{C}[x, y]$. Given ψ as in (a), assume that $\tilde{t} \mapsto (u_1(\tilde{t}), u_2(\tilde{t}))$ is a representative of ψ, with u_1, u_2 convergent series. Since ψ is assumed to be non-constant, if $u_1 = 0$, then C and the y axis share infinitely many points, against the choice of the coordinates and 1.1.3. Then we may write $u_1 = \tilde{t}^m w$ where m is a positive integer, w is a convergent series in $\tilde{t}$ and $w(0) \neq 0$. The implicit function theorem and the latter inequality guarantee the existence of a convergent series v such that $v^m = w$ (the reader may also use Exercise 2.4 below). Since, clearly, $v(0) \neq 0$ we may take $t = \tilde{t}v$ as a new local coordinate around 0 in $\mathbb{C}$ in such a way that, using it, the representative of ψ now has the form $t \mapsto (t^m, u(t))$ with u a convergent power series, $u(0) = 0$.

Since the image of ψ is contained in $|C|$, we have $f(t^m, u(t)) = 0$ for all t in a neighbourhood of 0. Using the factorization (2.11) at the beginning of this section, for some j we have $g_{s_j}(t^m, u(t)) = 0$ for all t close enough to 0. In view of the definition of g_{s_j}, there is $\varepsilon \in \mathbb{C}$, $\varepsilon^{\nu_j} = 1$, such that $u(t) = \sigma_\varepsilon(s_j)(t^m)$ as series. In other words,

$$u = \sum_{i>0} a_i^j \varepsilon^i t^{mi/\nu_j}.$$

Since u is an integral power series and $\nu_j = \nu(s_j)$, ν_j divides m, say $m = \tilde{m}\nu_j$. Then ψ is represented by

$$t \longmapsto (t^m, \sum_{i>0} a_i^j \varepsilon^i t^{\tilde{m}i}).$$

Since by the hypothesis φ_j is represented by

$$t \longmapsto (t^{\nu_j}, \sum_{i>0} a_i^j t^i)$$

it is clear that it suffices to take $\bar{\psi}$ represented by

$$t \longmapsto \varepsilon t^{\tilde{m}}.$$

To close, it remains to prove the uniqueness of j and $\bar{\psi}$ above. Assume we have $\psi = \varphi_\ell \circ \delta$ for an analytic germ $\delta : (\mathbb{C}, 0) \to (\mathbb{C}, 0)$ represented by

$$t \longmapsto \alpha(t), \quad \alpha \in \mathbb{C}\{t\}.$$

$\varphi_\ell \circ \delta$ is thus represented by the map

$$t \longmapsto (\alpha(t)^{\nu_\ell}, \sum_{h>0} a_h^\ell \alpha(t)^h).$$

By comparing with the above representative of ψ, we first get $\alpha(t)^{\nu_\ell} = t^m$, for all t close enough to 0: it follows that $m/\nu_\ell = \tilde{m}\nu_j/\nu_\ell$ is an integer and $\alpha(t) = \eta t^{\tilde{m}\nu_j/\nu_\ell}$ for a certain ν_ℓ-root of unity η. Equating the second components gives

$$\sum_{i>0} a_i^j \varepsilon^i t^{\tilde{m}i} = \sum_{h>0} a_h^\ell \eta^h t^{h\tilde{m}\nu_j/\nu_\ell}. \qquad (2.14)$$

All exponents effectively appearing in the left-hand side being multiples of $\tilde{m}$, the same occurs on the right-hand side and therefore $h\nu_j/\nu_\ell$ is an integer for all h for which $a_h^\ell \neq 0$. Using that these h and ν_ℓ share no divisor, it turns out that ν_j is a multiple of ν_ℓ, say $\nu_j = n\nu_\ell$. Then (2.14) reads

$$\sum_{i>0} a_i^j \varepsilon^i t^{\tilde{m}i} = \sum_{h>0} a_h^\ell \eta^h t^{\tilde{m}nh}.$$

Comparing again both sides shows that n divides all i for which $a_i^j \neq 0$. Since n divides ν_j too, this yields $n = 1$. Then $\nu_j = \nu_\ell$ and so

$$\sum_{i>0} a_i^j \varepsilon^i t^{\tilde{m}i} = \sum_{h>0} a_h^\ell \eta^h t^{\tilde{m}h}.$$

It follows that, for all i,

$$a_i^j \varepsilon^i = a_i^\ell \eta^i \qquad (2.15)$$

with $\varepsilon^{\nu_j} = \eta^{\nu_j} = 1$. Therefore s_j and s_ℓ are conjugate series, and hence equal. Thus $\ell = j$, and $a_i^j = a_i^\ell$ for all i, and then the equalities (2.15) give $\sigma_\varepsilon(s_j) = \sigma_\eta(s_j)$ and so $\varepsilon = \eta$ by 2.1.3. It follows $\delta = \bar{\psi}$, thus ending the proof.   ◇

On one hand, the analytic germs φ_j, $j = 1, \ldots, k$, of 2.4.7(a), are uniquely determined by C and O up to isomorphism on the source by 2.4.7(b). On the other, by 2.4.7(c), once a local chart and suitable coordinates have been chosen, they correspond one to one to the conjugacy classes of Puiseux series, which in turn are in one to one correspondence with the branches of C at O, as defined at the beginning of this section. More precisely, for $j = 1, \ldots, k$, the representative of the germ φ_j given in 2.3.10(c) is just the parameterization (2.13) of the corresponding branch γ_j, which may in turn be taken to be the germ of the image of that parameterization.

Remark 2.4.9 The above shows that the branches of C at O are uniquely determined by C and O, the choices of the affine chart and the coordinates already used to define them being irrelevant.

It is usual to refer to each germ φ_j as the *uniformizing germ* of the corresponding branch γ_j. Its representatives are called *uniformizing maps* of γ_j: locally at O, they parameterize it. The reader may note that a Puiseux parameterization of a branch γ may be recovered from any uniformizing map $t \mapsto (u_1(t), u_2(t))$ of γ by just reparameterizing the latter with new parameter $u_1^{1/\nu}$, with $\nu = o_t(u_1)$, as done for the map ψ in the proof of 2.4.7. The next lemma shows that any uniformizing map becomes injective once restricted to a suitable neighbourhood of 0 in $\mathbb{C}$.

Lemma 2.4.10 *If $\varphi : U \to C$ is a uniformizing map of a branch of a curve C, then there is a smaller open neighbourhood $U' \subset U$ of 0 in $\mathbb{C}$ such that the restriction $\varphi_{|U'}$ is injective.*

PROOF: Up to a first reduction of U to a smaller neighbourhood, by 2.4.7(c) we may assume φ to be given by the rule $z \mapsto (z^\nu, s(z^\nu))$ where $s = \sum_{i>0} a_i x^{i/\nu}$ and $\nu = \nu(s)$. If z and z', $z \neq z'$, have the same image, then we have $z' = \varepsilon z$ with $\varepsilon^\nu = 1$, $\varepsilon \neq 1$. In particular, $z \neq 0$. Further, we should have

$$\sum_{i>0} a_i z^i = \sum_{i>0} a_i \varepsilon^i z^i,$$

that is,

$$\sum_{i>0} (a_i - \varepsilon^i a_i) z^i = 0.$$

The power series $\sum_{i>0} (a_i - \varepsilon^i a_i) t^i$, $\varepsilon^\nu = 1$, $\varepsilon \neq 1$, are not zero due to 2.1.3, hence they define analytic functions which have no zeros other than $t = 0$ in a neighbourhood $U' \subset U$ of 0 in $\mathbb{C}$: such a U' fulfills the claim   ◇

Remark 2.4.11 The polynomial x obviously has no y-root and therefore no Puiseux series comes associated to it. Hence the need of the hypothesis of the y-axis being not contained in the curve C, repeatedly used above; otherwise the local representation of C by Puiseux series and their corresponding uniformizing maps would not be complete.

Remark 2.4.12 As is quite usual – for simplicity – when working locally at a point O, we have been using affine coordinates with the origin at O. However, there is no problem in using coordinates x, y for which $O = (a, b)$ provided the line $x - a = 0$ is not contained in the curve C. Indeed, using coordinates $\bar{x} = x - a$ and $\bar{y} = y - b$, the point O becomes the origin. Then, for each Puiseux series $\bar{s} = \sum_{i>0} a_i \bar{x}^{i/\nu}$ associated to C at O using coordinates $\bar{x}, \bar{y}$,

$$s = b + \bar{s}(x - a) = b + \sum_{i>0} a_i (x - a)^{i/\nu}$$

is taken as a Puiseux series of C at O relative to the coordinates x, y. Then still $f(x, s) = 0$ if $f(x, y)$ is an equation of C. It is easy to get a factorization of f similar to (2.11) and parameterizations like (2.12) and (2.13) – with $x = t^{\nu_i} + a$ for the latter. If $\nu = \nu(s) = \nu(\bar{s})$, s gives rise to the uniformizing map $t \mapsto (a + t^\nu, s(t^\nu))$.

2.5   Branches and Irreducible Components

Let us begin by recalling some quite easy algebraic facts. Assume that K is a field of characteristic zero and s an element algebraic over K, that is, a root, in an extension of K, of a polynomial $Q \in K[y]$. Then $\{P \in K[y] \mid P(s) = 0\}$ is a principal non-zero prime ideal; any basis of it is an irreducible polynomial M called a *minimal polynomial* of s.

Remark 2.5.1 Not only does $P(s) = 0$ if and only if $M|P$, by the above definition of M, but also the multiplicity of M as an irreducible factor of P equals the multiplicity of s as a root of P. Indeed, s is a simple root of M, as otherwise it would be a root of dM/dy which is non-zero and has degree less that $\deg M$, against the definition of M. If $P = M^r\bar{P}$, M not dividing $\bar{P}$, then $\bar{P}(s) \neq 0$ again by the definition of M, and so the multiplicity of s as a root of P is r.

We will deal with the case $K = \mathbb{C}(x)$, the field of complex rational fractions in x, and need to relate divisibility in the rings of polynomials $\mathbb{C}(x)[y]$ and $\mathbb{C}[x, y]$. A polynomial $F \in \mathbb{C}[x, y]$ is said to be *primitive as a polynomial in y* – we will say just primitive – if and only if, viewed as an element of $\mathbb{C}[x][y]$, $F = \sum_{i=0}^{d} a_i(x)y^i$, its coefficients $a_i(x)$, $i = 0, \ldots, d$, are coprime. We have:

Lemma 2.5.2 *A primitive $F \in \mathbb{C}[x, y]$ divides $G \in \mathbb{C}[x, y]$ in $\mathbb{C}(x)[y]$ if and only if F divides G in $\mathbb{C}[x, y]$.*

PROOF: If $G = AF$, $A \in \mathbb{C}(x)[y]$, then write $A = A'/P$, with $A' \in \mathbb{C}[x, y]$ and $P \in \mathbb{C}[x]$, to get $PG = A'F$. If $\deg P = 0$, we are done. Otherwise take a root a of P. Then

$$0 = P(a)G(a, y) = A'(a, y)F(a, y)$$

as polynomials in $\mathbb{C}[y]$. Necessarily $F(a, y) \neq 0$, because F is primitive and therefore its coefficients cannot have a common root. a is thus a common root of the coefficients of A' as a polynomial in y and therefore $x - a$ divides A' in $\mathbb{C}[x, y]$. After dividing both sides of $PG = A'F$ by $x - a$, it is enough to use induction on $\deg P$. The converse is obvious. ◇

Lemma 2.5.3 *A primitive $F \in \mathbb{C}[x, y]$ is irreducible in $\mathbb{C}(x)[y]$ if and only if it is irreducible in $\mathbb{C}[x, y]$.*

PROOF: The polynomial $F = \sum_{i=0}^{d} a_i(x)y^i$ being primitive, it has no non-trivial factor $P \in \mathbb{C}[x]$, as otherwise P would divide all $(\partial^i F/\partial y^i)_{x,0}$ and therefore all the coefficients $a_i(x)$, $i = 0, \ldots, d$. Therefore, if $F = F_1 F_2$ is a non-trivial factorization in $\mathbb{C}[x, y]$, then both F_1 and F_2 have positive degree in y and so $F = F_1 F_2$ is also a non-trivial factorization in $\mathbb{C}(x)[y]$. Conversely, a factorization $F = F_1 F_2$, $F_i \in \mathbb{C}(x)[y]$, $0 < \deg_y F_i < \deg_y F$, $i = 1, 2$, may be easily modified to $F = F_1' F_2'$, where F_2' is a primitive element of $\mathbb{C}[x, y]$ and $\deg_y F_2' = \deg_y F_2$. Then 2.5.2 applies. ◇

Back to geometry, the next proposition also applies to affine curves with obvious modifications. Its second claim shows in particular that the multiplicities of the branches, introduced at the beginning of Section 2.3, do not depend on the choices of the affine chart and the coordinates.

Proposition 2.5.4 *Let C be a curve of $\mathbb{P}_2$, O a point of C and γ a branch of C at O. Then:*

(a) *γ is a branch of one and only one irreducible component C_1 of C.*

(b) *If after choosing an affine chart containing O and coordinates x, y on it, with the y-axis not contained in C, f is an equation of the affine part of C and s any of the – conjugate – Puiseux series of C corresponding to γ, the multiplicity of s as a y-root of f – that is, the multiplicity of γ as a branch of C – equals the multiplicity of C_1 as an irreducible component of C.*

PROOF: Take the affine chart $\mathbb{A}_2$, coordinates x, y, and the Puiseux series s as described in part (b). Choose a minimal polynomial f_1 of s belonging to $\mathbb{C}[x, y]$ and primitive, which clearly is always possible. Since f has the root s, f_1 is irreducible and divides f in $\mathbb{C}(x)[y]$. By 2.5.2 and 2.5.3, the same occurs in $\mathbb{C}[x, y]$; therefore $C_1^0 : f_1 = 0$ is an irreducible component of the affine part C^0 of C. Furthermore, it has γ as one of its branches at O because $f_1(x, s) = 0$.

If $f = f_1^r f'$, with $f_1 \nmid f'$, then $f'(x, s) \neq 0$, as otherwise, arguing as above, f_1 would divide f'. It follows that no other irreducible component of C^0 has branch γ. Again by 2.5.2 and 2.5.3, f_1 is an irreducible factor of multiplicity r of f in $\mathbb{C}(x)[y]$. Therefore, the multiplicity r of C^0 as an irreducible component of C^0 is the multiplicity of s as a y-root of f by 2.5.1.

We have seen that C^0 has a unique irreducible component C_1^0 with branch γ and also that the multiplicities of C_1^0 and γ as an irreducible component and as a branch of C^0, respectively, do agree. By 1.2.7 the same holds for C and the irreducible component of C whose affine part is C_1^0. $\diamond$

In particular

Corollary 2.5.5 *If a curve C is reduced, then all its branches have multiplicity one in C.*

Corollary 2.5.6 *If two irreducible plane curves share a branch, then they are equal.*

PROOF: If C_1 and C_2, both irreducible, are different and share a branch, then $C_1 + C_2$ contradicts 2.5.4. $\diamond$

Remark 2.5.7 Assume that γ is a branch of a curve of $\mathbb{P}_2$. Due to 2.5.4 and 2.5.6, there is a unique irreducible curve C_γ of $\mathbb{P}_2$ which has γ as a branch, and all the curves which have the branch γ have C_γ as an irreducible component.

2.6 Intersection Multiplicity

In this section we define the intersection multiplicity of two plane curves at a point and prove its basic properties. The intersection multiplicity is a very important local invariant which will be widely used in the sequel. We will deal with projective curves; the interested reader may easily give a similar definition for affine curves and prove similar results.

Fix a point $p \in \mathbb{P}_2$ and let γ be a branch at p of a curve C' of $\mathbb{P}_2$. If F/G is a rational function on $\mathbb{P}_2$ defined at p, then F/G is analytic in a neighbourhood of p and therefore its germ $(F/G)_p$ at p may be composed with the uniformizing germ φ of γ to give a germ of complex analytic function $(F/G)_p \circ \varphi$ at $0 \in \mathbb{C}$. The infinitesimal order at 0 of (any representative of) $(F/G)_p \circ \varphi$ will be called the *order of F/G along γ*, denoted by $o_\gamma(F/G)$: it is either a non-negative integer or ∞, the latter possibility occurring if and only if $(F/G)_p \circ \varphi = 0$. It is clear from the definition that $o_\gamma((F/G)(F'/G')) = o_\gamma(F/G) + o_\gamma(F'/G')$ for any other rational function F'/G' defined at p, and also that $o_\gamma(F/G) = 0$ if and only if $F(p) \neq 0$. Therefore, if G' is homogeneous of the same degree as F and $G'(p) \neq 0$,

$$o_\gamma(F/G') = o_\gamma(F/G) + o_\gamma(G/G') = o_\gamma(F/G)$$

and so $o_\gamma(F/G')$ depends only on the numerator F: we will put $o_\gamma(F) = o_\gamma(F/G)$ for any homogeneous G for which $\deg G = \deg F$ and $G(p) \neq 0$; we will call it the *order of F along γ*.

Assume now to have given the branch γ of C' at p, as above, and another curve $C : F = 0$ of $\mathbb{P}_2$. Then $o_\gamma(F)$ is called the *intersection multiplicity* of the curve C and the branch γ, denoted $[C \cdot \gamma]$ in the sequel. It is either a non-negative integer or ∞. Next is a more effective expression of it:

Proposition 2.6.1 *Let $p \in \mathbb{P}_2$ be a point, C and C' curves of $\mathbb{P}_2$ and γ a branch of C' at p. Assume that projective coordinates x_0, x_1, x_2 have been taken such that $p = [1,0,0]$ and the line $x_1 = 0$ is not contained in C', and take $x = x_1/x_0$, $y = x_2/x_0$ as coordinates on the 0-th affine chart $\mathbb{A}_2 : x_0 \neq 0$. If f is an equation of the affine part of C', s a Puiseux series of γ and $\nu = \nu(s)$, then*

$$[C \cdot \gamma]_p = o_t f(t^\nu, s(t^\nu)).$$

PROOF: Up to a non-zero scalar, in $\mathbb{A}_2$ we have

$$f(x, y) = F(1, x, y) = F(1, x_1/x_0, x_2/x_0) = \frac{F(x_0, x_1, x_2)}{x_0^d},$$

$d = \deg F$. Therefore, by its definition, $o_\gamma F$ may be computed using f. Then the claim follows from the definition of $o_\gamma f$ and 2.4.7(c). ◇

Proposition 2.6.2 *If p is a point and C, C_1, C_2, C' are curves, all of $\mathbb{P}_2$, and γ is a branch of C' at p, then*

(a) $[(C_1 + C_2) \cdot \gamma] = [C_1 \cdot \gamma] + [C_2 \cdot \gamma]$.

(b) $[C \cdot \gamma] = 0$ *if and only if the point p does not belong to C.*

(c) $[C \cdot \gamma] = \infty$ *if and only if γ is a branch of C.*

(d) *If C_1, C_2 and C have the same degree and equations F_1, F_2 and $F_1 + F_2$, respectively, then*

$$[C \cdot \gamma] \geq \min([C_1 \cdot \gamma], [C_2 \cdot \gamma])$$

and equality holds if $[C_1 \cdot \gamma] \neq [C_2 \cdot \gamma]$.

PROOF: Direct from 2.6.1 using 2.1.1   ◇

Remark 2.6.3 An easy computation shows that all the lines ℓ through the origin O of γ have $[\ell \cdot \gamma] = e_O(\gamma)$, with the only exception of the tangent line ℓ_0 to γ, whose intersection multiplicity with γ is higher. This proves that neither the multiplicity of γ nor the tangent line to γ depend on the coordinates used in their definition in Section 2.4. The tangent cone to γ is also independent of the choice of the coordinates, as it is $e_O(\gamma)\ell_0$. The positive integer $[\ell_0 \cdot \gamma] - e_O(\gamma)$ is called the *class* of γ.

The intersection multiplicity of a curve and a branch will play an important role in Chapter 4. At present, we are more interested in the intersection multiplicity of two curves, defined next:

The *intersection multiplicity* – or *intersection number* – of the curves C and C' at p is defined by either of the equivalent equalities

$$[C \cdot C']_p = \sum_\gamma [C \cdot \gamma] = \sum_\gamma o_\gamma(F)$$

with the summations extended to all branches γ of C' at p, each repeated as many times as indicated by its multiplicity as a branch of C', and still $C : F = 0$. The intersection multiplicity $[C \cdot C']_p$ is thus either a non-negative integer or ∞.

Remark 2.6.4 If the curve C' is reduced, then, by 2.5.5, one may ignore the multiplicities of the branches involved in the definition and just take the summation extended to all the branches of C at p.

It follows at once from 2.6.1:

Proposition 2.6.5 *Let p be a point and C and C' curves of $\mathbb{P}_2$. Assume that projective coordinates x_0, x_1, x_2 have been taken such that $p = [1, 0, 0]$ and the line $x_1 = 0$ is not contained in C', and take $x = x_1/x_0$, $y = x_2/x_0$ as coordinates on $\mathbb{A}_2 : x_0 \neq 0$. Let f and g be equations of the affine parts* *of C and C', respectively. If $s'_1, \ldots, s'_r$ are Puiseux series of C' at p, one per conjugacy class and repeated according to their multiplicities as y-roots of g, and $\nu_i = \nu(s'_i)$, then we have*

$$[C \cdot C']_p = \sum_{i=1,\ldots,r} o_{t_i} f(t_i^{\nu_i}, s'_i(t_i^{\nu_i})).$$

Remark 2.6.4 still applies: if C' is reduced, then the multiplicities of the Puiseux series as y-roots are all equal to 1 and may be ignored.

The reader may use 2.6.5 to check that the intersection multiplicity introduced here does agree, for $n = 2$, with the intersection multiplicity of a line and a hypersurface as defined in Section 1.1.

The most expressive presentation of the intersection multiplicity is the following:

Theorem 2.6.6 (Halphen's formula; Halphen, 1874) *Let p be a point, and C and C' be curves of $\mathbb{P}_2$. Assume that projective coordinates x_0, x_1, x_2 have been taken such that $p = [1, 0, 0]$ and the line $x_1 = 0$ is not contained in C or C', and take $x = x_1/x_0$, $y = x_2/x_0$ as coordinates on $\mathbb{A}_2 : x_0 \neq 0$. Then*

$$[C \cdot C']_p = \sum_{s, s'} o_x(s - s'), \tag{2.16}$$

where the summation runs over the Puiseux series s of C at p and the Puiseux series s' of C' at p, in both cases including all conjugates and each series repeated according to its multiplicity as a root of the corresponding equation.

PROOF: Take the notations as in 2.6.5. For each $i = 1, \ldots, r$,

$$o_{t_i} f(t_i^{\nu_i}, s'_i(t_i^{\nu_i})) = \nu_i o_x f(x, s'_i(x)).$$

Furthermore, since f is invariant under conjugation, clearly

$$o_x f(x, \sigma_\varepsilon(s'_i)(x)) = o_x f(x, s'_i(x))$$

for any ν_i-th root of unity ε. Therefore, the number of conjugates of s'_i being ν_i,

$$\nu_i o_x f(x, s'_i(x)) = \sum_{\varepsilon^{\nu_i} = 1} o_x f(x, \sigma_\varepsilon(s'_i)(x)).$$

All together, the equality of 2.6.5 may be written

$$[C \cdot C']_p = \sum_{s'} o_x f(x, s'(x)),$$

the summation running over all the Puiseux series of C' at p, including conjugates and repeating the series according to multiplicities.

On the other hand, making explicit the Puiseux series in the factorization (2.4.7) shown at the beginning of Section 2.4,

$$f = \prod_s (y - s(x)) \tilde{f}(x, y),$$

where the summation runs over all Puiseux series s of C at p, including conjugates, repeated according to multiplicities. Since $\tilde{f}(0,0) \neq 0$, the factor $\tilde{f}$ is irrelevant and the last two displayed equalities together give

$$[C \cdot C']_p = \sum_{s'} o_x \prod_s (s'(x) - s(x)),$$

and hence the claim.   ◇

Again, the multiplicities of Puiseux series as roots may be safely ignored in 2.6.6 if both curves C and C' are reduced.

Formal properties of the intersection multiplicity of curves at a point are:

Proposition 2.6.7 *If p is a point and C, C_1, C_2, C' curves of $\mathbb{P}_2$, then*

(a) $[C \cdot C']_p = [C' \cdot C]_p$.

(b) $[(C_1 + C_2) \cdot C']_p = [C_1 \cdot C']_p + [C_2 \cdot C']_p$.

(c) $[C \cdot C']_p = 0$ *if and only if the point p does not belong to one of the curves C, C'.*

(d) $[C \cdot C']_p = \infty$ *if and only if the curves C, C' share an irreducible component that goes through p.*

PROOF: Claim (a) is clear from Halphen's formula 2.6.6. Claim (b) follows from 2.6.2(a). For claim (c), note that if the intersection multiplicity is zero, then either the summation in the definition (2.16) is empty and then $p \notin C'$, or 2.6.2(b) applies to at least one branch of C' at p and then $p \notin C$; the converse is clear. For claim (d), if $[C \cdot C']_p = \infty$, then by 2.6.2(c) (or 2.6.6), C and C' share a branch γ with origin at p; as a consequence, the irreducible components of C and C' with branch γ (2.5.4) do agree by 2.5.6. Again, the converse is clear.   ◇

Halphen's formula easily gives a lower bound for the intersection multiplicity in terms of the multiplicities of the curves:

Corollary 2.6.8 *If as above p is a point and C and C' curves of $\mathbb{P}_2$ with respective multiplicities $e_p(C)$ and $e_p(C')$ at p, then*

$$[C \cdot C']_p \geq e_p(C) e_p(C')$$

and equality holds if and only if C and C' share no tangent at p.

PROOF: Throughout the proof, dots indicate terms of higher degree. Take a local chart and affine coordinates on it such that the origin is p and neither of the coordinate axes is tangent to C or C' (see Remark 1.3.10). Put $e = e_p(C)$ and $e' = e_p(C')$. By 1.3.2, there are no monomials of degree less than e in the equation f of C, $f = f_e + \ldots$ with f_e non-zero and homogeneous of degree e. The coordinate axes being not tangent to C, they have intersection multiplicity with C at p equal to e and therefore f has non-zero monomials in both x^e and y^e, which in particular ensures that the y-axis is not contained in C. The local Newton polygon of C thus has a single side Γ, which goes from $(0, e)$ to $(e, 0)$ and whose associated equation is $f_e(1, z)$. Since the lines tangent to C at p are the lines $ax - y = 0$ for which $f_e(1, a) = 0$, by performing the first step of the Newton–Puiseux algorithm in all possible ways, we see that the Puiseux series of C at p all have the form $s = ax + \dots$ where a is the slope of a line tangent to C at p, each of the slopes of the tangent lines occurring at least once. Furthermore, the number of these series, repeated according to multiplicities, is $e = h(N(f))$ by 2.3.6.

All the arguments above apply to C'. Therefore, the differences in the right-hand side of Halphen's formula are ee' in number and all have the form

$$s - s' = (a - b)x + \dots,$$

where a and b are slopes of lines tangent to C and C', respectively, at p and every pair of these slopes occurs. Then the claimed inequality is clear, and so is the characterization of the equality because the curves share a tangent if and only if $a = b$ for at least one pair of series s, s'.   ◇

Remark 2.6.9 By 2.6.8, $[C \cdot C']_p = 1$ if and only if $e_p(C) = e_p(C') = 1$ – that is, both curves are smooth at p – and C and C' have different tangents at p. If this is the case, it is said that the intersection of C and C' at p is *simple* or *transverse*, and then the curves are said to be *transverse at p*.

The reader may easily adapt 2.6.5 and 2.6.6 to the case in which p is not taken as the origin of the coordinates, say $p = (a, b)$, using 2.4.12. In particular, no change is required in Halphen's formula other than taking o_{x-a} instead of o_x.

2.7 Exercises

2.1 Determine partial sums of the y-roots of each of the polynomials below, one per conjugacy class, enough to show the polydromy order of each root and to distinguish it from the other roots.

(a) $-x^5 + x^7 + 2x^3 y + x^4 y - x^6 y - xy^2 - 2x^2 y^2 + y^3$.

(b) $-x^5 + x^6 + 2x^3 y + x^4 y - x^5 y - xy^2 - 2x^2 y^2 + y^3$.

(c) $-x^7 + x^8 + x^4 y + x^5 y - 2x^2 y^2 - x^3 y^2 + y^3$.

(d) $-x^3 - 3x^4 - 9x^5 - 13x^6 - 17x^7 - 11x^8 - 6x^9 + 3x^2 y + 6x^3 y + 15x^4 y + 12x^5 y + 11x^6 y - 3xy^2 - 3x^2 y^2 - 6x^3 y^2 + y^3$.

(e) $x^8 - 2x^9 - 3x^{10} - x^{11} - 4x^6 y + 4x^7 y + 4x^8 y + 6x^4 y^2 - 2x^5 y^2 - 4x^2 y^3 + y^4$.

(f) $x^4 - x^7 - xy + x^4 y + 2x^5 y - 2x^2 y^2 - x^3 y^2 + y^3$.

(g) $x^8 - 2x^{10} + x^{12} - 4x^{13} - x^{16} - 4x^6 y + 4x^8 y + 4x^{11} y + 6x^4 y^2 - 2x^6 y^2 - 4x^2 y^3 + y^4$.

(h) $x^8 - 2x^9 - x^{10} - 2x^{11} + x^{12} - 4x^6 y + 4x^7 y + 4x^8 y + 6x^4 y^2 + 2x^5 y^2 - 2x^6 y^2 - 4x^2 y^3 + y^4$.

(i) $x^4 - 5x^6 + 4x^8 - 4x^3 y + 10x^5 y + 6x^2 y^2 - 5x^4 y^2 - 4xy^3 + y^4$.

2.2 The converse being obvious, prove that any $u \in \mathbb{C}\{x\}$ with $u(0) \neq 0$ is invertible in $\mathbb{C}\{x\}$. *Hint:* reduce to the case $u(0) = 1$ and use the Newton–Puiseux algorithm and 2.3.11 on $(y+1)u - 1$.

2.3 Prove that the elements of $\mathbb{C}\langle\langle x\rangle\rangle$ (cf. Section 2.1) are the series $\sum_{i \geq r} a_i x^{i/n}$, $a_i \in \mathbb{C}$ and $n, r \in \mathbb{Z}$, $n > 0$ (*Laurent fractionary power series*).

2.4 For any positive integer m, prove that any $u \in \mathbb{C}\{x\}$ with $u(0) \neq 0$ has one m-th root in $\mathbb{C}\{x\}$. *Hint:* Reduce to the case $u(0) = 1$ and proceed as for Exercise 2.2, this time using $(y+1)^m - u$.

2.5 Prove that the field $\mathbb{C}\langle\langle x\rangle\rangle$ is algebraically closed (which is sometimes also called *Puiseux's theorem*) following the steps below:

(1) Prove that for any positive integer n we have $\mathbb{C}\langle\langle x^{1/n}\rangle\rangle = \mathbb{C}\langle\langle x\rangle\rangle$, after which it suffices to prove that any positive-degree polynomial $P(x, y) \in \mathbb{C}[[x]][y]$ with no factor x has a root in $\mathbb{C}\langle\langle x\rangle\rangle$.

(2) If $\deg P(0, y) > 0$ take $a \in \mathbb{C}$ to be a root of $P(0, y)$, the new variable $\bar{y} = y - a$ and use Puiseux's theorem 2.2.7.

(3) If, otherwise, $\deg P(0, y) = 0$, take $d = \deg_y P$ and $P' = y^d P(x, 1/y)$, and use 2.2.7 again.

2.6 Continuity of (some) algebraic functions. For any complex number z and any real number $\varepsilon > 0$, write $B(z, \varepsilon)$ for the open ball with centre z and radius ε. Assume that $P \in \mathbb{C}[x][y]$ is monic and has degree $d > 0$ as a polynomial in y. For any complex value $\bar{x}$ of x denote by $\Lambda(\bar{x})$ the set of roots of $P(\bar{x}, y)$ and write $\Lambda = \Lambda(0)$. Prove that for any real $\varepsilon > 0$ there is a $\delta > 0$ such that if $|\bar{x}| < \delta$, then $\Lambda(\bar{x})$ is contained in the union $\bigcup_{z \in \Lambda} B(z, \varepsilon)$ and, furthermore, $B(z, \varepsilon) \cap \Lambda(\bar{x}) \neq \emptyset$ for any $z \in \Lambda$. A similar claim holds if the leading coefficient of P has positive degree in x, but then some roots converge to infinity, see [3, Ex. 1.9].

2.7 Prove that the Newton–Puiseux algorithm still works if $\mathbb{C}$ is replaced with an arbitrary algebraically closed field of characteristic zero, but it does not give any root of $y^p + y^{p+1} + x$ if the characteristic of the base field is p. Find the point at which the proof of 2.2.5 fails if $\mathbb{C}$ is replaced with a field of positive characteristic.

2.8 In $\mathbb{P}_2$, consider the conic $C : x_0 x_1 + x_0 x_2 - x_1^2 = 0$. Prove that the rule

$$u \longmapsto [1 - u, u^2, -u + \sum_{i \geq 0} u^{4+i}]$$

defines a germ of analytic map

$$\psi : (\mathbb{C}, 0) \longrightarrow C \subset \mathbb{P}_2.$$

Prove that ψ is not a uniformizing germ of a branch of C, but factors $\psi = \varphi \circ \bar{\psi}$, where $\bar{\psi} : (\mathbb{C}, 0) \to (\mathbb{C}, 0)$ is analytic and φ, given by

$$t \longmapsto [1, t, -t + t^2],$$

is a uniformizing germ of the only branch of C at $[1, 0, 0]$. Make $\bar{\psi}$ explicit.

2.9 Compute the intersection multiplicities at the origin of the following pairs of curves of $\mathbb{A}_2$, using Halphen's formula 2.6.6:

(1) $C_1 : y^3 - x^4 y = 0$, $C_1' : y^2 - y^3 - 4x^4 + 4x^4 y = 0$.

(2) $C_2 : y^3 - x^4 = 0$, $C_2' : y^3 - 2x^4 = 0$.

(3) $C_3' : (y^2 - x^3)^2 + x^5 y = 0$, $C_3' : (y^2 - x^3)^2 + 2x^2 y^3 = 0$.

2.10 Assume that γ is a branch of a curve at $(0,0) \in \mathbb{A}_2$, of multiplicity $n > 1$ and whose Puiseux series $\sum_{i>0} a_i x^{i/n}$ has highest characteristic exponent m/n. Prove that for each $j \geq 0$, there is an algebraic curve C_j of $\mathbb{A}_2$ whose only branch at $(0,0)$ has Puiseux series $\sum_{i=1}^{m+j} a_i x^{i/n} + bx^{(m+j+1)/n}$, $b \neq a_{m+j+1}$. Prove that $[C_j \cdot \gamma] = [C_{j-1} \cdot \gamma] + 1$ for all $j > 0$.

2.11 Let γ be a branch of a curve of $\mathbb{A}_2$.

(1) Prove that the subset of $\mathbb{N}$

$$\Gamma = \{[C \cdot \gamma] \mid C \text{ a curve of } \mathbb{A}_2\}$$

is a semigroup. It is called the *semigroup* of γ.

(2) Prove that $\Gamma = \mathbb{N}$ if and only if γ is smooth.

(3) Prove that $\mathbb{N} - \Gamma$ is a finite set.

Hint: Use Exercise 2.10 for claim (3).

2.12 Fulton's axioms for intersection multiplicity, see [12], 3.3. Fix a projective plane $\mathbb{P}_2$ and a point $p \in \mathbb{P}_2$. Denote by $\mathfrak{C}$ the set of all curves of $\mathbb{P}_2$.

(1) Prove that there is at most one map

$$\mathfrak{C}^2 \longrightarrow \mathbb{N} \cup \{\infty\}$$
$$(C, C') \longmapsto |C \cdot C'|_p$$

satisfying the following conditions for any $C, C', C'' \in \mathfrak{C}$:

(a) $|C \cdot C'|_p = \infty$ if and only if C and C' share an irreducible component going through p.

(b) $|C \cdot C'|_p = 0$ if and only if $p \notin C \cap C'$.

(c) $|C \cdot C'|_p = |C' \cdot C|_p$.

(d) $|C \cdot C'|_p \geq e_p(C)e_p(C')$ and equality holds if and only if C and C' share no tangent at C.

(e) $|C \cdot (C' + C'')|_p = |C \cdot C'|_p + |C \cdot C''|_p$.

(f) $|C \cdot C'|_p = |C \cdot C''|_p$ if C'' is defined by any equation of the form $G + AF = 0$, where F is an equation of C, G is an equation of C' and A is any homogeneous polynomial of degree $\deg G - \deg F$ in the projective coordinates.

Hint: fix coordinates such that $p = [1,0,0]$, let F and G be equations of C and C', respectively, and prove that $|C \cdot C'|_p$ is well determined using double induction on $|C \cdot C'|_p$ and $\min\{\deg_{x_1} F(x_0, x_1, 0), \deg_{x_1} G(x_0, x_1, 0)\}$, condition (f) being used to decrease the latter.

(2) Prove that the intersection multiplicity map $(C, C') \mapsto [C \cdot C']_p$ satisfies the conditions (a) to (f) above, and therefore is the only map satisfying them.

(3) Following the proof of (1), describe a rational algorithm which computes $[C \cdot C']_p$ from homogeneous equations of C and C'. Use it to compute the intersection multiplicity of $C : x_1^2 - x_0x_2 + 2x_2^2 = 0$ and $C' : x_1^3 - x_1x_2^2 - x_0^2x_2 = 0$ at $[1, 0, 0]$.

Chapter 3

Projective Properties of Plane Curves

3.1 Sylvester's Resultant

This section is devoted to introducing and proving the main properties of the resultant, in the form due to Sylvester. Roughly speaking the resultant of two polynomials is a polynomial expression in the coefficients of the polynomials that is zero if and only if either the polynomials have a common root or they both have degree strictly less than prescribed.

Assume that A is a ring with no zero divisors; denote by K its fraction field and by $\bar{K}$ the algebraic closure of K. Assume given two polynomials written in the form

$$
\begin{aligned}
P &= a_0 y^n + a_1 y^{n-1} + \cdots + a_{n-1} y + a_n, \\
Q &= b_0 y^m + b_1 y^{m-1} + \cdots + b_{m-1} y + b_m,
\end{aligned}
\tag{3.1}
$$

both belonging to $A[y]$ and with $n, m > 0$. We do not assume $a_0 \neq 0$ or $b_0 \neq 0$ and therefore we have just $\deg P \leq n$ and $\deg Q \leq m$. We will refer to n and m as being the *formal degrees* of the polynomials. The usual degrees $\deg P, \deg Q$ will be called *effective degrees* if some confusion may occur.

The *resultant* (or *Sylvester's resultant*) of the polynomials P, Q, taken with respective formal degrees n, m, is defined as being the element of A

$$
R(P,Q) = \begin{vmatrix}
a_0 & a_1 & \cdots & a_{n-1} & a_n & 0 & \cdots & 0 \\
0 & a_0 & a_1 & \cdots & a_{n-1} & a_n & \cdots & 0 \\
\vdots & & & & \vdots & & & \vdots \\
0 & \cdots & 0 & a_0 & a_1 & \cdots & a_{n-1} & a_n \\
b_0 & b_1 & \cdots & b_{m-1} & b_m & 0 & \cdots & 0 \\
0 & b_0 & b_1 & \cdots & b_{m-1} & b_m & \cdots & 0 \\
\vdots & & & & \vdots & & & \vdots \\
0 & \cdots & 0 & b_0 & b_1 & \cdots & b_{m-1} & b_m
\end{vmatrix},
\tag{3.2}
$$

© Springer Nature Switzerland AG 2019
E. Casas-Alvero, *Algebraic Curves, the Brill and Noether Way*, Universitext,
https://doi.org/10.1007/978-3-030-29016-0_3

the $(n+m) \times (n+m)$ determinant having m rows with coefficients of P and n rows with coefficients of Q. The reader may note that the resultant depends not only on P and Q, but also on the formal degrees n, m, which are not determined by P and Q. In the sequel, when no mention of formal degrees is made, they are taken equal to the effective degrees.

The main property of the resultant is as follows:

Theorem 3.1.1 (Sylvester 1853) *If $P, Q \in A[y]$ are taken with respective formal degrees n, m, then the following conditions are equivalent:*

(i) $R(P, Q) = 0$.

(ii)   *Either P and Q have a common root in $\bar{K}$ or $\deg P < n$ and $\deg Q < m$.*

(iii)   *Either P and Q have a common factor in $K[y]$ or $\deg P < n$ and $\deg Q < m$.*

PROOF: The equivalence of (ii) and (iii) is clear: a minimal polynomial (see the beginning of Section 2.5) of a common root is a common factor and any root of a common factor is a common root. We will prove that (ii) and (iii) are also equivalent to the following claim:

(iv)   *There exist non-zero polynomials $M, N \in K[y]$, with $\deg M < m$ and $\deg N < n$, such that $MP + NQ = 0$.*

Indeed, assume (iii): if $\deg P < n$ and $\deg Q < m$, just take $M = Q$ and $N = -P$; if F is a common factor of P and Q, take $M = Q/F$ and $N = -P/F$. For the converse we will prove (ii). If $\deg P < n$ and $\deg Q < m$ there is nothing to do. Otherwise, the arguments being the same in both cases, assume $\deg P = n$; then $P = a_0 \prod_{j=1}^{h} (y - \alpha_j)^{r_j}$, where $\alpha_1, \ldots, \alpha_h$ are the roots of P (in $\bar{K}$) and $r_1 + \cdots + r_h = n$. Since we assume $MP + NQ = 0$, the polynomial $\prod_{j=1}^{h} (y - \alpha_j)^{r_j}$ divides NQ, and then at least one of its factors divides Q because $\deg N < n$. The corresponding root is thus a common root.

Now taking
$$N = z_0 y^{n-1} + \cdots + z_{n-1},$$
$$M = t_0 y^{m-1} + \cdots + t_{m-1},$$

the equality $MP + NQ = 0$ translates into a system of $n + m$ linear and homogeneous equations in $z_0, \ldots, z_{n-1}, t_0, \ldots, t_{m-1}$ which has a non-trivial solution if and only if claim (iv) is satisfied. The reader may take care of checking that the determinant of such a system of equations is $R(P, Q)$, which proves the equivalence of (i) and (iv) and hence ends the proof.   ◇

Using two homogeneous variables allows us to replace the two possibilities of 3.1.1(ii) with a single one. Indeed, assume given homogeneous polynomials

$$\bar{P} = a_0 y_1^n + a_1 y_0 y_1^{n-1} + \cdots + a_{n-1} y_0^{n-1} y_1 + a_n y_0^n,$$
$$\bar{Q} = b_0 y_1^m + b_1 y_0 y_1^{m-1} + \cdots + a_{m-1} y_0^{m-1} y_1 + y_0^m b_m,$$

with coefficients in A and positive degrees and take

$$
R(\bar{P}, \bar{Q}) =
\begin{vmatrix}
a_0 & a_1 & \cdots & a_{n-1} & a_n & 0 & \cdots & 0 \\
0 & a_0 & a_1 & \cdots & a_{n-1} & a_n & \cdots & 0 \\
\vdots & & & & \vdots & & & \vdots \\
0 & \cdots & 0 & a_0 & a_1 & \cdots & a_{n-1} & a_n \\
b_0 & b_1 & \cdots & b_{m-1} & b_m & 0 & \cdots & 0 \\
0 & b_0 & b_1 & \cdots & b_{m-1} & b_m & \cdots & 0 \\
\vdots & & & & \vdots & & & \vdots \\
0 & \cdots & 0 & b_0 & b_1 & \cdots & b_{m-1} & b_m
\end{vmatrix},
$$

which is also called the resultant of $\bar{P}, \bar{Q}$. Then we have

Corollary 3.1.2 $R(\bar{P}, \bar{Q}) = 0$ *if and only if there is* $(\alpha, \beta) \in \bar{K}^2 - \{(0,0)\}$ *such that* $\bar{P}(\alpha, \beta) = \bar{Q}(\alpha, \beta) = 0$.

PROOF: Take P, Q as in (3.1) above. After an easy computation, we have $\bar{P}(\alpha, \beta) = \bar{Q}(\alpha, \beta) = 0$ with $\alpha \neq 0$ if and only if $P(\beta/\alpha) = Q(\beta/\alpha) = 0$, while $\bar{P}(0, \beta) = \bar{Q}(0, \beta) = 0$, $\beta \neq 0$, if and only if $a_0 = b_0 = 0$. Since $R(\bar{P}, \bar{Q}) = R(P, Q)$, the claim follows from 3.1.1. ◇

The properties of the resultant we will need in the sequel are proved below. In all of them P, Q are given by the equalities (3.1) above.

Proposition 3.1.3 *The resultant* $R(P, Q)$ *belongs to the ideal generated by* P, Q *in* $A[y]$.

PROOF: Successively add to the last column of the determinant in (3.2) the first column multiplied by y^{m+n-1}, then the second column multiplied by y^{m+n-2} and so on, until adding the last but one column multiplied by y. This obviously does not modify the value of the determinant and turns its last column into a new one all whose entries are either multiples of P or multiples of Q. Taking the development by this new last column proves the claim. ◇

Remark 3.1.4 According to 3.1.3, there are $N, M \in A[y]$ such that $R(P, Q) = MP + NQ$. The computation made in the proof of 3.1.3 proves that they may be taken with $\deg M < m$ and $\deg N < n$.

For the next proposition, take the coefficients a_i, b_j, $i = 0 \ldots, n$, $j = 0, \ldots, m$, of P and Q, as being free variables. The *weight* of a non-zero monomial $c a_{i_1} \cdots a_{i_r} b_{j_1} \cdots b_{j_k}$, $c \in \mathbb{C}$, is taken to be the sum $i_1 + \cdots + i_r + j_1 + \cdots + j_k$ of the indices of the variables appearing in it. A polynomial in the a_i, b_j is called *isobaric of weight* m if and only if all its non-zero monomials have weight m.

Proposition 3.1.5 *The resultant $R(P, Q)$ is:*

– *homogeneous of degree n as a polynomial in $a_0, \ldots, a_n$,*

– *homogeneous of degree m as a polynomial in $b_0, \ldots, b_m$, and*

– *isobaric of weight nm as a polynomial in $a_0, \ldots, a_n, b_0, \ldots, b_m$.*

PROOF: The determinant on the right of (3.2) is the sum of all products of $n + m$ non-zero entries taken from different rows and columns, each product taken with a suitable sign. After this, the first and second claims are clear because the determinant has m rows whose non-zero entries are coefficients of P and n rows whose non-zero entries are coefficients of Q. For the last claim, consider any of the above products, say

$$a_{i_1} \cdots a_{i_m} b_{j_1} \cdots b_{j_n}$$

where each a_{i_ℓ} has been picked from the ℓ-th row and each b_{j_k} from the $(m + k)$-th row. Then, by the form of the determinant, a_{i_ℓ} comes from the $(i_\ell + \ell)$-th column and b_{j_k} from the $(j_k + k)$-th column, for $\ell = 1, \ldots, m$ and $k = 1, \ldots, n$. Since these columns all need to be different,

$$\{i_1 + 1, \ldots, i_m + m, j_1 + 1, \ldots, j_n + n\} = \{1, \ldots, n + m\}$$

and therefore, by adding up the elements on each side,

$$\sum_{\ell=1}^{m} i_\ell + \frac{m(m+1)}{2} + \sum_{k=1}^{n} j_k + \frac{n(n+1)}{2} = \frac{(n+m)(n+m+1)}{2}.$$

This results in

$$\sum_{\ell=1}^{m} i_\ell + \sum_{k=1}^{n} j_k = nm,$$

as claimed.   ⬦

Next we will get an expression for the resultant in terms of the roots of the polynomials. We will make use of the following:

Lemma 3.1.6 *Assume that $F \in K[z_1, \ldots, z_h]$ becomes zero when the variable z_i is replaced with z_j, $i \neq j$. Then $z_i - z_j$ divides F in $K[z_1, \ldots, z_h]$.*

PROOF: Up to reordering the variables assume $i = 1$ and $j = 2$ for simpler notations. Taking both F and $z_1 - z_2$ as polynomials in the single variable z_1 and coefficients in $K[z_2, \ldots, z_h]$, we can perform the Euclidean division of F by $z_1 - z_2$ in $K[z_2, \ldots, z_h][z_1]$, because the divisor is monic. It will give

$$F = (z_1 - z_2)F' + G,$$

with $F', G \in K[z_1, \ldots, z_h]$ and $\deg_{z_1} G = 0$. Replacing z_1 with z_2 in the above equality gives $0 = G$, because G does not contain z_1, and hence $F = (z_1 - z_2)F'$, as claimed.   ⬦

Theorem 3.1.7 *Assume that P and Q as above have $a_0 \neq 0$ and $b_0 \neq 0$, that the roots of P are $s_1, \ldots, s_n$ and that the roots of Q are $s'_1, \ldots, s'_m$, these roots being in both cases repeated according to multiplicities. Then*

$$R(P, Q) = a_0^m b_0^n \prod_{\substack{i=1,\ldots,n \\ j=1,\ldots,m}} (s_i - s'_j). \tag{3.3}$$

PROOF: Since dividing both sides of the equality (3.3) by $a_0^m b_0^n$ gives the equivalent equality

$$R(\frac{1}{a_0}P, \frac{1}{b_0}Q) = \prod_{\substack{i=1,\ldots,n \\ j=1,\ldots,m}} (s_i - s'_j),$$

we are allowed to assume the supplementary hypothesis $a_0 = b_0 = 1$.

Take free variables over $K(y)$, $S_1, \ldots, S_n, S'_1, \ldots, S'_m$, and write

$$K[S, S'] = K[S_1, \ldots, S_n, S'_1, \ldots, S'_m].$$

Consider the polynomials

$$\mathcal{P} = \prod_{i=1}^{n} (y - S_i),$$

$$\mathcal{Q} = \prod_{i=1}^{m} (y - S'_i).$$

If they are written as polynomials in y with coefficients in $K[S, S']$,

$$\mathcal{P} = \sum_{i=0}^{n} A_i y^{n-i},$$

$$\mathcal{Q} = \sum_{i=0}^{m} B_i y^{m-i},$$

each coefficient A_i is a homogeneous polynomial of degree i in $S_1, \ldots, S_n$ with coefficients in K, and so is each B_i in $S'_1, \ldots, S'_m$.

Take $\mathcal{R} = R(\mathcal{P}, \mathcal{Q})$ to be the resultant of $\mathcal{P}, \mathcal{Q}$ still taken as polynomials in y.

Replacing S_i with S'_j induces a ring-homomorphism

$$\Psi : K[S, S'] \longrightarrow K[S, S'],$$

and its obvious extension

$$K[S, S'][y] \longrightarrow K[S, S'][y]$$

still denoted Ψ. Since still $\deg_y \Psi(\mathcal{P}) = n$ and $\deg_y \Psi(\mathcal{Q}) = m$ and the resultant is a polynomial function of the coefficients of the polynomials,

$$\Psi(R(\mathcal{P}, \mathcal{Q})) = R(\Psi(\mathcal{P}), \Psi(\mathcal{Q})).$$

Now, $R(\Psi(\mathcal{P}), \Psi(\mathcal{Q})) = 0$ because the polynomials share the root S'_j (3.1.1). This shows that $R(\mathcal{P}, \mathcal{Q})$, which is an element of $K[S, S']$, becomes zero when any of the S_i, $i = 1, \ldots, n$, is replaced with any of the S'_j, $j = 1, \ldots, m$. By Lemma 3.1.6, any $S_i - S'_j$, $i = 1, \ldots, n$, $j = 1, \ldots, m$, divides $R(\mathcal{P}, \mathcal{Q})$; these polynomials being coprime, its product also divides $R(\mathcal{P}, \mathcal{Q})$, that is,

$$R(\mathcal{P}, \mathcal{Q}) = \prod_{\substack{i=1,\ldots,n \\ j=1,\ldots,m}} (S_i - S'_j)M \tag{3.4}$$

with $M \in K[S, S']$.

We already know the coefficients A_i and B_i, of $\mathcal{P}$ and $\mathcal{Q}$, to be homogeneous of degree i in $S_1, \ldots, S_n$ and $S'_1, \ldots, S'_m$, respectively. By the third claim of 3.1.5, $R(\mathcal{P}, \mathcal{Q})$ is homogeneous of degree nm in $S_1, \ldots, S_n$, $S'_1, \ldots, S'_m$. Back to (3.4), we see that $M \in K$.

The reader may compare the product of the elements on the principal diagonal of the determinant giving $R(\mathcal{P}, \mathcal{Q})$, which is the only monomial involving no S_i, with the corresponding monomial on the right of (3.4) to see that $M = 1$ and therefore

$$R(\mathcal{P}, \mathcal{Q}) = \prod_{\substack{i=1,\ldots,n \\ j=1,\ldots,m}} (S_i - S'_j), \tag{3.5}$$

which is a sort of transcendent version of the equality we want to prove.

Now, consider the specialization ring-homomorphism

$$\chi : K[S, S'] \longrightarrow \bar{K},$$

obtained by replacing S_i with s_i and S'_j with s'_j, $i = 1, \ldots, n$, $j = 1, \ldots, m$, and its extension to polynomials in y, still denoted χ. Once χ is applied to both sides of (3.5), we have

$$\chi(R(\mathcal{P}, \mathcal{Q})) = \prod_{\substack{i=1,\ldots,n \\ j=1,\ldots,m}} (s_i - s'_j).$$

On the other hand, clearly, $\chi(\mathcal{P}) = P$ and $\chi(\mathcal{Q}) = Q$. Since the degrees in y of $\mathcal{P}$ and $\mathcal{Q}$ are preserved by χ, again, as for Ψ above,

$$\chi(R(\mathcal{P}, \mathcal{Q})) = R(\chi(\mathcal{P}), \chi(\mathcal{Q})) = R(P, Q)$$

and the proof is complete.   ◇

3.2 Plane Curves and their Points

We will deal with curves $C : F = 0$ of $\mathbb{P}_2$, $F \in \mathbb{C}[x_0, x_1, x_2]$ and their affine parts. As done before, we will usually fix our attention on the 0-th affine chart $x_0 \neq 0$, taken with affine coordinates $x = x_1/x_0, y = x_2/x_0$. The corresponding affine part of C is then $C_0 : f = 0$, $f(x,y) = F(1,x,y)$. If the line $\ell_0 : x_0 = 0$ is not an irreducible component of C and therefore x_0 is not a factor of F, then $d = \deg C = \deg C_0$. Assume further that f is written as the sum of its homogeneous parts,

$$f = f_0 + \cdots + f_d,$$

with f_i homogeneous of degree i; then $f_d \neq 0$ and the intersection $C \cap \ell_0$ is the group of points defined by

$$F(0, x_1, x_2) = f_d(x_1, x_2).$$

Remark 3.2.1 Notations being as above, in the sequel we will often choose the projective coordinates such that the third vertex of the reference, $[0, 0, 1]$, does not belong to C, which in particular ensures that ℓ_0 is not an irreducible component of C. The condition $[0,0,1] \notin C$ is clearly equivalent to the monomial of F in x_2^d, $d = \deg F$, being non-zero, and also to the monomial in y^d of f being non-zero. After such a choice, $x_0 = 0$ not being a component of C, the irreducible components of the affine part C_0 are the affine parts of the irreducible components of C (by 1.2.7).

In the sequel, for any two affine or projective curves C, C' of the same plane, we will write $C \cap C' = |C| \cap |C'|$; we will call it the intersection of C and C' and its points the intersection points of C and C'.

Let us begin by proving an almost obvious fact:

Lemma 3.2.2 *Any algebraic curve C, of either $\mathbb{A}_2$ or $\mathbb{P}_2$, contains infinitely many points.*

PROOF: For the affine case, assume the curve to be $C : f = 0$. Since f is non-constant, up to swapping the affine coordinates, let us assume that ordered as a polynomial in y it is

$$f(x,y) = a_0(x)y^d + \cdots + a_d(x)$$

with $d > 0$ and $a_0(x) \neq 0$. Then, for infinitely many $z \in \mathbb{C}$, $a_0(z) \neq 0$, the polynomial in y, $f(z,y)$, has at least one root and therefore there is a point of C with abscissa z. In the projective case we know from 1.1.5 that C has some point, and therefore some non-empty affine part to which the affine statement applies. $\diamond$

The key to the remaining results in this section is the following:

Proposition 3.2.3 *Two curves C, C' of $\mathbb{P}_2$ either have a common irreducible component, and then they share infinitely many points, or they intersect in finitely many points.*

PROOF: Choose the projective coordinates x_0, x_1, x_2 such that $[0, 0, 1]$ does not belong to C or C' and, as usual, take $x = x_1/x_0$ and $y = x_2/x_0$ as affine coordinates on the 0-th affine chart. Since the line $x_0 = 0$ is not contained in

C or C', it is not an irreducible component of either curve and therefore, by 1.2.8, C and C' share an irreducible component if and only if so do their 0-th affine parts C_0 and C'_0. On the other hand, the intersection of $x_0 = 0$ with either curve being finite, the intersection of C and C' is finite if and only if so is the intersection of their 0-th affine parts. Therefore, it will be enough to prove the claim for C_0 and C'_0, which we will do next.

Let f and g to be affine equations of C_0 and C'_0, $d = \deg f$ and $d' = \deg f'$. The choice of the reference guarantees that $d = \deg_y f$ and $d' = \deg_y f'$ (see 3.2.1). Consider the resultant $R_y(f, g)$ of f and g taken as polynomials in y of degrees d and d', $R(f, g) \in \mathbb{C}[x]$.

Assume that $R_y(f, g) = 0$, as a polynomial in x. Then, by 3.1.1, f and g share a factor h, of positive degree in y, in $\mathbb{C}(x)[y]$. Write $h = uh'/v$ with $u, v \in \mathbb{C}[x]$ and $h' \in \mathbb{C}[x][y]$ primitive. Then, h' still has positive degree in y and divides f and g in $\mathbb{C}(x)[y]$. Using 2.5.2, $h' = 0$ is a curve any of whose irreducible components is shared by C_0 and C'_0. The set $C_0 \cap C'_0$ is then infinite due to 3.2.2.

Assume now $R_y(f, g) \neq 0$, as a polynomial in x. Fix any $a \in \mathbb{C}$. By the choice of the reference (again, see 3.2.1) $f(a, y)$ has effective degree d and $g(a, y)$ has effective degree d'; therefore, the replacement of x with a being a ring-homomorphism

$$\mathbb{C}[x, y] \longrightarrow \mathbb{C}[x]$$
$$P(x, y) \longmapsto P(a, y),$$

we have $R_y(f, g)_{|x=a} = R(f(a, y), g(a, y))$. Since, by 3.1.1, the latter is 0 if and only if $f(a, y)$ and $g(a, y)$ share a root in $\mathbb{C}$ ($\mathbb{C}$ is algebraically closed), it follows that $R_y(f, g)(a) = 0$ if and only if there is a $b \in \mathbb{C}$ such that $f(a, b) = g(a, b) = 0$: the abscissae of the intersection points of C_0 and C'_0 are the roots of $R_y(f, g)$. Since there are finitely many roots of $R_y(f, g)$ and, for any a, there are at most d roots of $f(a, y)$, there are finitely many points in $C_0 \cap C'_0$. In particular, C_0 and C'_0 share no irreducible component in this case.   ⋄

Remark 3.2.4 The proof of 3.2.3 shows how to use the resultant to detect if the curves C and C' – both still assumed to be missing the point $[0, 0, 1]$ – share an irreducible component and, this being not the case, how to compute the intersection points of their 0-th affine parts. There is a shared irreducible component if and only if $R_y(f, g) = 0$. Otherwise, the abscissae of the intersection points are the roots a of $R_y(f, g)$ and, for each such a, the roots of $\gcd(f(a, y), g(a, y))$ are the ordinates of the intersection points with abscissa a.

The intersection of two curves C, C' of $\mathbb{P}_2$ is the union of all the intersections $C_i \cap C'_j$ where C_i is an irreducible component of C and C'_j an irreducible component of C'. Therefore $C \cap C'$ consists of all the points of the irreducible components shared by C and C', together with the points of the pairwise intersections of the non-shared irreducible components of C and C'. There are finitely many of the latter by 3.2.3. From the points of $C \cap C'$, those which do not belong to any shared irreducible component are called *isolated intersection points* of C and C'. They are of course finitely many. Repeating the argument and including the obvious case $r = 1$, we have:

Corollary 3.2.5 *If* $C_1, \dots, C_r$, $r \geq 1$, *are curves of* $\mathbb{P}_2$, *the points of* $C_1 \cap \cdots \cap C_r$ *are the points belonging to one of the irreducible components shared by* $C_1, \dots, C_r$ *(if there is one) plus finitely many points.*

Still from 3.2.3, we have:

Corollary 3.2.6 *If infinitely many points of an irreducible curve* C *of* $\mathbb{P}_2$ *belong to another curve* C' – *which in particular occurs if* $|C| \subset |C'|$ – *then* C *is an irreducible component of* C'. *If* C' *is irreducible, then* $C = C'$.

PROOF: By 3.2.3 C and C' share an irreducible component which, C being irreducible, by 1.1.2, necessarily equals C; it also equals C' if the latter is irreducible too. ◇

The next corollary is the particular case of Hilbert's Nullstelensatz already mentioned in Section 1.1.

Corollary 3.2.7 *Two curves* C, C' *of* $\mathbb{P}_2$ *have the same points if and only if they have the same irreducible components, that is,* $C_{red} = C'_{red}$.

PROOF: If $|C| = |C'|$, then the points of any irreducible component C_1 of C all belong to C'; therefore, by 3.2.6, C_1 is an irreducible component of C'. The same argument applies to the irreducible components of C'. The converse is clear. ◇

Remark 3.2.8 Due to 3.2.7, the reduced curves – in particular the irreducible curves – are determined by their sets of points. In the sequel we will make no distinction between irreducible curves and their sets of points.

3.3 Bézout's Theorem

Assume that $f \in \mathbb{C}\{x\}[y]$ and has no factor x. In Section 2.3 we have found all the y-roots s of f with $s(0) = 0$. Now we will drop the latter condition and look for all the y-roots of f. Thus assume that a fractionary power series s in x satisfies $f(x, s) = 0$. On one hand, the polynomial $f(0, y)$ is not zero, as otherwise f would have a factor x. On the other, $f(x, s) = 0$ obviously implies $f(0, s(0)) = 0$, and so the independent term $s(0)$ of s is a root of $f(0, y)$. Therefore, if $\deg f(0, y) = 0$, no fractionary power series is a y-root of f. Otherwise, take $b \in \mathbb{C}$ such that $f(0, b) = 0$. If $\bar{y} = y - b$ and $\bar{f}(x, \bar{y}) = f(x, \bar{y} + b)$, then it is clear that $\bar{f}(0, 0) = 0$ and therefore $h(N(\bar{f})) > 0$. More precisely, $h(N(\bar{f}))$ is the multiplicity of 0 as a root of $\bar{f}(0, \bar{y})$, which in turn equals the multiplicity of b as a root of $f(0, y)$. If

$h = h(N(\bar{f}))$, by 2.3.4 and 2.3.6, there are h convergent fractionary power series $\bar{s}_1, \ldots, \bar{s}_h$ – possibly repeated – such that

$$\bar{f}(x, \bar{y}) = \prod_{i=1}^{h} (\bar{y} - \bar{s}_i) \bar{f}'(x, \bar{y}) \tag{3.6}$$

with $\bar{f}'(x, \bar{y}) \in \mathbb{C}\{x\}[y]$ and $\bar{f}'(0,0) \neq 0$.

Taking $s_i = b + \bar{s}_i$ and making in (3.6) the substitution $\bar{y} = y - b$ we get

$$f(x, y) = \prod_{i=1}^{h} (y - s_i) f'(x, y)$$

where h is the multiplicity of b as a root of $f(0, y)$, $f'(x, y) \in \mathbb{C}\{x\}[y]$ and $\bar{f}'(0, b) \neq 0$.

Clearly, $f(0, y) = (y - b)^h f'(0, y)$. Therefore, the roots of $f'(0, y)$ are the roots of $f(0, y)$ other than b, with the same multiplicities. If $\deg f'(0, y) = 0$, then no other fractionary power series is a y-root of f. Otherwise $\deg f'(0, y) = \deg f(0, y) - h$ and we use induction on $\deg f(0, y)$ to get:

Proposition 3.3.1 *Assume that $f \in \mathbb{C}\{x\}[y]$ and has no factor x. Let $b_1, \ldots b_\ell$ be the roots of $f(0, y)$ and $h_1, \ldots h_\ell$ their respective multiplicities. Then for each $j = 1, \ldots, \ell$ there are convergent fractionary power series $s_{j,i}$, $i = 1, \ldots, h_j$, possibly repeated, such that $s_{j,i}(0) = b_j$ and*

$$f(x, y) = \prod_{j=1}^{\ell} \prod_{i=1}^{h_j} (y - s_{j,i}) \hat{f}(x, y), \tag{3.7}$$

where $\hat{f}(0, y)$ has degree zero.

Remark 3.3.2 For a more geometric view, the reader may consider the analytic curve $\xi : f = 0$, which is defined in a suitable neighbourhood of the y axis because f is a polynomial in y. The equality (3.7) shows its Puiseux series at each of the points in which it intersects the y-axis. It is worth comparing with the similar equality (2.10) in Puiseux's theorem 2.3.4: in the latter only the Puiseux series at the origin are explicit, the other factors appearing in (3.7) are hidden in the factor $\tilde{f}$ in (2.10). While $\tilde{f}$ is irrelevant locally at the origin, it is not so if the whole of ξ is considered. Even the factor $\hat{f}$ in (3.7) may be relevant, as it gives the asymptotic part of ξ, see Exercise 3.1.

Back to the algebraic case, 3.3.1 gives:

Proposition 3.3.3 *Assume that the algebraic curve $C : F = 0$ does not contain $[0, 0, 1]$. Then the y-roots of the equation of its 0-th affine part $f(x, y) = F(1, x, y)$ are the Puiseux series of C at the points of C lying on the y-axis.*

PROOF: Proposition 3.3.1 applies to f and the Puiseux series appearing in the factorization (3.7) are the Puiseux series of C at the points of the y-axis. Thus, we need only to show that the last factor $\hat{f}$ has no y-roots. To this end we will compare the degrees in y of both sides of (3.7). On one hand, if $d = \deg C$, then, by 3.2.1, $d = \deg_y f = \deg f(0, y)$. On the other, the h_j, $j = 1, \ldots, \ell$ are the multiplicities of the roots of $f(0, y)$ and therefore

$$\deg_y \prod_{j=1}^{\ell} \prod_{i=1}^{h_j} (y - s_{j,i}) = \sum_{j=1}^{\ell} h_j = \deg f(0, y) = d.$$

It follows that $\deg_y \hat{f} = 0$ and hence $\hat{f}$ has no y-root. ◇

The next proposition is an important step to Bézout's theorem; it shows the relationship between the resultant and the intersection multiplicity, and is far more precise than just saying that the abscissae of the intersection points are the roots of the resultant (see Remark 3.2.4).

Proposition 3.3.4 *Let* $C : F = 0$ *and* $C' : G = 0$ *be curves of* $\mathbb{P}_2$ *of degrees* d *and* d'. *Assume that they share no irreducible component and that neither of them contains the point* $[0, 0, 1]$. *Take* $f(x, y) = F(1, x, y)$ *and* $g(x, y) = G(1, x, y)$ *as equations of their 0-th affine parts and* $R_y(f, g)$ *the resultant of* f *and* g *as polynomials in* y *of degrees* d *and* d', *respectively. For any* $a \in \mathbb{C}$ *call* L_a *the line* $x - a = 0$. *Then the multiplicity* μ_a *of* a *as a root of* $R_y(f, g)$ *is*

$$\mu_a = \sum_{p \in L_a} [C \cdot C']_p.$$

PROOF: After replacement of x with $\bar{x} + a$, it is not restrictive to assume $a = 0$. Then, by 3.2.1, still $d = \deg_y f$ and $d' = \deg_y g$, from which, by 3.1.7 and 3.3.3,

$$\mu_0 = o_x R_y(f, g) = \sum_{s, s'} o_x (s - s'),$$

where in the summation s runs over the Puiseux series of C, and s' over the Puiseux series of C', both at the points of the y-axis L_0 and repeated according to multiplicities. Now, if s and s' are Puiseux series at different points, their independent terms are different, because they equal the ordinates of the points, and therefore $o_x(s - s') = 0$. Hence, the above equality may be written

$$\mu_0 = \sum_{p \in L_0} \sum_{s_p, s'_p} o_x(s_p - s'_p),$$

where in the second summation s_p and s'_p run over the Puiseux series of C and C', respectively, at p, both repeated according to multiplicities. Using Halphen's formula 2.6.6 ends the proof. ◇

Here is the main result in this section. It is named after the French mathematician Bézout, who gave a version of it in 1779. Many other versions appear in the old literature since at least Cramer (1750); most are obscure regarding the case of non-transverse intersections or just do not cover it:

Theorem 3.3.5 (Bézout's theorem) *If C and C' are curves of $\mathbb{P}_2$, of degrees d and d' and sharing no irreducible component, then*

$$dd' = \sum_{p \in \mathbb{P}_2} [C \cdot C']_p.$$

PROOF: The curves sharing no irreducible component, their intersection is finite by 3.2.3. Therefore we may choose a projective reference such that $[0, 0, 1]$ does not belong to C or C' and, furthermore, the line $L : x_0 = 0$ misses all the points in $C \cap C'$. Take $\mathbb{A}_2 = \mathbb{P}_2 - L$ and all notations as in 3.3.4. Since there are no intersection points on L,

$$\sum_{p \in \mathbb{P}_2} [C \cdot C_p] = \sum_{p \in \mathbb{A}_2} [C \cdot C']_p = \sum_{a \in \mathbb{C}} \sum_{p \in L_a} [C \cdot C']_p = \deg R_y(f, g),$$

the last equality due to 3.3.4.

To complete the proof we will prove that $\deg R_y(f, g) = dd'$. For the remainder of the proof, dots ... at the end of an expression indicate terms of lower degree. Write

$$f = a_0 y^d + a_1 y^{d-1} + \cdots + a_{d-1} y + a_d,$$
$$g = b_0 y^{d'} + b_1 y^{d'-1} + \cdots + a_{d'-1} y + a_{d'},$$

and

$$a_i = \alpha_i x^i + \dots, \quad i = 0, \dots, d,$$
$$b_i = \beta_i x^i + \dots, \quad i = 0, \dots, d'.$$

Splitting the columns of the resultant according to the above,

$R(f, g) =$

$$
\begin{vmatrix}
\alpha_0 & \alpha_1 x & \cdots & \alpha_{d-1} x^{d-1} & \alpha_d x^d & 0 & \cdots & 0 \\
0 & \alpha_0 & \alpha_1 x & \cdots & \alpha_{d-1} x^{d-1} & \alpha_d x^d & \cdots & 0 \\
\vdots & & & & \vdots & & & \vdots \\
0 & \cdots & 0 & \alpha_0 & \alpha_1 x & \cdots & \alpha_{d-1} x^{d-1} & \alpha_d x^d \\
\beta_0 & \beta_1 x & \cdots & \beta_{d'-1} x^{d'-1} & \beta_{d'} x^{d'} & 0 & \cdots & 0 \\
0 & \beta_0 & \beta_1 x & \cdots & \beta_{d'-1} x^{d'-1} & \beta_{d'} x^{d'} & \cdots & 0 \\
\vdots & & & & \vdots & & & \vdots \\
0 & \cdots & 0 & \beta_0 & \beta_1 x & \cdots & \beta_{d'-1} x^{d'-1} & \beta_{d'} x^{d'}
\end{vmatrix} + \dots.
$$

By 3.1.5, the determinant shown above is a homogeneous polynomial of degree dd' in the single variable x, hence a monomial $cx^{dd'}$. Since its coefficient c results from replacing x with 1,

$$
R(f,g) = \begin{vmatrix}
\alpha_0 & \alpha_1 & \cdots & \alpha_{d-1} & \alpha_d & 0 & \cdots & 0 \\
0 & \alpha_0 & \alpha_1 x^1 & \cdots & \alpha_{d-1} & \alpha_d & \cdots & 0 \\
\vdots & & & & \vdots & & & \vdots \\
0 & \cdots & 0 & \alpha_0 & \alpha_1 & \cdots & \alpha_{d-1} & \alpha_d \\
\beta_0 & \beta_1 & \cdots & \beta_{d'-1} & \beta_{d'} & 0 & \cdots & 0 \\
0 & \beta_0 & \beta_1 x & \cdots & \beta_{d'-1} & \beta_{d'} & \cdots & 0 \\
\vdots & & & & \vdots & & & \vdots \\
0 & \cdots & 0 & \beta_0 & \beta_1 x & \cdots & \beta_{d'-1} & \beta_{d'}
\end{vmatrix} x^{dd'} + \cdots .
$$

$$(3.8)$$

Now we will examine the intersections of C and C' with L. By recovering F and G from f and g, it is easy to see that they may be written

$$F = \alpha_0 x_2^d + \alpha_1 x_1 x_2^{d-1} + \cdots + \alpha_{d-1} x_1^{d-1} x_2 + \alpha_d x_1^d + x_0 F',$$
$$G = \beta_0 x_2^{d'} + \beta_1 x_1 x_2^{d'-1} + \cdots + \beta_{d'-1} x_1^{d'-1} x_2 + \beta_{d'} x_1^{d'} + x_0 G',$$

where F' and G' are homogeneous polynomials. The groups of points $C \cap L$ and $C' \cap L$ thus have equations

$$\alpha_0 x_2^d + \alpha_1 x_1 x_2^{d-1} + \cdots + \alpha_{d-1} x_1^{d-1} x_2 + \alpha_d x_1^d$$

and

$$\beta_0 x_2^{d'} + \beta_1 x_1 x_2^{d'-1} + \cdots + \beta_{d'-1} x_1^{d'-1} x_2 + \beta_{d'} x_1^{d'}.$$

Since we have assumed $C \cap C' \cap L = \emptyset$, by 3.1.2, the resultant of these equations is not zero; it being just the determinant appearing in (3.8) above, the proof is complete. $\diamond$

Obviously, by 2.6.9, Bézout's theorem gives the true number of intersection points if all the intersections are transverse:

Corollary 3.3.6 *In the hypothesis of 3.3.5, if C and C' are transverse at all their intersection points, then the number of points in $C \cap C'$ is dd'.*

The following are other direct corollaries of Bézout's theorem.

Corollary 3.3.7 *Any two curves of $\mathbb{P}_2$ have at least one intersection point*

Corollary 3.3.8 *Any reducible curve of $\mathbb{P}_2$ has at least one singular point.*

PROOF: A reducible curve C is composed of two curves C_1, C_2, $C = C_1 + C_2$, which share a point p by 3.3.7. Then p is a singular point of C by 1.3.3. $\diamond$

Corollary 3.3.9 *If C and C' are curves of $\mathbb{P}_2$, of degrees d and d', and there is a finite set $T \subset \mathbb{P}_2$ such that*

$$dd' < \sum_{p \in T} [C \cdot C']_p,$$

then C and C' share an irreducible component. In particular, if one of them is irreducible, then it is contained in the other.

3.4   Pencils of Curves

According to the definition in Section 1.5, once coordinates have been chosen on a projective plane $\mathbb{P}_2$, a *pencil of plane curves* of $\mathbb{P}_2$ is any family of curves of the form

$$\mathcal{P} = \{C_{\lambda_0,\lambda_1} : \lambda_0 F_0 + \lambda_1 F_1 = 0 \mid (\lambda_0, \lambda_1) \in \mathbb{C}^2 - \{0\}\},$$

where F_0, F_1 are linearly independent homogeneous polynomials of the same degree $d > 0$ in the coordinates. We will say just 'pencil' if no confusion may occur. Note that all curves in $\mathcal{P}$ have degree d. The complex numbers λ_0, λ_1 are usually referred to as the *parameters* of C_{λ_0,λ_1}; they determine the curve and are determined by it up to a non-zero complex factor, as in fact they are projective coordinates of C_{λ_0,λ_1}. The curves $C_{1,0}, C_{0,1}$ span $\mathcal{P}$ and any two different curves of $\mathcal{P}$ may be also taken as generators: doing this causes a linear and invertible substitution of the parameters.

Pencils are nice families of curves, very easy to deal with. In this section we will see some of their properties and their use in applications of the Bézout theorem.

Next is a fundamental property which is a particular case of 1.5.3. Anyway, its proof is straightforward after taking the curves as generators:

Lemma 3.4.1 *If $p \in \mathbb{P}_2$ belongs to two different curves of a pencil $\mathcal{P}$, then it belongs to all curves of $\mathcal{P}$.*

The points that belong to two – and therefore to any – of the curves of a pencil $\mathcal{P}$ are called the *base points* of $\mathcal{P}$.

The pencil $\mathcal{P}$ being as above, if $D = \gcd(F_0, F_1)$ has positive degree, all the points of the fixed part $D = 0$ of $\mathcal{P}$ are base points of $\mathcal{P}$. If, otherwise, $D = 1$ (that is, $\mathcal{P}$ has no fixed part), then $\mathcal{P}$ has finitely many base points by 3.2.3. More precisely, by Bézout's theorem 3.3.5, a pencil $\mathcal{P}$ with no fixed part and whose curves have degree d has at most d^2 base points.

Two other basic properties of pencils are:

Proposition 3.4.2 *If p is not a base point of a pencil $\mathcal{P}$, then there is one and only one curve of $\mathcal{P}$ through p*

PROOF: This is a particular case of 1.5.3. For a direct argument, if $p = [a_0, a_1, a_2]$, then the equation in λ, μ

$$\lambda F_0(a_0, a_1, a_2) + \mu F_1(a_0, a_1, a_2) = 0$$

is not trivial because p is not a base point. $\diamond$

Proposition 3.4.3 *Assume that $\mathcal{P}$ is a pencil of curves, C_0, C_1 are two different curves of $\mathcal{P}$ and γ is a branch of curve. Then*

(a) *If $[C_i \cdot \gamma] \geq \delta$ for $i = 0, 1$, then $[C \cdot \gamma] \geq \delta$ for any $C \in \mathcal{P}$.*

(b) *If still $[C_i \cdot \gamma] \geq \delta$ for $i = 0, 1$ and one of the inequalities is an equality, then $[C \cdot \gamma] = \delta$ for all $C \in \mathcal{P}$ but for a single curve $C' \in \mathcal{P}$ for which $[C' \cdot \gamma] > \delta$.*

PROOF: Take an affine chart and affine coordinates on it so that $p = (0,0)$. Let f_0 and f_1 be affine equations of C_0 and C_1 and $t \mapsto (u(t), v(t))$ a uniformizing map of γ. An affine equation of an arbitrary $C \in \mathcal{P}$ being $\lambda f_0 + \mu f_1 = 0$, we have

$$[C \cdot \gamma] = o_t(\lambda f_0(u(t), v(t)) + \mu f_1(u(t), v(t))).$$

Since $[C_i \cdot \gamma] = o_t f_i(u(t), v(t))$, $i = 0, 1$, claim (a) is clear. For claim (b), the argument being the same in both cases, assume we have $[C_0 \cdot \gamma] = \delta$. Then the initial form of $f_0(u(t), v(t))$ has degree δ, and the same occurs with all $\lambda f_0(u(t), v(t)) + \mu f_1(u(t), v(t))$, but for a single ratio $\lambda : \mu$ for which the part of degree δ cancels. ◇

Remark 3.4.4 By adding up the intersection multiplicities with the branches of E at p, claim (a) of 3.4.3 still holds if the branch γ is replaced with an arbitrary curve E and the intersection multiplicities are all taken at a fixed $p \in \mathbb{P}_2$.

The next proposition is often used in the applications:

Proposition 3.4.5 *Assume that a curve E has $C_0 \cap E = C_1 \cap E = \{p_1, \ldots, p_r\}$ and $[C_0 \cdot E]_{p_i} \leq [C_1 \cdot E]_{p_i}$, $i = 1, \ldots, r$, for two different curves C_0, C_1 of a pencil $\mathcal{P}$. Then there is a curve in $\mathcal{P}$ sharing an irreducible component with E.*

PROOF: Assume that the curves of $\mathcal{P}$ have degree d and that E has degree d'. Take any $q \in E$, $q \neq p_i$, $i = 1, \ldots, r$. By 3.4.2 there is a $C \in \mathcal{P}$ going through q. Then, by 3.4.4, $[C \cdot E]_{p_i} \geq [C_0 \cdot E]_{p_i}$, $i = 1, \ldots, r$. Using Bézout's theorem 3.3.5,

$$\sum_{i=1}^{r} [C \cdot E]_{p_i} + [C \cdot E]_q > \sum_{i=1}^{r} [C_0 \cdot E]_{p_i} = dd'$$

and then, by 3.3.9, C and E share an irreducible component. ◇

Remark 3.4.6 If the curve E of 3.4.4 is in addition assumed to be irreducible, then there is one curve in $\mathcal{P}$ containing it.

Here is one application dating from the first half of the 19th century:

Corollary 3.4.7 (Gergonne 1827) *Assume that C_0 and C_1 are curves of degree d of $\mathbb{P}_2$ that intersect in different points p_i, $i = 1, \ldots d^2$. If md of these points lie on an irreducible curve E of degree $m < d$, then the remaining $d(d - m)$ points lie on a curve of degree $d - m$.*

PROOF: Up to renumbering the intersection points, assume that E contains $p_1, \ldots, p_{dm}$. Then E misses all the remaining points $p_{dm+1}, \ldots, p_{d^2}$, as otherwise (3.3.9) E would be contained in both C_0 and C_1 which by hypothesis intersect in finitely many points. By 3.4.6, there is a curve C in the pencil spanned by C_0 and C_1 that contains E, say $C = E + E'$; then E' has degree $d - m$ and necessarily contains $p_{dm+1}, \ldots, p_{d^2}$ by 3.4.1.   ◇

We close this section with a description of the local behaviour of the curves of a pencil at the base points. We delay to the forthcoming Section 4.8 an important result concerning the singular points of the curves of a pencil (Bertini's theorem 4.8.10).

Proposition 3.4.8 *Let p be a base point of a pencil of curves $\mathcal{P}$. Then either*

(a) *all curves of $\mathcal{P}$ have the same multiplicity at p and no two of them have the same tangent cone at p, or*

(b) *all curves of $\mathcal{P}$ have the same multiplicity and tangent cone at p, except a single one which has higher multiplicity at p.*

PROOF: Fix an affine chart of $\mathbb{P}_2$ and affine coordinates x, y in it, so that $p = (0,0)$. Then the affine equations of the curves of $\mathcal{P}$ will have the form $\lambda f + \mu g = 0$, $(\lambda, \mu) \in \mathbb{C}^2 - \{0,0\}$, where $f, g \in \mathbb{C}[x,y]$ are linearly independent and have $f(0,0) = g(0,0) = 0$. Assume that the initial forms of f and g are f_e and g_r, with positive degrees e and r, respectively. Up to swapping over f and g, assume $e \leq r$ and take g_e to be the form of degree e of g: $g_e = g_r$ if $e = r$, and $g_e = 0$ otherwise.

If f_e and g_e are linearly independent, then for all pairs $(\lambda, \mu) \neq (0,0)$, the initial form of $\lambda f + \mu g$ is $\lambda f_e + \mu g_e$ and two of these initial forms are proportional if and only if so are the corresponding pairs of parameters (λ, μ). We are thus in case (a).

Otherwise, the initial forms of the $\lambda f + \mu g$ are $\lambda f_e + \mu g_e$, all proportional, but for a single ratio $\lambda : \mu$ which gives $\lambda f_e + \mu g_e = 0$ and therefore causes $\lambda f + \mu g$ to have an initial form of higher degree. This is case (b).   ◇

3.5 Singular Points, Tangent Lines and Polar Curves

The local structure of a plane algebraic curve at singular point may be rather complicated (see for instance [3] or [22]); fortunately, for developing the intrinsic geometry on a curve there is no need to deal with all types of singular points, but only with the simplest ones; they will be presented next. Recall first from Section 1.3 that if p is a point of an algebraic curve C of $\mathbb{P}_2$, and an affine chart and coordinates on it have been taken in such a way that p is the origin of coordinates, then the equation of the affine part of C has the form

$$f = f_e + f_{e+1} + \cdots + f_d$$

where each f_i is homogeneous of degree i, $f_e \neq 0$ and $e = e_p(C)$ is the multiplicity of p. The tangent cone to C at p, $TC_p(C)$, is the projective closure of $f_e = 0$; its irreducible components are the tangent lines to C at p (1.3.10). The multiplicity of each tangent line as an irreducible component of $TC_p(C)$ will be called just the *multiplicity* of the tangent; of course it is the multiplicity of the corresponding linear factor of f_e.

Remark 3.5.1 The sum of the multiplicities of the tangents to C at p is the multiplicity $e_p(C)$ of C at p.

In the descriptions that follow, an affine chart and affine coordinates have been taken as above, still $p = (0,0)$, $f = f_e + f_{e+1} + \cdots + f_d$ is an equation of the affine part of C, $e = e_p(C)$ and dots $\ldots$ at the end of an expression indicate terms of higher degree.

3.5.2 Simple points: Simple points p of a curve C have been defined as being those with $e_p(C) = 1$. By 3.5.1 above, if p is a simple point of C, then C has a single – necessarily simple – tangent line T at p, and of course $[C \cdot T]_p \geq 2$. Simple points p of C with $[C \cdot T]_p > 2$ are called *flexes* of C, $[C \cdot T]_p - 2$ then being called the *order* of the flex. The flexes with order one are called *ordinary flexes*.

With the conventions set above, the point p is a simple point of C if and only if $e = 1$, or, equivalently, at least one of the points $(1,0)$ and $(0,1)$ is a vertex of the local Newton polygon $N(f)$ of f.

If neither of the coordinate axes is T, then $o_x f(x,0) = o_y f(0,y) = 1$ and then $N(f)$ has the segment with ends $(0,1)$ and $(1,0)$ as its only side. In particular, $h(N(f)) = 1$ and, by 2.3.6, C has a unique Puiseux series at p, necessarily simple as a y-root of f. By the Newton–Puiseux algorithm this series has the form $s = ax + \ldots$, $a \neq 0$; the series being unique, we have $\nu(s) = 1$ and a is the slope of the tangent line T (2.4.6).

Assume that the coordinates have been chosen such that T is $T : y = 0$, in which case $o_x f(x,0) > 1$ and $o_y f(0,y) = 1$; still $h(N(f)) = 1$ and C has a single Puiseux series s at p, which is a simple y-root of f. Then either $N(f)$ has $(0,1)$ as its only vertex (if $o_x f(x,0) = \infty$), or the other vertex is $(r,0)$, where $r = o_x f(x,0) = [C \cdot T]_p \geq 2$. In the first case T is the only, necessarily simple, irreducible component of C through p and $s = 0$. In the second case $s = ax^r + \ldots$, $a \neq 0$, $\nu(s) = 1$ and $r = [C \cdot T]_p$. In particular, p is a flex if and only if $r > 2$, and then it has order $r - 2$.

3.5.3 Nodes and ordinary singularities: *Ordinary singular points – or ordinary singularities –* of a curve C are the points of multiplicity $e \geq 1$ at which C has e different – necessarily simple – tangents. Note that, in spite of the name, the ordinary singularities of multiplicity one are the non-singular points. Ordinary singularities of multiplicity e are called *ordinary e-fold points*, and *nodes* if $e = 2$. If, with the above conventions, the coordinates are taken such that neither of the coordinate axes is tangent to C at p, then p is an ordinary singularity if and only if $N(f)$ has a single side Γ, going from $(0, e)$ to $(e, 0)$, and furthermore the associated equation $F_\Gamma = f_e(1, z)$ has no multiple roots. Again due to 2.3.6, the number of Puiseux series of C at p, counted with multiplicities, is e. The Newton–Puiseux algorithm provides at least one Puiseux series $s_i = a_i x + \ldots$ for each root a_i of F_Γ, $i = 1, \ldots, e$. Since $a_i \neq a_j$ for $i \neq j$, the series s_i, $i = 1, \ldots, e$, are pairwise different; hence they are all the Puiseux series of C at p and each is simple as a y-root of f. Due to their initial terms, s_i and s_j are conjugate if and only if they are equal, which shows that $\nu(s_i) = 1$ for $i = 1, \ldots, e$. Note that the slopes of the tangents at p are a_i, $i = 1, \ldots, e$, just because $F_\Gamma = f_e(1, z)$. It follows that a curve has e different branches at each of its ordinary e-fold points and all these branches are smooth.

3.5.4 Ordinary cusps: An *ordinary cusp* of a curve C is a double point p of C at which C has a single tangent T and furthermore $[C \cdot T]_p = 3$. With the conventions as above, assume that the coordinates are taken such that T is the first coordinate axis. Then $o_x f(x, 0) = 3$ due to the last condition of the definition, while $o_y f(0, y) = 2$ because the y-axis is not tangent to C. If the coefficient corresponding to the point $(1, 1)$ is non-zero, then the C has two different tangent lines at p, against the definition. The local Newton polygon thus has a single side, with ends $(0, 2)$ and $(3, 0)$. The number of Puiseux series, counted with multiplicities, is two, again by 2.3.6. Since the Newton–Puiseux algorithm provides two series $s = ax^{3/2} + \ldots$ and $s' = -ax^{3/2} + \ldots$, $a \neq 0$, these are the only ones, they have $\nu(s) = \nu(s') = 2$ and are conjugate to each other. Therefore, at any ordinary cusp a curve has a single branch.

The names *cusp* and *unibranched point* apply to all the singular points at which the curve has a single branch. The curves $y^2 - x^{2k+1} = 0$, k a positive integer, have different types of cusps of multiplicity two at the origin. The ordinary cusps are particularly simple due to the condition $[C \cdot T]_p = 3$.

The next proposition provides a useful form of the equation of the tangent line at a simple point:

Proposition 3.5.5 *Assume that p is a point of a curve $C : F = 0$ of $\mathbb{P}_2$. Then the equation in the homogeneous coordinates x_0, x_1, x_2*

$$\left(\frac{\partial F}{\partial x_0}\right)_p x_0 + \left(\frac{\partial F}{\partial x_1}\right)_p x_1 + \left(\frac{\partial F}{\partial x_2}\right)_p x_2 = 0 \qquad (3.9)$$

is an identity $0 = 0$ if and only if p is a singular point of C. Otherwise it is an equation of the tangent line to C at p.

PROOF: The first claim is straightforward from 1.3.12. Thus assume (3.9) to be non-trivial and $p = [a_0, a_1, a_2]$ to be a simple point of C. If $d = \deg C$, by

Euler's formula,

$$0 = dF(a_0, a_1, a_2) = \left(\frac{\partial F}{\partial x_0}\right)_p a_0 + \left(\frac{\partial F}{\partial x_1}\right)_p a_1 + \left(\frac{\partial F}{\partial x_2}\right)_p a_2 = 0$$

and therefore p belongs to the line ℓ defined by (3.9). Take any $q = [b_0, b_1, b_2] \neq p$. The line ℓ_q spanned by p and q has parametric equations

$$(x_0, x_1, x_2) = \lambda(a_0, a_1, a_2) + \mu(b_0, b_1, b_2)$$

and therefore its intersection with C is given by

$$G(\lambda, \mu) = F(\lambda a_0 + \mu b_0, \lambda a_1 + \mu b_1, \lambda a_2 + \mu b_2).$$

Now, G has the factor μ because $p \in C$, and ℓ_q is tangent to C at p if and only if the factor μ has multiplicity at least two. This is in turn equivalent to

$$0 = \left(\frac{\partial G}{\partial \mu}\right)_{(1,0)} = \left(\frac{\partial F}{\partial x_0}\right)_p b_0 + \left(\frac{\partial F}{\partial x_1}\right)_p b_1 + \left(\frac{\partial F}{\partial x_2}\right)_p b_2$$

and therefore to $q \in \ell$. We have thus seen that $p \in \ell$ and that a point $q \neq p$ spans with p a line tangent to C at p if and only if $q \in \ell$; this proves that ℓ is the tangent to C at p ◇

Coordinates x_0, x_1, x_2 being fixed in $\mathbb{P}_2$, let $C : F = 0$ be a curve of degree $d > 1$ and $q = [a_0, a_1, a_2]$ a point, both of $\mathbb{P}_2$. Consider the polynomial

$$\partial_q F = a_0 \frac{\partial F}{\partial x_0} + a_1 \frac{\partial F}{\partial x_1} + a_2 \frac{\partial F}{\partial x_2}. \tag{3.10}$$

If it is non-zero, it defines a curve $\mathcal{P}_q(C)$ of $\mathbb{P}_2$, of degree $d-1$, which is called the *polar of C relative to q* (and sometimes also the *polar of q relative to C*). If $\partial_q F = 0$, then we say that $\mathcal{P}_q(C)$ is *undefined* (old texts say *undetermined*). Even in the case when $\partial_q F = 0$, we will refer to $\partial_q F$ as an equation of the polar.

Obviously the definition of polar curve is independent of the non-zero factors up to which the equation of C and the coordinates of q are determined. Its independence of the choice of the projective coordinates needs to be checked:

Lemma 3.5.6 *Neither the fact of it being defined, nor the polar curve itself, depend on the coordinates used in its definition.*

PROOF: Assume we have new coordinates y_0, y_1, y_2 related to the old ones by the equalities

$$x_i = \sum_{j=0}^{2} c_i^j y_j, \quad i = 0, 1, 2;$$

in particular q has new coordinates b_0, b_1, b_2 satisfying

$$a_i = \sum_{j=0}^{2} c_i^j b_j, \quad i = 0, 1, 2.$$

For any polynomial $G \in \mathbb{C}[x_0, x_1, x_2]$, we will write

$$\widehat{G} = G(\sum_{j=0}^{2} c_0^j y_j, \sum_{j=0}^{2} c_1^j y_j, \sum_{j=0}^{2} c_2^j y_j).$$

Then, on one hand, changing coordinates in the equation (3.10) gives

$$\widehat{\partial_q F} = a_0 \frac{\widehat{\partial F}}{\partial x_0} + a_1 \frac{\widehat{\partial F}}{\partial x_1} + a_2 \frac{\widehat{\partial F}}{\partial x_2} \tag{3.11}$$

as an equation of the polar in the new coordinates. On the other hand, computing an equation of the polar using the new coordinates gives

$$\sum_{j=0}^{2} b_j \frac{\partial \widehat{F}}{\partial y_j} = \sum_{j=0}^{2} b_j \sum_{i=0}^{2} \frac{\widehat{\partial F}}{\partial x_i} c_i^j = \sum_{i=0}^{2} \left(\sum_{j=0}^{2} b_j c_i^j \right) \frac{\widehat{\partial F}}{\partial x_i} = \sum_{i=0}^{2} a_i \frac{\widehat{\partial F}}{\partial x_i} = \widehat{\partial_q F},$$

from which the claim directly follows.   ◇

Remark 3.5.7 If there are no restrictions to the choice of the coordinates, they may be taken such that $q = [0, 0, 1]$, after which an equation of the polar is just $\partial F / \partial x_2$.

Remark 3.5.8 The derivatives $\partial F / \partial x_i$, $i = 0, 1, 2$, are linearly dependent if and only if there is an undefined polar. Therefore, if all polars of C are defined, then they describe a net called the *net of polars* of C. Furthermore, mapping q to $\mathcal{P}_q(C)$ is a projectivity between $\mathbb{P}_2$ and the net of polars of C.

The cases in which a polar is undefined are very special, namely:

Proposition 3.5.9 *The polar $\mathcal{P}_q(C)$ is undefined if and only if all irreducible components of C are lines through q. In particular, any irreducible curve of degree $d > 1$ has all its polars defined.*

PROOF: By 3.5.6 and 3.5.7, it is not restrictive to assume that $q = [0, 0, 1]$ and the equation of the polar is $\partial F / \partial x_2$. Then the equation is zero if and only if the equation F of C is a homogeneous polynomial in x_0, x_1 only. Since such a polynomial is a product of linear factors, the claim follows.   ◇

Remark 3.5.10 Take the affine chart $x_0 \neq 0$ and, as usual, the equation $f(x, y) = F(1, x, y)$ of the affine part of C. The obvious equalities

$$\frac{\partial F}{\partial x_1}(1, x, y) = \frac{\partial f}{\partial x}, \quad \frac{\partial F}{\partial x_2}(1, x, y) = \frac{\partial f}{\partial y}$$

show that the affine part of the polar $\mathcal{P}_q(C)$ relative to $q = [0, a, b]$ has equation

$$a \frac{\partial f}{\partial x} + b \frac{\partial f}{\partial y}.$$

In particular, the affine parts of the polars relative to $[0, 1, 0]$ and $[0, 0, 1]$ have equations $\partial f / \partial x$ and $\partial f / \partial y$, respectively.

Polars will become useful due to the following result:

Proposition 3.5.11 *Let C be a curve of $\mathbb{P}_2$, $\deg C > 1$ and assume $\mathcal{P}_q(C)$ defined. Then a point $p \in \mathbb{P}_2$ belongs to $\mathcal{P}_q(C) \cap C$ if and only if either*

(a) *p is a singular point of C, or*

(b) *p is a simple point of C and the tangent line to C at p goes through q.*

PROOF: The notations being as above, a point $p \in C$ belongs to $\mathcal{P}_q(C)$ if and only if

$$\left(\frac{\partial F}{\partial x_0}\right)_p a_0 + \left(\frac{\partial F}{\partial x_1}\right)_p a_1 + \left(\frac{\partial F}{\partial x_2}\right)_p a_2 = 0. \tag{3.12}$$

If either p is a singular point of C, or p is a simple point of C and the tangent at p goes through q, (3.12) above is satisfied due to 3.5.5. Conversely, if (3.12) holds, then either $(\partial F/\partial x_i)_p = 0$ for $i = 0, 1, 2$ and then p is a singular point of C by 1.3.12, or one of the derivatives is not zero, p is then a simple point of C and, again by 1.3.12, (3.12) just says that the tangent to C at p goes through q. ◇

Not only do the polar curves $\mathcal{P}_q(C)$ go through the singular points of C, but their multiplicities there may be bounded from below:

Proposition 3.5.12 *If defined, any polar curve $\mathcal{P}_q(C)$ has multiplicity $e_p(\mathcal{P}_q(C)) \geq e_p(C) - 1$ at each point p of C. Furthermore, $e_p(\mathcal{P}_q(C)) = e_p(C)$ if $p = q$.*

PROOF: If $e = e_p(C)$, take the coordinates such that $p = [1, 0, 0]$ and, if $q \neq p$, $q = [0, 0, 1]$. Then an equation of C has the form

$$F = x_0^{d-e} f_e + \cdots + f_d$$

with $f_i \in \mathbb{C}[x_1, x_2]$ homogeneous of degree i for $i = e, \ldots, d$, $f_e \neq 0$. The claim follows by just computing either $\partial F/\partial x_2$ if $p \neq q$, or $\partial F/\partial x_0$ if $p = q$. Notice that in the latter case $e = d$ has to be excluded, because it gives rise to an undefined polar. ◇

Lemma 3.5.13 *If C is irreducible and $\deg C > 1$, then, for any point $q \in \mathbb{P}_2$, $\mathcal{P}_q(C)$ is defined and $\mathcal{P}_q(C) \cap C$ is a finite set.*

PROOF: The polar $\mathcal{P}_q(C)$ is defined due to 3.5.9. The curve C is not an irreducible component of $\mathcal{P}_q(C)$ because $\deg C > \deg \mathcal{P}_q(C)$. Then $\mathcal{P}_q(C)$ and C share no irreducible component and therefore $\mathcal{P}_q(C) \cap C$ is a finite set by 3.2.3. ◇

A very important fact easily follows from 3.5.11:

Corollary 3.5.14 *An irreducible curve of $\mathbb{P}_2$ has finitely many singular points.*

PROOF: The case of a line being obvious, assume $\deg C > 1$ and take any $q \in \mathbb{P}_2$. Then $\mathcal{P}_q(C)$ is defined by 3.5.9 and 3.5.11 applies: $\mathcal{P}_q(C) \cap C$ contains all singular points of C and is a finite set by 3.5.13. $\diamond$

Remark 3.5.15 Actually, any reduced curve C of $\mathbb{P}_2$ has finitely many singular points, as by 1.3.3 these are either points shared by two different irreducible components of C, or singular points of an irreducible component of C. There are finitely many of the former by 3.2.3, while there are finitely many of the latter by 3.5.14 above.

Corollary 3.5.16 *If q is a point and C a curve, both of $\mathbb{P}_2$, then there are finitely many lines through q tangent to C.*

PROOF: Assume first C is irreducible. The case of a line being obvious, assume also $\deg C > 1$; then $\mathcal{P}_q(C)$ is defined due to 3.5.9. By 3.5.13, there are finitely many lines obtained joining q and the points of $\mathcal{P}_q(C) \cap C$ other than q. By 3.5.11, among these lines there are:

(a) all the lines through q that are tangent to C at simple points other than q, and

(b) all the lines joining q and a singular point other than q, and so, in particular, all the lines through q tangent to C at a singular point other than q.

Since there are finitely many lines tangent to C at q (none if $q \notin C$), the claim for C irreducible follows. The same claim then holds for arbitrary curves due to 1.3.3(c). $\diamond$

Polar curves allow us to detect the multiple irreducible components of a curve:

Proposition 3.5.17 *If a curve C, $\deg C > 1$, has a multiple irreducible component C', then C' is an irreducible component of any polar curve of C. Conversely, if C and one of its polar curves $\mathcal{P}_q(C)$, $q \notin C$, share an irreducible component C', then C' is a multiple irreducible component of C.*

PROOF: If C has a multiple irreducible component C', then all points of C' are singular points of C, due to 1.3.3; hence, by 3.5.11, they belong to any polar curve $\mathcal{P}_q(C)$ of C and therefore C' is an irreducible component of $\mathcal{P}_q(C)$ by 3.2.6. Conversely, if $q \notin C$ and $\mathcal{P}_q(C)$ and C share an irreducible component C', then any point of C' is either a singular point of C, or a simple point of C the tangent at which goes through q (by 3.5.11). There are finitely many of the latter because there are finitely many tangents to C through q by 3.5.16, and none of these tangents is contained in C since $q \notin C$. All but finitely many points of C' are thus singular points of C. Since there are finitely many singular points of C' (3.5.14), and hence finitely many intersection points of C' and the other irreducible components of C, there are infinitely many non-singular points of C' that do not belong to another irreducible component of C and are singular points of C. If p is any of these points and r is the multiplicity of C' as an irreducible component of C, then $r = re_p(C') = e_p(C) > 1$; hence, C' is a multiple component of C, as wanted. ◇

In particular, a characterization of the non-reduced curves follows:

Corollary 3.5.18 *Let $C : F = 0$ be a curve of $\mathbb{P}_2$, assume that the coordinates are taken such that $[0,0,1] \notin C$ and take $f(x,y) = F(1,x,y)$ as an equation of the 0-th affine part of C. Then the curve C has a multiple irreducible component if and only if the resultant $R_y(f, \partial f/\partial y)$ is the polynomial zero.*

PROOF: The case of a line being obvious, assume $\deg C > 1$. Since $q = [1,0,0]$ does not belong to C, $\mathcal{P}_q(C)$ is defined and, by Euler's formula, q does not belong to $\mathcal{P}_q(C)$ either. Then 3.2.4 applies to C and $\mathcal{P}_q(C)$, and the claim follows from it and 3.5.17.   ◇

In algebraic geometry, *enumerative problems* are the problems whose goal is to compute the number of certain objects rather than the objects themselves. Computing how many tangent lines to a curve C may be drawn from a point is an example of an enumerative problem; the next proposition contains a classical – partial – solution to it.

Proposition 3.5.19 (Plücker's first formula, Plücker, 1835) *Let C be an irreducible curve of degree d of $\mathbb{P}_2$ whose only singularities are δ nodes and κ ordinary cusps. If $q \in \mathbb{P}_2$ does not belong to C or to any of the tangents to C at its singular points, then the number of tangent lines to C through q is*

$$d(d-1) - 2\delta - 3\kappa, \tag{3.13}$$

each tangent line T counted as many times as $\sum_p([C \cdot T]_p - 1)$, with the summation extended to all contact points of T.

Note that the hypothesis of 3.5.19 excludes the points q which either belong to C, or belong to one of the tangents to C at a singular point, there being finitely many such tangents. Excluding these points is often done, in a less precise formulation, by saying that the claim holds *for q in general position*. Note also that all contact points considered in 3.5.19 are simple points, and that, according to the claim, a tangent counts once unless either it is tangent at a flex (*flex tangent*) or it has two or more contact points (*multiple tangent*).

PROOF OF 3.5.19: The case of a line being obvious, we assume $\deg C > 1$. Then the polar $\mathcal{P}_q(C)$ is defined and shares no irreducible component with C by 3.5.13; therefore Bézout's theorem 3.3.5 applies and gives

$$\sum_{\substack{p \text{ a smooth} \\ \text{point of } C}} [C \cdot \mathcal{P}_q(C)]_p = d(d-1) - \sum_{\substack{p \text{ a node} \\ \text{of } C}} [C \cdot \mathcal{P}_q(C)]_p - \sum_{\substack{p \text{ a cusp} \\ \text{of } C}} [C \cdot \mathcal{P}_q(C)]_p.$$

$$\tag{3.14}$$

Note that the summands on the left are zero unless p is a contact point of a tangent through q, by 3.5.11. The proof will be complete after computing the intersection multiplicities involved. We will do that in three separate lemmas; for further use, the second one applies to any ordinary singularities and not only to nodes. In their proofs, dots $\dots$ mean terms of higher degree.

Lemma 3.5.20 *If p is a smooth point of a curve C of $\mathbb{P}_2$, $\deg C > 1$, $q \neq p$ and the line pq is not contained in C, then the polar $\mathcal{P}_q(C)$ is defined and $[C \cdot \mathcal{P}_q(C)]_p = [C \cdot pq]_p - 1$.*

PROOF: Take a reference such that $p = [1,0,0]$, $q = [0,1,0]$ and the line spanned by p and $[0,0,1]$ is not tangent to C. Write $r = [C \cdot pq]_p$, which is finite by the hypothesis. In the affine chart $x_0 \neq 0$, $p = (0,0)$ and, by 3.5.2, an equation of the affine part of C has the form

$$f = a_{0,1}y + a_{r,0}x^r + \sum_{i+rj>r} a_{i,j}x^iy^j, \quad a_{0,1}, a_{r,0} \neq 0.$$

Its local Newton polygon having a single side, going from $(0,1)$ to $(0,r)$, there is a unique, necessarily simple, Puiseux series s of C at p, $\nu(s) = 1$ and

$$s = -\frac{a_{r,0}}{a_{0,1}}x^r + \dots$$

On the other hand, by 3.5.10, the affine part of the polar has equation

$$\frac{\partial f}{\partial x} = ra_{r,0}x^{r-1} + \sum_{i+rj>r} ia_{i,j}x^{i-1}y^j, \quad a_{r,0} \neq 0.$$

Then it is straightforward to check that $o_x(\partial f/\partial x)(x,s) = r-1$, which, using 2.6.1, ends the proof.   ◇

Lemma 3.5.21 *Assume that p is an ordinary singularity of multiplicity e of a curve C of $\mathbb{P}_2$ and q a point that does not belong to any of the tangents to C at p. Then for each branch γ of C at p, $[\mathcal{P}_q(C) \cdot \gamma] = e - 1$ and hence $[\mathcal{P}_q(C) \cdot C]_p = e(e-1)$. In particular, in case of p being a node of C, $[\mathcal{P}_q(C) \cdot C]_p = 2$.*

PROOF: Take a reference such that $p = [1,0,0]$, $q = [0,0,1]$ and the line spanned by p and $[0,1,0]$ is not tangent to C at p, and again the affine chart $x_0 \neq 0$ and affine coordinates $x = x_1/x_0$, $y = x_2/x_0$ on it. Then, as seen in 3.5.3, on one hand an equation of the affine part of C is $f = f_e + \dots + f_d$ where each $f_i \in \mathbb{C}[x,y]$ is homogeneous of degree i, $f_e \neq 0$ and $f_e(1,z)$ has roots $a_1, \dots, a_e$, all different and therefore simple. On the other hand, the Puiseux series of C at p, all simple, are $s_i = a_ix + \dots$, $\nu(s_i) = 1$, $i = 1, \dots, e$. By 3.5.10, the affine part of the polar has equation $\partial f/\partial y$ and

$$\frac{\partial f}{\partial y}(x,s_i) = \frac{\partial f_e}{\partial y}(x,s_i) + \dots = \frac{\partial f_e}{\partial y}(x,a_ix) + \dots = \frac{\partial f_e}{\partial y}(1,a_i)x^{e-1} + \dots$$

Since $(\partial f_e/\partial y)(1,a_i) \neq 0$ because a_i is a simple root of $f_e(1,z)$, we have $o_x(\partial f/\partial y)(x,s_i) = e - 1$; then 2.6.1 gives the claim.   ◇

Lemma 3.5.22 *If p is an ordinary cusp of a curve C of $\mathbb{P}_2$ and q a point that does not belong to the tangent to C at p, then $[C \cdot \mathcal{P}_q(C)]_p = 3$.*

PROOF: Take a reference such that $p = [1,0,0]$, $q = [0,0,1]$ and the line spanned by p and $[0,1,0]$ is the tangent to C at p. We are in the conditions of 3.5.4: the equation of the affine part of C in the affine chart $x_0 \neq 0$ has the form

$$f = a_{0,2}y^2 + a_{3,0}x^3 + \sum_{2i+3j>6} a_{i,j}x^i y^j, \quad a_{0,2}, a_{3,0} \neq 0$$

and C has at p two, necessarily simple, conjugate Puiseux series, which have the form $s = \pm ax^{3/2} + \dots$, $\nu(s) = 2$. An equation of the affine part of the polar being

$$\frac{\partial f}{\partial y} = 2a_{0,2}y + \sum_{2i+3j>6} ja_{i,j}x^i y^{j-1}, \quad a_{0,2} \neq 0,$$

the claim follows after substituting in it the Puiseux parameterization

$$x = t^2, \quad y = at^3 + \dots$$

of the only branch of C at p, once again using 2.6.1.   ◇   ◇

The version of the first Plücker formula in 3.5.19 is the most usual in the literature. An obvious extension allowing ordinary singularities follows from 3.5.21. Other extensions are presented in exercises 3.16 and 3.17. The integers 2 and 3 appearing in the version of 3.5.19 are called the *contributions* – to the class – of each node and cusp, respectively. The contributions to the class of arbitrary singularities may also be computed after a deeper analysis of the singularities; in which respect, the reader may see [3, Section 6.3].

The method used to solve the above enumerative problem is quite representative. The first step is to present the objects to be counted – or closely related ones, above we have counted contact points rather than tangent lines – as part of a finite intersection. The other elements of this intersection – in our case the singular points – are named *improper solutions*, while the solutions we want to count are called *proper solutions*. Then, intersection theory – Bézout's theorem here – allows us to count, with multiplicities, proper and improper solutions together. The last step is to determine the multiplicities of the improper solutions, in order to subtract them from the whole number of solutions, and also compute the multiplicities of the proper solutions in order to give a precise meaning to the final formula. For a second example of an enumerative problem (Plücker's second formula), see Exercise 3.18. The full set Plücker's formulas is in Exercise 4.35.

To close this section, the next proposition bounds the number of singular points of an irreducible plane curve in terms of its degree. It is important by itself and will be useful later on.

Proposition 3.5.23 *If C is an irreducible curve of $\mathbb{P}_2$, of degree d, and, as before, $e_p(C)$ denotes the multiplicity of the point p on C, then*

$$(d-1)(d-2) \geq \sum_{\substack{p \text{ a singular} \\ \text{point of } C}} e_p(C)(e_p(C)-1). \qquad (3.15)$$

Obviously the summation in (3.15) may be extended to all the points of C as well.

PROOF OF 3.5.23: Write $e_p = e_p(C)$. Let C' be a polar of C. As already argued, since C is irreducible and $\deg C' = d - 1 < \deg C$, the curves C and C' share no irreducible component. Since (from 3.5.12) we know C' to have multiplicity at least $e_p - 1$ at each point p of C, Bézout's theorem and 2.6.8 give

$$d(d-1) \geq \sum_{p \in C} e_p(e_p - 1). \qquad (3.16)$$

In Section 1.5 we found the dimension k of the projective space $\mathcal{H}_{d-1}(\mathbb{P}_2)$, of the curves of degree $d - 1$ of $\mathbb{P}_2$; it is

$$k = \frac{d(d+1)}{2} - 1 = \frac{d(d-1)}{2} + d - 1 \geq \sum_{p \in C} \frac{e_p(e_p - 1)}{2}.$$

So, we may choose $k - \sum_{p \in C} e_p(e_p - 1)/2$ simple points of C. For each of these points, the curves through it describe a hypersurface of $\mathcal{H}_{d-1}(\mathbb{P}_2)$, while for each singular point p of C, the curves having multiplicity at least $e_p - 1$ at p describe a linear variety of codimension $e_p(e_p - 1)/2$ (see Example 1.5.2). The sum of the codimensions of all these linear varieties equals the dimension of $\mathcal{H}_{d-1}(\mathbb{P}_2)$, hence their intersection is non-empty and therefore there is a curve C'' of degree $d - 1$ going through all the selected simple points and having a point of multiplicity at least $e_p - 1$ at each singular point p of C. As for C' above, C'' cannot share an irreducible component with C. Then, using again Bézout's theorem and 2.6.8,

$$d(d-1) \geq k - \sum_{p \in C} \frac{e_p(e_p - 1)}{2} + \sum_{p \in C} e_p(e_p - 1),$$

which gives the wanted inequality.   ◇

Remark 3.5.24 If C is still an irreducible curve of $\mathbb{P}_2$, then

$$\sigma(C) = \frac{(d-1)(d-2)}{2} - \sum_{\substack{p \text{ a singular} \\ \text{point of } C}} \frac{e_p(C)(e_p(C)-1)}{2},$$

which obviously is an integer, is called the *deficiency*, and sometimes also the *apparent genus*, of C. We have seen in 3.5.23 that it is always non-negative.

3.6   On Some Rings and Ideals

In this section we will study some rings and ideals of rational functions. We will start by recalling the basic facts about localization of entire rings. For more generality and details the reader may see for instance [15, II.3].

Throughout this section A denotes a commutative and entire (that is, with no zero divisors) ring, and K its field of quotients. Recall from algebra that a proper ideal $\mathfrak{p} \neq A$ of A is called a *prime ideal* if and only if, for any $a, b \in A$, if $ab \in \mathfrak{p}$ then either $a \in \mathfrak{p}$ or $b \in \mathfrak{p}$. Recall also that an ideal $\mathfrak{m} \neq A$ of A is called *maximal* if and only if no ideal of A, other that A and $\mathfrak{m}$ itself, contains it. Any maximal ideal is prime, see for instance [15, II.2]. The next elementary lemma will be useful later on.

Lemma 3.6.1 *If a prime ideal $\mathfrak{p}$ contains the product II' of ideals I and I', then either $I \subset \mathfrak{p}$ or $I' \subset \mathfrak{p}$.*

PROOF: Assume for instance $I \not\subset \mathfrak{p}$ and take $a \in I - \mathfrak{p}$. Then for any $b \in I'$ we have $ab \in II' \subset \mathfrak{p}$. Since $a \notin \mathfrak{p}$ and $\mathfrak{p}$ is prime, $b \in \mathfrak{p}$.   ⋄

A subset $S \subset A$ is called a multiplicative system if and only if $1 \in S$, $0 \notin S$ and S is closed under multiplication, that is, $ab \in S$ for any $a, b \in S$. If S is a multiplicative system, then the subset of K of all the quotients of the form a/b with $b \in S$ is obviously a subring of K which is called the *localized ring* or the *ring of fractions of A by S*, denoted $S^{-1}A$. Clearly $A \subset S^{-1}A \subset K$, $A = S^{-1}A$ if $S = \{1\}$ and $S^{-1}A = K$ if $S = A - \{0\}$. The reader may notice that all $b \in S$ become invertible in $S^{-1}A$, as the quotient $1/b$ belongs to $S^{-1}A$.

Remark 3.6.2 If $\varphi : A \to A'$ is a morphism of rings such that all the images $\varphi(b)$, $b \in S$, are invertible, then the reader may easily see that φ uniquely extends to a morphism $\tilde{\varphi} : S^{-1}A \to A'$ given by the rule $a/b \mapsto \varphi(a)\varphi(b)^{-1}$.

Assume that $\mathfrak{p} \subset A$ is a prime ideal. Then the complement $S = A - \mathfrak{p}$ of $\mathfrak{p}$ is clearly a multiplicative system: we will write $S^{-1}A = A_{\mathfrak{p}}$ and call $A_{\mathfrak{p}}$ the *localized ring of A at $\mathfrak{p}$*.

It is straightforward to check that for any ideal I of A,

$$S^{-1}I = \{a/b \in S^{-1}A \mid a \in I\}$$

is an ideal of $S^{-1}A$. Clearly, it is generated in $S^{-1}A$ by the elements of I. If $I \cap S \neq \emptyset$, then $S^{-1}I = S^{-1}A$ because it contains an invertible element. If $S = A - \mathfrak{p}$, $\mathfrak{p}$ a prime ideal of A, we will write $S^{-1}I = IA_{\mathfrak{p}}$. Prime ideals of A and $S^{-1}A$ are closely related:

Lemma 3.6.3 *Mapping $\mathfrak{p}$ to $S^{-1}\mathfrak{p}$ is a bijection between the set of the prime ideals of A disjoint from S and the set of all the prime ideals of $S^{-1}A$. The inverse bijection is $\mathfrak{q} \mapsto \mathfrak{q} \cap A$. Both bijections preserve inclusions.*

PROOF: Let $\mathfrak{p}$ be a prime ideal of A disjoint from S. If $(a/b)(a'/b') = aa'/bb' \in S^{-1}\mathfrak{p}$, then, by the definition of $S^{-1}\mathfrak{p}$, there is a $b'' \in S$ such that $b''aa' \in \mathfrak{p}$. S and $\mathfrak{p}$ being disjoint and the latter prime, this yields $aa' \in \mathfrak{p}$ and hence either $a \in \mathfrak{p}$ and therefore $a/b \in S^{-1}\mathfrak{p}$, or $a' \in \mathfrak{p}$ and $a'/b' \in S^{-1}\mathfrak{p}$. This proves that $S^{-1}\mathfrak{p}$ is prime. For the injectivity, assume that prime ideals $\mathfrak{p}, \mathfrak{p}'$ of A are disjoint from S and have $S^{-1}\mathfrak{p} = S^{-1}\mathfrak{p}'$. Then for any $a \in \mathfrak{p}$, $a \in S^{-1}\mathfrak{p}'$ and therefore there is a $b \in S$ such that $ba \in \mathfrak{p}'$; since $\mathfrak{p}$ is prime and $b \notin \mathfrak{p}'$, we have $a \in \mathfrak{p}'$ and so $\mathfrak{p} \subset \mathfrak{p}'$. Reversing the roles of $\mathfrak{p}$ and $\mathfrak{p}'$ gives $\mathfrak{p} = \mathfrak{p}'$. For the exhaustivity, we leave to the reader to check that if $\mathfrak{q}$ is a prime ideal of $S^{-1}A$, then $\mathfrak{p} = A \cap \mathfrak{q}$ is a prime ideal of A and $S^{-1}\mathfrak{p} = \mathfrak{q}$. The rest of the claim is clear. $\diamond$

As a non-trivial example, let $p \in \mathbb{P}_n$ and assume that $X_1, \ldots, X_n$ are affine coordinates on an affine chart $\mathbb{A}_n$ containing p, in such a way that $\mathcal{A}(\mathbb{A}_n) = \mathbb{C}[X_1, \ldots, X_n]$. Then

$$\mathfrak{m}_p = \{f \in \mathbb{C}[X_1, \ldots, X_n] \mid f(p) = 0\}$$

is obviously a prime ideal of $\mathbb{C}[X_1, \ldots, X_n]$: it is called the *ideal of* – or *corresponding to* – p; it is written $\mathfrak{m}_{\mathbb{A}_n,p}$ if a reference to the affine chart $\mathbb{A}_n$ is needed. We have seen in Section 1.4 that

$$\mathcal{O}_{\mathbb{P}_n,p} = \mathbb{C}[X_1, \ldots, X_n]_{\mathfrak{m}_p}$$

and

$$\mathfrak{M}_{\mathbb{P}_n,p} = \mathfrak{m}_p \mathcal{O}_{\mathbb{P}_n,p}. \tag{3.17}$$

Regarding the ideals $\mathfrak{m}_p$ we have:

Lemma 3.6.4 *If $p = (b_1, \ldots, b_n)$, then the ideal $\mathfrak{m}_p$ is generated by the polynomials $X_1 - b_1, \ldots, X_n - b_n$. Furthermore, $\mathfrak{m}_p$ is a maximal ideal and if $p \neq q$, then $\mathfrak{m}_p \neq \mathfrak{m}_q$.*

PROOF: Clearly, all the $X_i - b_i$ are zero at p and therefore belong to $\mathfrak{m}_p$. Assume $f \in \mathfrak{m}_p$. By putting $X_i = (X_i - b_i) + b_i$, $i = 1, \ldots n$, and operating, f may be written as a polynomial in $X_i - b_i$, $i = 1, \ldots n$, which will have the form

$$f = c + (X_1 - b_1)h_1 + \cdots + (X_n - b_n)h_n$$

with $c \in \mathbb{C}$ and $h_i \in \mathbb{C}[X_1, \ldots, X_n]$, $i = 1, \ldots n$. If $f \in \mathfrak{m}_p$, then we have $f(p) = 0$ and hence $c = 0$, which gives

$$f = (X_1 - b_1)h_1 + \cdots + (X_n - b_n)h_n,$$

as wanted. Regarding the second claim, assume now that f, written as above, does not belong to $\mathfrak{m}_p$. Then c is a non-zero complex number and therefore an invertible element of $\mathbb{C}[X_1, \ldots, X_n]$. Since

$$c = f - (X_1 - b_1)h_1 - \cdots - (X_n - b_n)h_n \in (f) + \mathfrak{m}_p$$

we have $(f) + \mathfrak{m}_p = (1)$ and hence no proper ideal strictly contains $\mathfrak{m}_p$.

To close, if $p = (b_1, \ldots, b_n) \neq q = (c_1, \ldots, c_n)$, then $b_i \neq c_i$ for some i and $X_i - b_i \notin \mathfrak{m}_q$. $\diamond$

Remark 3.6.5 Due to the equality (3.17), the polynomials $X_1 - b_1, \ldots, X_n - b_n$ generate the ideal $\mathfrak{M}_{\mathbb{P}_n, p}$ in $\mathcal{O}_{\mathbb{P}_n, p}$.

Back to the plane, the next proposition provides in particular other generators of $\mathfrak{M}_{\mathbb{P}_2, p}$:

Proposition 3.6.6 *Assume that C, C' are curves of $\mathbb{P}_2$, p an isolated intersection point of C and C', $\mathbb{A}_2$ an affine chart of $\mathbb{P}_2$ containing p and $C_0 : f = 0$ and $C'_0 : g = 0$ the corresponding affine parts of C and C'. If (f, g) denotes the ideal generated by f and g in the local ring $\mathcal{O}_{\mathbb{P}_2, p}$, then*

(a) $\mathfrak{M}_{\mathbb{P}_2, p}^r \subset (f, g)$ *for some positive integer r.*

Furthermore, if C and C' are transverse at p, then

(b) $\mathfrak{M}_{\mathbb{P}_2, p} = (f, g)$.

PROOF: First of all, we will prove that we may reduce ourselves to the case in which all the irreducible components of C or C' go through p. To this end, if C has some irreducible component missing p, write $C = \tilde{C} + \hat{C}$, where all the irreducible components of $\tilde{C}$ go through p and all the irreducible components of $\hat{C}$ are missing p. Otherwise take $\tilde{C} = C$. Then $f = f' \hat{f}$ where f' and $\hat{f}$ are equations of the affine parts of $\tilde{C}$ and $\hat{C}$, respectively, $\hat{f} = 1$ if $\tilde{C} = C$. Since $p \notin \hat{C}$, in all cases $\hat{f}(p) \neq 0$ and therefore $\hat{f}$ is invertible in $\mathcal{O}_{\mathbb{P}_2, p}$. Proceed similarly with C' to get $C' = \tilde{C}' + \hat{C}'$, all the irreducible components of $\tilde{C}$ going through p and all the irreducible components of $\hat{C}$ missing p, or just $\tilde{C}' = C'$ when there are no components missing p. Then $g = g' \hat{g}$, where g' is an equation of the affine part of $\tilde{C}'$ and $\hat{g}$ is also invertible in $\mathcal{O}_{\mathbb{P}_2, p}$. It is clear that if p is an isolated intersection of C and C', then it is an isolated intersection of $\tilde{C}$ and $\tilde{C}'$ too. Also, if C and C' are transverse at p, so are $\tilde{C}$ and $\tilde{C}'$. On the other hand $(f, g) = (f', g')$ because the generators differ in invertible factors.

Now we will proceed, as allowed, under the supplementary hypothesis that all the irreducible components of either curve are going through p. It follows from 3.6.5 that the affine equations of any two different lines through p may be taken as generators of $\mathfrak{M}_{\mathbb{P}_2, p}$. Then the claims may be equivalently written

(a') There are two different lines ℓ_1 and ℓ_2 through p whose affine equations $h_1 = 0$ and $h_2 = 0$ satisfy the relations

$$h_1^i h_2^{r-i} = A_i f + B_i g, \quad A_i, B_i \in \mathcal{O}_{\mathbb{P}_2, p}, \quad i = 0, \ldots, r.$$

(b') There are two different lines ℓ_1 and ℓ_2 through p whose affine equations $h_1 = 0$ and $h_2 = 0$ satisfy the relations

$$h_i = A_i f + B_i g, \quad A_i, B_i \in \mathcal{O}_{\mathbb{P}_2, p}, \quad i = 1, 2.$$

In this form it is clear that the claims do not depend on the choice of the affine coordinates on $\mathbb{A}_2$, as the algebraic relations involved still hold after a linear substitution of the variables.

Now we will change the affine coordinates. The curves C, C' do not have the line $\ell_0 = \mathbb{P}_2 - \mathbb{A}_2$ as an irreducible component, because $p \notin \ell_0$; therefore each intersects ℓ_0 in finitely many points. Furthermore, since p is assumed to be an isolated intersection and all the irreducible components of either C or C' are going through p, C and C' share no irreducible component and therefore $C \cap C'$ is finite. Then we take a projective reference with its second and third vertices on ℓ_0, neither belonging to C or C' or to any of the finitely many lines joining two different points of $C \cap C'$. If the corresponding coordinates are x_0, x_1, x_2, then $\mathbb{A}_2$ is still the affine chart $x_0 \neq 0$. As usual, take $x = x_1/x_0$ and $y = x_2/x_0$ as affine coordinates on $\mathbb{A}_2$. Assume that $p = (a, b)$ and let $f, g \in \mathbb{C}[x, y]$ be equations of C and C'. On one hand the resultant $R_y(f, g)$ belongs to (f, g) by 3.1.3. On the other, by 3.2.4, $R_y(f, g) = (x - a)^{r_1} P$ with $r_1 > 0$ and $P \in \mathbb{C}[x]$, $P(a) \neq 0$. Since P is then invertible in $\mathcal{O}_{\mathbb{P}_2, p}$, it follows that $(x - a)^{r_1} \in (f, g)$. The same argument swapping over the coordinates shows that $(y - a)^{r_2} \in (f, g)$ for some $r_2 > 0$. Then for claim (a') it suffices to take $h_1 = x - a$, $h_2 = y - b$ and $r = r_1 + r_2$.

For claim (b'), note first that, by the choice of the coordinates, p is the only point of $C \cap C'$ with abscissa a, and also the only point of $C \cap C'$ with ordinate b. Then, using 3.3.4 twice gives $r_1 = r_2 = [C \cdot C']_p = 1$ and (b') follows.   ◇

Next is a further fact related to localization which will be useful later on. Note that, any maximal ideal being a prime ideal, the localized ring $A_{\mathfrak{m}}$ is defined for any maximal ideal $\mathfrak{m}$ of A .

Lemma 3.6.7 *For any entire ring A we have*

$$A = \bigcap A_{\mathfrak{m}},$$

the intersection running over all the maximal ideals $\mathfrak{m}$ of A.

PROOF: For any $z \in K$ the set $(A : z) = \{b \in A \mid bz \in A\}$, of all possible denominators of z, is an ideal of A. By the definition of $A_{\mathfrak{m}}$, if $z \in A_{\mathfrak{m}}$, then $z = a/b$, $b \notin \mathfrak{m}$ and therefore there is a $b \notin \mathfrak{m}$ for which $bz \in A$. Thus if $z \in \bigcap A_{\mathfrak{m}}$, then for any maximal ideal $\mathfrak{m}$ there is a $b \notin \mathfrak{m}$ for which $bz \in A$ and therefore no maximal ideal contains the ideal $(A : z)$. It follows that $(A : z) = A$, hence $1 \in (A : z)$ and $z = 1z \in A$. The other inclusion is clear.
◇

Back to rational functions, the next proposition describes the prime ideals of $\mathbb{C}[x, y]$ – seen as the affine ring of an affine chart – as being the ideal zero, the ideals associated to irreducible curves and the ideals associated to points:

Proposition 3.6.8 *If $\mathbb{A}_2$ is an affine chart of $\mathbb{P}_2$ – or just an affine plane – with coordinates x, y, then the prime ideals of $\mathcal{A}(\mathbb{A}_2) = \mathbb{C}[x, y]$ are*

(a) *the ideal 0,*

(b) *the ideals associated to irreducible curves, namely the principal ideals (f) with f irreducible, and*

(c) *the ideals $\mathfrak{m}_p = (x - a, y - b)$ for $p = (a, b) \in \mathbb{A}_2$.*

PROOF: Let $\mathfrak{p}$ be a prime ideal of $\mathbb{C}[x, y]$. If $\mathfrak{p} = 0$ we are done. Otherwise take $f' \in \mathfrak{p}$, $f' \neq 0$. Since $\mathfrak{p}$ is a prime ideal, one of the irreducible factors f of f' belongs to $\mathfrak{p}$, and therefore $(f) \subset \mathfrak{p}$. Again if $(f) = \mathfrak{p}$ we are done. Otherwise take $g \in \mathfrak{p} - (f)$. Since f and g share no factor, the resultant $R_y(f, g) \in \mathbb{C}[x]$ is not zero and, due to 3.1.3, it belongs to $\mathfrak{p}$. Then $R_y(f, g)$ has positive degree, as otherwise it would be $\mathfrak{p} = (1)$, which is excluded. By splitting $R_y(f, g)$ into linear factors and using again that $\mathfrak{p}$ is prime it follows that one of the factors, say $x - a$ for certain $a \in \mathbb{C}$, belongs to $\mathfrak{p}$. A similar argument with $R_x(f, g)$ proves that for certain $b \in \mathbb{C}$, $y - b \in \mathfrak{p}$ and therefore $(x - a, y - b) \subset \mathfrak{p}$. If $p = (a, b)$, by 3.6.4, $\mathfrak{m}_p = (x - a, y - b)$ and it is a maximal ideal. We have thus seen that $\mathfrak{m}_p \subset \mathfrak{p}$, which gives $\mathfrak{m}_p = \mathfrak{p}$ and the proof is complete. $\diamond$

Remark 3.6.9 Directly from the definitions, for any $f \in \mathbb{C}[x, y]$, $f \neq 0$, a point p belongs to the curve $f = 0$ if and only if $(f) \subset \mathfrak{m}_p$. The other inclusions between non-zero prime ideals of $\mathbb{C}[x, y]$ are the obvious equalities: $\mathfrak{m}_p \subset \mathfrak{m}_q$ if and only if $\mathfrak{m}_p = \mathfrak{m}_q$, because $\mathfrak{m}_p$ is maximal, and, for f and f' irreducible, $(f) \subset (f')$ if and only if $(f) = (f')$, because f is irreducible.

Remark 3.6.10 It follows in particular from 3.6.4 and 3.6.9 that the maximal ideals of $\mathbb{C}[x, y]$ are the ideals of points $\mathfrak{m}_p$, $p \in \mathbb{A}_2$.

Remark 3.6.11 From 3.6.7 and 3.6.10,

$$\mathbb{C}[x, y] = \bigcap_{p \in \mathbb{A}_2} \mathbb{C}[x, y]_{\mathfrak{m}_p} = \bigcap_{p \in \mathbb{A}_2} \mathcal{O}_{\mathbb{P}_2, p}$$

and therefore a rational function on $\mathbb{P}_2$ is regular at all points of $\mathbb{A}_2$ if and only if it belongs to the affine ring $\mathbb{C}[x, y]$.

Remark 3.6.12 Since for $p \in \mathbb{A}_2$, $\mathcal{O}_{\mathbb{P}_2, p} = \mathbb{C}[x, y]_{\mathfrak{m}_p}$, it follows from 3.6.8 and 3.6.3 that the prime ideals of $\mathcal{O}_{\mathbb{P}_2, p}$ besides (0) and $\mathfrak{M}_{\mathbb{P}_2, p}$ are the ideals $f\mathcal{O}_{\mathbb{P}_2, p}$, for $f = 0$ an equation of the affine part of an irreducible curve C through p.

The prime ideal $f\mathcal{O}_{\mathbb{P}_2, p}$ above is called the *ideal of $\mathcal{O}_{\mathbb{P}_2, p}$ associated to C*, or also the *ideal of C in $\mathcal{O}_{\mathbb{P}_2, p}$*. It will be clear after 3.7.8(g) below that it does not depend on the choice of the affine part of C containing p.

Proposition 3.6.13 *Mapping each irreducible curve of $\mathbb{P}_2$ through p to the ideal of $\mathcal{O}_{\mathbb{P}_2,p}$ associated to it is a bijection between the set of all irreducible curves of $\mathbb{P}_2$ through p and the set of all prime ideals of $\mathcal{O}_{\mathbb{P}_2,p}$ other than the maximal and the zero ideals.*

PROOF: The exhaustivity has been seen in 3.6.12. If the affine equation f' of the affine part of an irreducible curve C' through p generates the same ideal as f, $f\mathcal{O}_{\mathbb{P}_2,p} = f'\mathcal{O}_{\mathbb{P}_2,p}$, then $f = f'b/b$, with b/b' invertible in $\mathcal{O}_{\mathbb{P}_2,p}$, that is, $fb' = f'b$ with $b, b' \in \mathbb{C}[x, y]$, $b(p) \neq 0$, $b'(p) \neq 0$. The last conditions ensure that neither f divides b', nor f' divides b. Both f and f' being irreducible, we have $f = cf'$, $c \in \mathbb{C} - \{0\}$, which yields the equality of the affine parts of C and C'; then $C = C'$ because both curves are assumed to be irreducible. $\diamond$

The following is an important fact regarding the polynomials vanishing at infinitely many points of an irreducible curve, given in projective and affine versions:

Proposition 3.6.14 *Assume that $C : F = 0$ is an irreducible curve of $\mathbb{P}_2$. If x_0, x_1, x_2 are homogeneous coordinates on $\mathbb{P}_2$, then:*

(a) *If a homogeneous polynomial $G \in \mathbb{C}[x_0, x_1, x_2]$ vanishes at infinitely many points of C, then $G \in (F)$ and therefore G vanishes at all points of C.*

(b) *Assume that x, y are affine coordinates on an affine chart of $\mathbb{P}_2$ and that $C_0 : f(x, y) = 0$ is the corresponding affine part of C. If a polynomial $g \in \mathbb{C}[x, y]$ vanishes at infinitely many points of C_0, then $g \in (f)$ and in particular g vanishes at all points of C_0.*

PROOF: Claim (a) is obvious if $G = 0$. Otherwise the curve $C' : G = 0$ does not have a finite intersection with C. Using 3.2.3, C and C' share an irreducible component which, C being irreducible, needs to be C itself; hence, F divides G.

For claim (b), one may assume that $x = x_1/x_0$ and $y = x_2/x_0$ (by 1.2.1), and so also that $f(x, y) = F(1, x, y)$. After this,

$$G = x_0^{\deg g} g(x_1/x_0, x_2/x_0)$$

is homogeneous and vanishes at infinitely many points of C. Then, by claim (a), G is a multiple of F, which causes $g = G(1, x, y)$ to be a multiple of f. $\diamond$

3.7 Rational Functions on an Irreducible Plane Curve

In this section we will introduce the rational functions on an irreducible curve C of $\mathbb{P}_2$ as well as some rings they compose. The reader may note that all definitions are independent of any choice of coordinates.

Roughly speaking the rational functions on C are the restrictions to C of the rational functions on $\mathbb{P}_2$. To be precise, let us first point out the following:

Lemma 3.7.1 *If a rational function on $\mathbb{P}_2$ is defined at a point p of an irreducible curve C, then it is defined at all but finitely many points of C.*

PROOF: After taking homogeneous coordinates, assume the rational function to be G_1/G_2, with G_1 and G_2 homogeneous of the same degree and $G_2(p) \neq 0$. Then just apply 3.6.14(a) to G_2. ◇

We are of course not interested in restricting to C functions that are defined at no point of C. Therefore we will restrict ourselves to considering the subset $\mathcal{R}_C$ of $\mathbb{C}(\mathbb{P}_2)$ composed of the functions which are defined at at least one point of C; by 3.7.1, each of these functions is defined at all but finitely many points of C. Hence it is obvious that $\mathcal{R}_C$ is a subring of $\mathbb{C}(C)$ containing $\mathbb{C}$. After fixing homogeneous coordinates x_0, x_1, x_2, if C has equation F, then, by 3.6.14(a), the elements of $\mathcal{R}_C$ are the rational functions that may be written as a quotient G_1/G_2, where $G_1, G_2 \in \mathbb{C}[x_0, x_1, x_2]$ are homogeneous, $\deg G_1 = \deg G_2$ and $G_2 \notin (F)$.

We are interested in making no distinction between restrictions of two rational functions to C if these restrictions patch together to give a map defined on a larger subset of C. To this end let us consider the subset of all the functions in $\mathcal{R}_C$ that are defined and vanish at all but finitely many points of C. It is straightforward to check that these functions describe a proper ideal $\mathcal{I}_C$ of $\mathcal{R}_C$. In fact, after taking coordinates and representing the rational functions in $\mathcal{R}_C$ as quotients G_1/G_2 of homogeneous polynomials of the same degree, $G_2 \notin (F)$, $G_1/G_2 \in \mathcal{I}_C$ if and only if $G_1 \in (F)$, by 3.6.14(a). Regarding $\mathcal{I}_C$ we have:

Lemma 3.7.2 *The ideal $\mathcal{I}_C$ is the only maximal ideal of $\mathcal{R}_C$.*

PROOF: By definition of $\mathcal{I}_C$, an arbitrary $\Phi \in \mathcal{R}_C - \mathcal{I}_C$ cannot be zero at all the points of C at which it is defined, because these points are all but finitely many points of C. Therefore Φ is defined and non-zero at some $p \in C$. As a consequence, $1/\Phi$ is defined at p, it thus belongs to $\mathcal{R}_C$ and Φ is invertible in $\mathcal{R}_C$. It follows that any proper ideal of $\mathcal{R}_C$ is contained in $\mathcal{I}_C$, and hence the claim. ◇

The quotient $\mathcal{R}_C/\mathcal{I}_C$ is thus a field: it will be called the *field of rational functions on C*, denoted $\mathbb{C}(C)$ in the sequel. We will identify $\mathbb{C} \subset \mathcal{R}_C$ with its isomorphic image by the morphism onto the quotient, after which $\mathbb{C}(C)$ is a $\mathbb{C}$-algebra. The elements of $\mathbb{C}(C)$ will be called *rational functions on C*. The next proposition will allow us to see them as true functions, each defined at all but finitely many points of C.

Proposition 3.7.3 *If $\Phi, \Psi \in \mathcal{R}_C$, then the following conditions are equivalent:*

(i) $\Phi - \Psi \in \mathcal{I}_C$.

(ii) Φ *and* Ψ *are defined and agree at infinitely many points of C.*

(iii) $\Phi(p) = \Psi(p)$ *for all $p \in C$ at which both Φ and Ψ are defined.*

PROOF: The implications (i) $\Rightarrow$ (ii) and (iii) $\Rightarrow$ (i) are clear. We will prove that (ii) $\Rightarrow$ (iii). Take homogeneous coordinates, let F be an equation of C and assume that the rational functions Φ and Ψ are represented as quotients of coprime homogeneous polynomials of the same degree whose denominators are not multiples of F, say $\Phi = G_1/G_2$ and $\Psi = H_1/H_2$. If condition (ii) is satisfied, then the homogeneous polynomial $G_1 H_2 - G_2 H_1$ vanishes at infinitely many points of C and therefore, by 3.6.14(a), it vanishes at all points of C. Then

$$\Phi(p) = G_1(p)/G_2(p) = H_1(p)/H_2(p) = \Psi(p)$$

provided $G_2(p) \neq 0$ and $H_2(p) \neq 0$, which are the conditions for Φ and Ψ to be defined at p.   ◇

Corollary 3.7.4 *Assume $\chi \in \mathbb{C}(C)$ and take U to be the set of points of C at which at least one representative of χ is defined. Then:*

(a) *$|C| - U$ is finite and mapping each $p \in U$ to the image of p by any representative of χ defined at p is a well-defined map $\tilde{\chi} : U \to \mathbb{C}$ which in turn determines χ.*

(b) *If $\chi = a \in \mathbb{C}$, then $U = C$ and $\tilde{\chi}$ is the constant map with value a.*

(c) *If $\chi_1, \chi_2 \in \mathbb{C}(C)$, then*

$$(\widetilde{\chi_1 + \chi_2})(p) = \widetilde{\chi_1}(p) + \widetilde{\chi_2}(p) \quad and \quad (\widetilde{\chi_1 \chi_2})(p) = \widetilde{\chi_1}(p)\widetilde{\chi_2}(p)$$

for all $p \in C$ at which both $\widetilde{\chi_1}$ and $\widetilde{\chi_2}$ are defined.

PROOF: Any representative of χ being defined at all but finitely many points of C (3.7.1), it is clear that $|C| - U$ is finite. The map $\tilde{\chi}$ is then well defined due to 3.7.3. If the classes of $\Phi, \Psi \in \mathcal{R}_C$ give rise to the same map, then Φ and Ψ agree at all points of U at which both are defined and, there being infinitely many such points, by 3.7.3, $\Phi - \Psi \in \mathcal{I}_C$. This proves claim (a). Claims (b) and (c) directly follow from the definitions.   ◇

As allowed by 3.7.4, in the sequel we will not distinguish between $\chi \in \mathbb{C}(C)$ and the map $\tilde{\chi}$ it defines, and therefore will write $\chi(p)$ for $\tilde{\chi}(p)$. Often χ is referred to as being the *restriction to C* of any of its representatives, even if, as a map, χ may be defined at points at which the representative is not. The homomorphism onto the quotient $\mathcal{R}_C \to \mathbb{C}(C)$ thus maps rational functions on $\mathbb{P}^2$ to their restrictions to C: it will be called the *restriction homomorphism* in the sequel.

Example 3.7.5 The rational functions on $\mathbb{P}_2$

$$\frac{x_1 - x_2 - x_0}{x_1 + x_2 - x_0} \quad \text{and} \quad -\frac{x_1 + x_2 + x_0}{x_1 - x_2 + x_0}$$

represent the same rational function χ on the curve $x_1^2 + x_2^2 - x_0^2 = 0$. Both the points $(1,1,0)$ and $(1,-1,0)$ are regular points of χ, but each of them is an indetermination point of one of the representatives.

As for rational functions on $\mathbb{P}_n$, $\chi \in \mathbb{C}(C)$ is said to be *defined* – or to be *regular* – at the points at which some representative of it is defined; these points are in turn called *regular points* of χ. Following the usual conventions, the *zeros* of χ are the points at which χ is defined and takes value 0, while the *poles* of $\chi \neq 0$ are the zeros of $1/\chi$. The points which are neither regular points nor poles of χ are called *indetermination points* of χ.

Remark 3.7.6 According to 3.7.3 and the conventions set above, if $\chi_1, \chi_2 \in \mathbb{C}(C)$ are defined and take the same value at infinitely many points of C, then $\chi_1 = \chi_2$.

A rational function $\chi \in \mathbb{C}(C)$ is not expected to have a representative in $\mathcal{R}_C$ which is defined at all the points of C at which χ itself is defined. In other words, the set of regular points of χ may be larger than any of the sets of points of C at which a representative of χ is defined, hence the need of patching together restrictions of different representatives.

Fix $p \in C$. Since $\chi \in \mathbb{C}(C)$ is regular at p if and only if so is one of its representatives, the subset of $\mathbb{C}(C)$

$$\mathcal{O}_{C,p} = \{\chi \in \mathbb{C}(C) \mid \chi \text{ is regular at } p\}$$

is the image of $\mathcal{O}_{\mathbb{P}_2,p}$ by the restriction homomorphism $\mathcal{R}_C \to \mathbb{C}(C)$, and therefore may be identified with $\mathcal{O}_{\mathbb{P}_2,p}/(\mathcal{O}_{\mathbb{P}_2,p} \cap \mathcal{I}_C)$. It is thus a local ring (and a $\mathbb{C}$-algebra), called the *local ring of C at p*, and sometimes also the *local ring of p on C*. Its maximal ideal is the image of $\mathfrak{M}_{\mathbb{P}_2,p}$, hence the set of all the rational functions on C which have a zero at p; we will denote it by $\mathfrak{M}_{C,p}$.

A clearer view of the above, as well as some further information, will be obtained after taking coordinates. Assume that x_0, x_1, x_2 are homogeneous coordinates of $\mathbb{P}_2$ and that C has equation F. Take $d = \deg C = \deg F$. Identify the rational functions on $\mathbb{P}_2$ with their representations as quotients G_1/G_2, $G_1, G_2 \in \mathbb{C}[x_0, x_1, x_2]$ homogeneous of the same degree, $G_2 \neq 0$. We have already seen that the elements of $\mathcal{R}_C$ are the rational functions that may be written G_1/G_2 as above, with G_2 not a multiple of F. Regarding the ideal $\mathcal{I}_C$ we have:

Proposition 3.7.7

(a) *A rational function $G_1/G_2 \in \mathcal{R}_C$, with G_2 not a multiple of F, belongs to $\mathcal{I}_C$ if and only if G_1 is a multiple of F.*

(b) *The ideal $\mathcal{I}_C$ is principal, generated by any quotient F/G with G homogeneous of degree d and not a multiple of F.*

(c) *For any $p \in C$, $\mathcal{O}_{\mathbb{P}_2,p} \cap \mathcal{I}_C$ is principal, and is generated by any quotient F/G with G homogeneous of degree d, $G(p) \neq 0$.*

PROOF: Claim (a) is straightforward from 3.6.14 applied to G_1. Using it, any element of $\mathcal{I}_C$ has the form HF/G_2, with G_2 not a multiple of F and H homogeneous (by 1.1.1) of degree $\deg G_2 - d$. Then just write

$$\frac{HF}{G_2} = \frac{HG}{G_2}\frac{F}{G}.$$

For the last claim, if $G(p) \neq 0$, then, by claim (b) F/G is a basis of $\mathcal{I}_C$. Therefore, any element of $\mathcal{O}_{\mathbb{P}_2,p} \cap \mathcal{I}_C$ may be written

$$\frac{G_1}{G_2} = \frac{H_1}{H_2}\frac{F}{G},$$

with H_2 not a multiple of F and $G_2(p) \neq 0$. It follows that

$$G_1 H_2 G = G_2 H_1 F.$$

Now F is irreducible and does not divide H_2 or G, hence it divides G_1. Then proceeding as for claim (b) ends the proof.   ◇

Using affine coordinates provides an easier presentation. Assume that the projective coordinates have been taken such that the affine chart $\mathbb{A}_2 : x_0 \neq 0$ contains some point of C. As usual take $x = x_1/x_0$, $y = x_2/x_0$ as affine coordinates and $f = F(1, x, y)$ as an equation of the affine part C_0 of C. Note that f is the rational function $f = F/x_0^d$, $d = \deg C$. Therefore, according to 3.7.7, f generates the ideal $\mathcal{I}_C$, and also the ideal $\mathcal{O}_{\mathbb{P}_2,p} \cap \mathcal{I}_C$ of $\mathcal{O}_{\mathbb{P}_2,p}$ if $p \in C_0$. After this the next proposition needs no proof. It describes the above $\mathbb{C}$-algebras and ideals in terms of elements of $\mathbb{C}(x, y)$; some of the descriptions have appeared before and are included here for the sake of completeness.

Proposition 3.7.8 *Assume $p \in C_0$. Then*

(a) $\mathbb{C}(\mathbb{P}_2) = \mathbb{C}(x, y)$.

(b) $\mathcal{R}_C = \mathbb{C}[x, y]_{(f)}$.

(c) $\mathcal{I}_C = f\mathbb{C}[x, y]_{(f)}$.

(d) $\mathbb{C}(C) = \mathcal{R}_C/\mathcal{I}_C = \mathbb{C}[x, y]_{(f)}/f\mathbb{C}[x, y]_{(f)}$.

(e) $\mathcal{O}_{\mathbb{P}_2,p} = \mathbb{C}[x, y]_{\mathfrak{m}_p}$.

(f) $\mathfrak{M}_{\mathbb{P}_2,p} = \mathfrak{m}_p\mathbb{C}[x,y]_{\mathfrak{m}_p}$.

(g) $\mathcal{O}_{\mathbb{P}_2,p} \cap \mathcal{I}_C = f\mathbb{C}[x,y]_{\mathfrak{m}_p}$.

(h) $\mathcal{O}_{C,p} = \mathcal{O}_{\mathbb{P}_2,p}/f\mathcal{O}_{\mathbb{P}_2,p} = \mathbb{C}[x,y]_{\mathfrak{m}_p}/f\mathbb{C}[x,y]_{\mathfrak{m}_p}$.

(i) $\mathfrak{M}_{C,p} = \mathfrak{M}_{\mathbb{P}_2,p}/f\mathcal{O}_{\mathbb{P}_2,p} = \mathfrak{m}_p\mathbb{C}[x,y]_{\mathfrak{m}_p}/f\mathbb{C}[x,y]_{\mathfrak{m}_p}$.

Remark 3.7.9 Using 3.7.8(d), $\mathbb{C}(C)$ appears as the extension of $\mathbb{C}$ generated by the restrictions $\bar{x}, \bar{y}$ of the coordinate functions x, y: $\mathbb{C}(C) = \mathbb{C}(\bar{x}, \bar{y})$. The reader may note that, unlike x, y, the restrictions $\bar{x}, \bar{y}$ are not free over $\mathbb{C}$, but related by the equality $f(\bar{x}, \bar{y}) = 0$. To be more precise, the extension $\mathbb{C} \subset \mathbb{C}(C)$ has degree of transcendence one. Indeed, assume for instance that the equation f effectively depends on y: then, $\bar{x}$ is free over $\mathbb{C}$, because a non-trivial algebraic relation $P(\bar{x}) = 0$, $P \in \mathbb{C}[x] - \{0\}$, would imply P to be a multiple of f; the extension $\mathbb{C}[\bar{x}] \subset \mathbb{C}(C)$ is algebraic due to the relation $f(\bar{x}, \bar{y}) = 0$.

There are other $\mathbb{C}$-algebras associated to C that, unlike the above, depend on the choice of an affine chart of $\mathbb{P}_2$. Still assume to have chosen an affine chart $\mathbb{A}_2$ containing some point of C, with affine coordinates x, y as above, and call C_0 the corresponding affine part of C. Recall that the affine ring $\mathcal{A}(\mathbb{A}_2)$ of $\mathbb{A}_2$ has been introduced in Section 1.4 as the subring of $\mathbb{C}(\mathbb{P}_n)$ composed of the rational functions regular at all points of $\mathbb{A}_n$. The *affine ring* $\mathcal{A}(C_0)$ of C_0 is defined to be the image of $\mathcal{A}(\mathbb{A}_2)$ under the restriction homomorphism; its elements are thus the rational functions on C that have a representative regular at all points of $\mathbb{A}_2$, after which they are obviously regular at all points of C_0. We will see in 3.7.13 below that, conversely, any rational function on C, regular at all points of C_0, belongs to $\mathcal{A}(C_0)$.

Using affine representations, by 1.4.7, the affine ring $\mathcal{A}(C_0)$ appears as the subring

$$\mathcal{A}(C_0) = \mathbb{C}[\bar{x}, \bar{y}] \subset \mathbb{C}(\bar{x}, \bar{y}) = \mathbb{C}(C).$$

Remark 3.7.10 By 3.6.3, or by a direct check, we have

$$(f\mathbb{C}[x,y]_{(f)}) \cap \mathbb{C}[x,y] = f\mathbb{C}[x,y].$$

Therefore the restriction homomorphism induces a $\mathbb{C}$-algebra isomorphism

$$\mathbb{C}[x,y]/f\mathbb{C}[x,y] \to \mathbb{C}[\bar{x}, \bar{y}] = \mathcal{A}(C_0)$$

mapping the class of any $P(x,y) \in \mathbb{C}[x,y]$ to its restriction $P(\bar{x}, \bar{y})$. It is usual to identify $\mathbb{C}[x,y]/f\mathbb{C}[x,y]$ with $\mathcal{A}(C_0)$ through this isomorphism.

For each $p \in C_0$ take

$$\bar{\mathfrak{m}}_p = \{g \in \mathcal{A}(C_0) \mid g(p) = 0\}.$$

Clearly, $\bar{\mathfrak{m}}_p$ is an ideal of $\mathcal{A}(C_0)$, and is the image of $\mathfrak{m}_p$ under the restriction homomorphism. We have:

Proposition 3.7.11 *The maximal ideals of $\mathcal{A}(C_0)$ are the ideals $\bar{\mathfrak{m}}_p$, $p \in C_0$; they are the only non-zero prime ideals of $\mathcal{A}(C_0)$.*

PROOF: The affine ring $\mathcal{A}(C_0)$ being the homomorphic image of $\mathcal{A}(\mathbb{A}_2)$, its maximal (resp. prime) ideals are the images of the maximal (resp. prime) ideals of $\mathcal{A}(\mathbb{A}_2)$ containing the kernel of the restriction homomorphism, which in turn is generated by f by 3.7.10. By 3.6.8 and 3.6.9, the maximal ideals of $\mathcal{A}(\mathbb{A}_2)$ containing f are the ideals $\mathfrak{m}_p$ for which $p \in C_0$, hence the first claim. Again by 3.6.8, the only prime ideal of $\mathcal{A}(\mathbb{A}_2)$ containing f other than the $\mathfrak{m}_p$, $p \in C_0$, is the kernel $f\mathcal{A}(\mathbb{A}_2)$, after which the second claim is clear.   ◇

The above definitions directly provide presentations to be added to those of 3.7.8:

Proposition 3.7.12

(d') $\mathbb{C}(C) = \mathcal{A}(C_0)_{(0)}$.

(h') $\mathcal{O}_{C,p} = \mathcal{A}(C_0)_{\bar{\mathfrak{m}}_p}$.

Using 3.6.7 and 3.7.12(h'),

$$\mathcal{A}(C_0) = \bigcap_{p \in C_0} \mathcal{O}_{C,p}$$

and therefore:

Proposition 3.7.13 *The elements of the affine ring $\mathcal{A}(C_0)$ are the rational functions on C which are regular at all points of C_0.*

We close this section by showing an important property of the local rings of curves at their smooth points:

Proposition 3.7.14 *Assume that p is a smooth point of an irreducible curve C of $\mathbb{P}_2$, let C' be any curve of $\mathbb{P}_2$ transverse to C at p and, after fixing any affine chart $\mathbb{A}_2$ of $\mathbb{P}_2$ containing p, take h to be an equation of the affine part of C'. Then:*

(a) *The restriction $\bar{h}$ of h generates $\mathfrak{M}_{C,p}$ as an ideal of $\mathcal{O}_{C,p}$.*

(b) *If g is an equation of the affine part of a curve C'' of $\mathbb{P}_2$ and $\bar{g}$ the restriction of g to C, then, in the local ring $\mathcal{O}_{C,p}$,*

$$\bar{g} = u\bar{h}^r,$$

where $u \in \mathcal{O}_{C,p}$ is invertible and $r = [C \cdot C'']_p$.

PROOF: For claim (a) just note that 3.6.6(b) ensures that h and an equation f of the affine part of C generate $\mathfrak{M}_{\mathbb{P}_2,p}$ and then use 3.7.8(i). Claim (b) is obvious if $p \notin C''$ and therefore $\bar{g} \notin \bar{h}\mathcal{O}_{C,p} = \mathfrak{M}_{C,p}$, as then $\bar{g}$ itself is invertible and $[C \cdot C'']_p = 0$. Otherwise $\bar{g} \in \bar{h}\mathcal{O}_{C,p}$. Assume we have an equality

$$\bar{g} = u'\bar{h}^{r'}$$

with $u' \in \mathcal{O}_{C,p}$ and $r' > 0$, which is indeed the case for $r' = 1$. If u' is the restriction of a rational function $a' \in \mathcal{O}_{\mathbb{P}_2,p}$, then

$$g = a'h^{r'} + bf,$$

for some $b \in \mathcal{O}_{\mathbb{P}_2,p}$. If γ is the only branch of C at p, $o_\gamma f = \infty$ and therefore

$$[C.C'']_p = o_\gamma g = o_\gamma a' + r'o_\gamma h = o_\gamma a' + r'. \tag{3.18}$$

This proves that the exponent r' is bounded by $[C.C'']_p$. Since on the other hand r' can obviously be increased as far as u' is not invertible, as then $u' \in \bar{h}\mathcal{O}_{C,p}$, there are an invertible $u \in \mathcal{O}_{C,p}$ and a positive integer r for which

$$\bar{g} = u\bar{h}^r.$$

Furthermore, if $a \in \mathcal{O}_{\mathbb{P}_2,p}$ restricts to u, u being invertible, $a(p) \neq 0$ and so $o_\gamma a = 0$, after which computing as in (3.18) gives $[C.C'']_p = r$. $\diamond$

Remark 3.7.15 Using 3.7.14(b), it is straightforward to prove that – still assuming p to be a smooth point of C – all non-zero ideals of $\mathcal{O}_{C,p}$ are principal, generated by the powers of $\bar{h}$.

3.8   M. Noether's Fundamental Theorem

As before, assume we have fixed homogeneous coordinates x_0, x_1, x_2 on a projective plane $\mathbb{P}_2$. Assume that $C_1 : F = 0$, $C_2 : G = 0$ and $C_3 : H = 0$ are curves of $\mathbb{P}_2$, C_1 and C_2 sharing no irreducible component. In this section we will find sufficient conditions for the following:

Noether's condition: There are homogeneous polynomials A, B, of degrees $\deg A = \deg H - \deg F$, $\deg B = \deg H - \deg G$, such that $H = AF + BG$.

Since the condition still holds after a linear substitution of the variables, it is clear that it does not depend on the choice of the coordinates.

If Noether's condition is satisfied, then, obviously, all the intersection points of C_1 and C_2 belong to C_3. The converse is, however, far from being true, as in some sense, once Noether's condition is satisfied, the local behaviour of C_1 and C_2 at each common point p constrains the local behaviour of C_3 at p. For instance the reader may easily check, taking an affine chart and affine coordinates with origin at p, that if the Noether condition is satisfied and p is a simple point of C_1 and C_2 at which they have the same tangent line, then either p is a simple point of C_3 at which C_3 has the same tangent line as C_1 and C_2, or p is a singular point of C_3. Similarly, if p is a singular point of C_1 and C_2, then it must be a singular point of C_3 too.

Next we will see that Noether's condition may be reformulated in terms of local conditions at the points of $C_1 \cap C_2$. To this end we introduce affine and local versions of the Noether condition. Still take C_1, C_2 and C_3 as above and let $\mathbb{A}_2$ be an affine chart of $\mathbb{P}_2$, x, y affine coordinates on $\mathbb{A}_2$ and $f, g, h \in \mathbb{C}[x, y]$ equations of the affine parts of, respectively, C_1, C_2 and C_3.

Affine Noether's condition in $\mathbb{A}_2$: There are polynomials $a, b \in \mathbb{C}[x, y]$ such that $h = af + bg$.

If furthermore $p \in \mathbb{A}_2$, we set:

Local Noether's condition at p: There are $a_p, b_p \in \mathcal{O}_{\mathbb{P}_2, p}$ such that $h = a_p f + b_p g$.

Remark 3.8.1 The local condition at p may be equivalently stated, with no use of the affine chart or the affine equations, $I_3 \subset I_1 + I_2$, where I_1, I_2, I_3 are the ideals of $\mathcal{O}_{\mathbb{P}_2, p}$ associated to C_1, C_2 and C_3 respectively, as $I_1 = f\mathcal{O}_{\mathbb{P}_2, p}$, $I_2 = g\mathcal{O}_{\mathbb{P}_2, p}$ and $I_3 = h\mathcal{O}_{\mathbb{P}_2, p}$.

Reduction to a local setting is given by the next two propositions:

Proposition 3.8.2 *If the local Noether condition is satisfied at each point of $C_1 \cap C_2$ that belongs to $\mathbb{A}_2$, then the affine Noether condition is satisfied in $\mathbb{A}_2$.*

PROOF: Note first that the local Noether condition is trivially satisfied at any point $p \in \mathbb{A}_2 - C_1 \cap C_2$, as in the local ring $\mathcal{O}_{\mathbb{P}_2, p}$ of any such point either f or g is invertible and therefore $\mathcal{O}_{\mathbb{P}_2, p} = f\mathcal{O}_{\mathbb{P}_2, p} + g\mathcal{O}_{\mathbb{P}_2, p}$. Hence, for any $p \in \mathbb{A}_2$ there are polynomials $v_p, w_p, u_p \in \mathbb{C}[x, y]$, $u_p(p) \neq 0$, for which $h = (v_p/u_p)f + (w_p/u_p)g$. In other words, for any $p \in \mathbb{A}_2$, there is a $u_p \in \mathbb{C}[x, y]$, $u_p(p) \neq 0$, for which $u_p h$ belongs to the ideal (f, g) generated by f, g in $\mathbb{C}[x, y]$. Clearly

$$T = \{u \in \mathbb{C}[x, y] \mid uh \in (f, g)\}$$

is an ideal of $\mathbb{C}[x, y]$ and we have seen above that it is contained in no ideal $\mathfrak{m}_p$ for $p \in \mathbb{A}_2$. Since the latter are all the maximal ideals of $\mathbb{C}[x, y]$ (by 3.6.10) we have $T = \mathbb{C}[x, y]$; in particular $1 \in T$ and so $h = 1h \in (f, g)$, as claimed.   ◇

Proposition 3.8.3 *If the affine Noether condition in $\mathbb{A}_2$ is satisfied and all the points of $C_1 \cap C_2$ belong to $\mathbb{A}_2$, then the Noether condition is satisfied.*

For the proof of 3.8.3 we need:

Lemma 3.8.4 *If the curves $C_1 : F = 0$ and $C_2 : G = 0$ share no point on the line $x_0 = 0$, $L \in \mathbb{C}[x_0, x_1, x_2]$ and $x_0 L$ belongs to the ideal (F, G), generated by F, G in $\mathbb{C}[x_0, x_1, x_2]$, then L also belongs to (F, G).*

PROOF OF 3.8.4: Assume we have $M, N \in \mathbb{C}[x_0, x_1, x_2]$ for which

$$x_0 L = MF + NG. \tag{3.19}$$

Then

$$0 = M(0, x_1, x_2)F(0, x_1, x_2) + N(0, x_1, x_2)G(0, x_1, x_2). \tag{3.20}$$

Since by the hypothesis about the points of $C_1 \cap C_2$, the polynomials $F(0, x_1, x_2)$ and $G(0, x_1, x_2)$ share no irreducible factor, $F(0, x_1, x_2)$ divides $N(0, x_1, x_2)$ and $G(0, x_1, x_2)$ divides $M(0, x_1, x_2)$. Using (3.20), there is a $T \in \mathbb{C}[x_1, x_2]$ such that

$$N(0, x_1, x_2) = TF(0, x_1, x_2) \quad \text{and} \quad M(0, x_1, x_2) = -TG(0, x_1, x_2). \tag{3.21}$$

Then rewrite 3.19 in the form

$$x_0 L = (M + TG)F + (N - TF)G.$$

It is clear from (3.21) that both the coefficients $M + TG$ and $N - TF$ vanish when x_0 is replaced with 0; they both are thus multiples of x_0 and the claim follows. ◇

PROOF OF 3.8.3: As above, let C_1, C_2 and C_3 have, respectively, equations F, G and H. If their 0-th affine parts satisfy the affine Noether condition, then there are $a, b \in \mathbb{C}[x, y]$ such that

$$H(1, x, y) = aF(1, x, y) + bG(1, x, y)$$

or, equivalently

$$H(1, x_1/x_0, x_2/x_0) =$$
$$a(x_1/x_0, x_2/x_0)F(1, x_1/x_0, x_2/x_0) + b(x_1/x_0, x_2/x_0)G(1, x_1/x_0, x_2/x_0).$$

After multiplying by a suitable power of x_0 we get

$$x_0^r H(x_0, x_1, x_2) = A'(x_0, x_1, x_2)F(x_0, x_1, x_2) + B'(x_0, x_1, x_2)G(x_0, x_1, x_2)$$

with $A', B' \in \mathbb{C}[x_0, x_1, x_2]$ and $r \geq 0$. This in turn, after r applications of 3.8.4, gives

$$H = A''F + B''G, \tag{3.22}$$

where also $A'', B'' \in \mathbb{C}[x_0, x_1, x_2]$. By equating the homogeneous parts of degree $\deg H$ of the two sides of (3.22), we obtain

$$H = AF + BG,$$

where A and B are the homogenous parts of degrees $\deg H - \deg F$ and $\deg H - \deg G$ of A'' and B'', respectively. This completes the proof. ◇

Remark 3.8.5 The reader may note that converses of 3.8.2 and 3.8.3 hold true, namely:

- If the Noether condition is satisfied, then for any affine chart $\mathbb{A}_2$, the affine Noether condition is satisfied on $\mathbb{A}_2$.

- If the affine Noether condition is satisfied in $\mathbb{A}_2$, then the local Noether condition is satisfied at each point of $\mathbb{A}_2$.

Corollary 3.8.6 *If curves C_1, C_2, C_3 of $\mathbb{P}_2$, with C_1 and C_2 sharing no irreducible component, satisfy the local Noether condition at every point of $C_1 \cap C_2$, then they satisfy the Noether condition.*

PROOF: By the hypothesis, $C_1 \cap C_2$ is a finite set and therefore the projective coordinates x_0, x_1, x_2 may be taken with $x_0 = 0$ missing all points in $C_1 \cap C_2$. The 0-th affine chart then contains all the points of $C_1 \cap C_2$ and it suffices to apply 3.8.2 and 3.8.3.   ◇

The rest of this section is devoted to setting sufficient conditions for Noether's local condition at a point. The reader may note that 3.6.6 gives a set of sufficient conditions, as it ensures that the local Noether condition at p is satisfied if C_1 and C_2 are transverse at p and p belongs to C_3. Each of the next two propositions gives sufficient conditions that are more general. In both of them assume as above that C_1, C_2, C_3 are curves of $\mathbb{P}_2$ with C_1 and C_2 sharing no irreducible component. Still $e_p(C)$ denotes the multiplicity of the curve C at the point p.

Proposition 3.8.7 *If C_1 and C_2 share no tangent at a point p and $e_p(C_3) \geq e_p(C_1) + e_p(C_2) - 1$, then C_1, C_2, C_3 satisfy the local Noether condition at p.*

PROOF: Take a local chart, and affine coordinates x, y on it, such that p is the origin of coordinates. Call I the ideal generated by equations f and g of the affine parts of C_1 and C_2 in the local ring $\mathcal{O}_{\mathbb{P}_2,p}$ and put $e = e_p(C_1)$, $e' = e_p(C_2)$.

Since any equation h of the affine part of C_3 is a sum of homogeneous polynomials of degree not less than $e_p(C_3)$, it will suffice to prove that any homogeneous polynomial in x, y of degree $k \geq e + e' - 1$ belongs to I. In fact, since $x, y \in \mathfrak{M}_{\mathbb{P}_2,p}$, we have seen in 3.6.6 that any homogeneous polynomial in x, y of degree high enough belongs to I; after this we are allowed to use decreasing induction on k. Factoring the initial form of f as a product of (possibly repeated) linear forms $\ell_1, \ldots, \ell_e$ we have

$$f = \ell_1 \ldots \ell_e + f'$$

where all the terms of f' have degree higher than e. Similarly

$$g = t_1 \ldots t_{e'} + g'$$

with each t_i linear and all the terms of g' have degree higher than e'.

Take $\ell_r = \ell_e$ for $r > e$ and $M_i = \prod_{r=1}^{i} \ell_r$. Similarly, take $t_r = t_{e'}$ for $r > e'$ and $N_i = \prod_{r=1}^{i} t_r$. Note that since C and C' are assumed not to share a tangent line at p, M_i and N_j are coprime for any i, j. Next we will check that the products $M_i N_{k-i}$, $i = 0, \ldots k$, compose a basis of the vector space of all homogeneous polynomials of degree k. On one hand their number is $k + 1$, as due. On the other, if there is an equality

$$\lambda_0 M_0 N_k + \lambda_1 M_1 N_{k-1} + \cdots + \lambda_k M_k N_0 = 0,$$

$\lambda_i \in \mathbb{C}$, then all summands have the factor ℓ_1 but the first one, which has not. It follows that $\lambda_0 = 0$. By dropping the first summand, dividing by ℓ_1 and repeating the argument, it follows that $\lambda_1 = 0$, and so on, until we get $\lambda_i = 0$ for $i = 0, \ldots, k$ This proves the linear independence of the $M_i N_{k-i}$, $i = 0, \ldots k$.

After the above, it suffices to prove that each $M_i N_{k-i}$, $i = 0, \ldots k$, belongs to I. If $i \geq e$, then $M_i N_{k-i}$ is a multiple of $\ell_1 \ldots \ell_e$, say

$$M_i N_{k-i} = T\ell_1 \ldots \ell_e = Tf - Tf'$$

with $T \in \mathbb{C}[x, y]$ homogeneous of degree $k - e$. Clearly $Tf \in I$, and also $Tf' \in I$ because it is a sum of monomials of degree higher than $k - e + e = k$ and the induction hypothesis applies. This proves that $M_i N_{k-i}$ belongs to I.

If $i < e$, then, using $k \geq e + e' - 1$, we have $k - i \geq e'$; then $M_i N_{k-i}$ is a multiple of $t_1 \ldots t_{e'}$ and an argument similar to the one above shows that also in this case $M_i N_{k-i}$ belongs to I. $\diamond$

Proposition 3.8.8 *If C_1 is irreducible, has an ordinary singularity at p and for each branch γ of C_1 at p*

$$[C_3 \cdot \gamma] \geq [C_2 \cdot \gamma] + e_p(C_1) - 1,$$

then C_1, C_2, C_3 satisfy the local Noether condition at p.

PROOF: Completing the proof will take a while, as a number of auxiliary lemmas will we proved on the way. Fix an affine chart $\mathbb{A}_2$ of $\mathbb{P}_2$ containing p and affine coordinates x, y on it such that p is the origin of coordinates and neither of the coordinate axes is tangent to C_1 at p. Since in the sequel we will work exclusively with the affine parts of C_1, C_2, C_3 we will use the same notations for the curves and their affine parts, the latter thus being denoted C_1, C_2, C_3. Assume that $f, g, h \in \mathbb{C}[x, y]$ are equations of C_1, C_2, C_3, respectively, and take $e = e_p(C_1)$. If f is written as a sum of homogeneous polynomials

$$f = \sum_{i=e}^{d} f_i, \quad \deg f_i = i, \quad f_e \neq 0, \tag{3.23}$$

then its initial form f_e has no factor x or y, because C_1 is tangent to neither of the axes. Therefore, up to multiplying f by a non-zero constant factor, f_e

may be written as a product of distinct linear factors

$$f_e = \prod_{j=1}^{e} (y - a_j x), \quad a_j \neq 0, \quad a_j \neq a_i \text{ for } j \neq i,$$

because p is an ordinary singular point of C_1. The local Newton polygon $N(f)$ then has a single side Γ, with ends $(0, e)$ and $(e, 0)$ and associated equation $F_\Gamma(z) = f_e(1, z) = \prod_{j=1}^{e}(z - a_j)$. The Newton–Puiseux algorithm gives rise to at least e different Puiseux series of the form $s_j = a_j x + \ldots$, the dots representing terms of higher order. Since $h(N(f)) = e$, by 2.3.6, the series $s_1, \ldots, s_e$ have each multiplicity and polydromy order both equal to one, and they are all the Puiseux series of C_1 at p. We write γ_i for the branch of C_1 corresponding to s_i, $\gamma_1, \ldots, \gamma_e$ thus being the branches of C_1 at p.

Let us take a second affine plane $\mathbb{A}_2'$ with coordinates t, z and consider the map

$$\Phi : \mathbb{A}_2' \longrightarrow \mathbb{A}_2$$
$$(t, z) \longmapsto (t, tz).$$

As is clear, Φ maps the line $E : t = 0$ to $p = (0, 0)$ and its restriction to the complement of E is injective. Its inverse $\Psi = \Phi^{-1}$ is a correspondence through which each point $(x, y) \in \mathbb{A}_2$ with $x \neq 0$ has $(x, y/x)$ as its only correspondent, $(0, 0)$ corresponds with all the points on E and the points $(0, y)$, $y \neq 0$, have no correspondent. It is said that Ψ *blows up* the point p into the line E.

We will use Ψ to reduce our problem to the case of a smooth point. To this end we will first transform the curves by Ψ: The curve of $\mathbb{A}_2'$, $\hat{C}_1 : \hat{f}(t, z) = f(t, tz) = 0$ is called the total transform of C_1 (by Ψ); its points are clearly those mapped to points of C_1 by Φ. Similarly $\hat{C}_2 : \hat{g}(t, z) = g(t, tz) = 0$ and $\hat{C}_3 : \hat{h}(t, z) = h(t, tz) = 0$ are the total transforms of C_2 and C_3.

Using (3.23), we have

$$\hat{f} = \sum_{i=e}^{d} f_i(t, tz) = t^e \sum_{i=e}^{d} t^{i-e} f_i(1, z)) = t^e \tilde{f},$$

where

$$\tilde{f} = \sum_{i=e}^{d} t^{i-e} f_i(1, z)$$

clearly has no factor t. The line E is thus an irreducible component of multiplicity e of the total transform $\hat{C}_1$. By dropping it, we get the curve $\tilde{C}_1 : \tilde{f} = 0$, which is called the strict transform of C_1. The intersection of $\tilde{C}_1$ with the line E is given by

$$0 = \tilde{f}(0, z) = f_e(1, z) = \prod_{j=1}^{e}(z - a_j)$$

and therefore consists of the points $p_j = (0, a_j)$, $j = 1, \ldots, e$. Furthermore, for each j, $[\tilde{C}_1 \cdot E]_{p_j} = 1$ because the a_j are pairwise different. We have thus in particular proved:

Lemma 3.8.9 *For $j = 1, \ldots, e$, p_j is a non-singular point of $\tilde{C}_1$ and $\tilde{C}_1$ and E are transverse at p_j.*

The reader may note that the points of $\tilde{C}_1 \cap E$ correspond one to one to the tangents to C_1 at p, which in turn are in a one to one correspondence with the branches of C_1 at p: the point p_j corresponds to the tangent line $y - a_j x = 0$, whose slope is the second coordinate of p_j, this line corresponding in turn to the branch γ_j with Puiseux series $s_j = a_j x + \ldots$. Since

$$t^e \tilde{f}(t, s_j(t)/t) = \hat{f}(t, s_j(t)/t) = f(t, s_j(t)) = 0,$$

we have $\tilde{f}(t, s_j(t)/t) = 0$ and we see that the – necessarily unique – Puiseux series of $\tilde{C}_1$ at p_j is $s_j(t)/t$. We may compute thus,

$$[C_2 \cdot \gamma_j] = o_x g(x, s_j(x)) = o_t g(t, s_j(t)) = o_t \hat{g}(t, s_j(t)/t) = [\hat{C}_2 \cdot \tilde{C}_1]_{p_j}.$$

This and a similar computation with C_3 give:

Lemma 3.8.10 *For all $j = 1, \ldots, e$,*

$$[C_2 \cdot \gamma_j] = [\hat{C}_2 \cdot \tilde{C}_1]_{p_j} \quad and \quad [C_3 \cdot \gamma_j] = [\hat{C}_3 \cdot \tilde{C}_1]_{p_j}.$$

The last piece we need from the geometric side is:

Lemma 3.8.11 *The curve $\tilde{C}_1$ is irreducible.*

PROOF OF 3.8.11 Note first that given $P \in \mathbb{C}[t, z]$, any product $t^r P$ may be written, if r is high enough, as $t^r P = Q(t, zt)$ where $Q \in \mathbb{C}[x, y]$. Indeed, for any $r > \deg_z P$, the degree in t of any monomial of $t^r P$ is not less than its degree in z, and therefore $t^r P$ may be written as a polynomial in t, tz. In the particular case of $P = \tilde{f}$ we already know that $t^e \tilde{f} = f(t, tz)$.

Assume now we have a factorization

$$\tilde{f} = P_1 P_2 \tag{3.24}$$

with $P_1, P_2 \in \mathbb{C}[t, z]$. After multiplying by t^r with r high enough, and higher in particular than e, by the above we will have an equality

$$t^{r-e} f(t, tz) = t^r \tilde{f}(t, z) = Q_1(t, zt) Q_2(t, zt)$$

where $t^{r_i} P_i(t, z) = Q_i(t, tz)$, $i = 1, 2$, $r_1 + r_2 = r$. This in turn, after replacing t and z with x and y/x, gives

$$x^{r-e} f(x, y) = Q_1(x, y) Q_2(x, y).$$

Now, since f is irreducible, it divides one of the factors on the right; then the other, assume it to be $Q_1(x, y)$, is $Q_1(x, y) = ax^s$, with $a \in \mathbb{C}$ and $s \geq 0$. It follows that

$$t^{r_1} P_1(t, z) = Q_1(t, tz) = at^s$$

and so $s \geq r_1$ and $P_1 = at^{s-r_1}$. Since we already know $\tilde{f}$ to have no factor t, the equality (3.24) gives $s - r_1 = 0$, P_1 is constant and the irreducibility of $\tilde{C}_1$ is proved.   ◇

Continuing with the proof of 3.8.8, it is time to care about more algebraic aspects of the transformation Ψ. First, consider the pull-back ring homomorphism induced by Φ between the affine rings of $\mathbb{A}_2$ and $\mathbb{A}'_2$, namely,

$$\varphi : \mathbb{C}[x, y] \longrightarrow \mathbb{C}[t, z]$$
$$P = P(x, y) \longmapsto \hat{P} = P \circ \Phi = P(t, tz).$$

It is straightforward to check that φ is a monomorphism; we will thus identify each $P \in \mathbb{C}[x, y]$ with its image $\hat{P}$, and take $\mathbb{C}[x, y]$ as a subring of $\mathbb{C}[t, z]$. In particular, after the identification, $t = x$ and $y = xz$; hence $\mathbb{C}[t, z]$ may be viewed as the extension $\mathbb{C}[x, y][y/x]$ of $\mathbb{C}[x, y]$ in its field of fractions $\mathbb{C}(x, y)$.

We will use a bar ⁻ to indicate restriction to C_1 or $\tilde{C}_1$, the meaning being clear in each case. The affine rings of C_1 and $\tilde{C}_1$ are $\mathcal{A}(C_1) = \mathbb{C}[\bar{x}, \bar{y}]$ and $\mathcal{A}(\tilde{C}_1) = \mathbb{C}[\bar{x}, \bar{z}]$, and we know the kernels of the restriction homomorphisms

$$\mathbb{C}[x, y] \longrightarrow \mathbb{C}[\bar{x}, \bar{y}] \quad \text{and} \quad \mathbb{C}[x, z] \longrightarrow \mathbb{C}[\bar{x}, \bar{z}]$$

to be, respectively $f\mathbb{C}[x, y]$ and $\tilde{f}\mathbb{C}[x, z]$ (3.7.10). Since the (images under φ of the) multiples of f in $\mathbb{C}[x, y]$ obviously are multiples of $\tilde{f}$ in $\mathbb{C}[x, z]$, φ induces a morphism between the affine rings of $\mathcal{A}(C_1)$ and $\mathcal{A}(\tilde{C}_1)$

$$\bar{\varphi} : \mathbb{C}[\bar{x}, \bar{y}] \longrightarrow \mathbb{C}[\bar{x}, \bar{z}]$$
$$\overline{P(x, y)} \longmapsto \overline{P(x, xz)}.$$

In fact, $\bar{\varphi}$ is a monomorphism too. For, assume that $P(x, xz)$ is a multiple of $\tilde{f}$, say

$$P(x, xz) = Q\tilde{f}, \quad Q \in \mathbb{C}[x, z].$$

Arguing as in the proof of 3.8.11, multiplication by a suitable power of x provides an equality

$$x^r P(x, y) = Q'f, \quad Q' \in \mathbb{C}[x, y].$$

Since f has no factor x (otherwise $f = x$ against the choice of the coordinates) x^r divides Q', P is a multiple of f in $\mathbb{C}[x, y]$ and therefore $\bar{P} = 0$ as wanted.

Once we have seen that $\bar{\varphi}$ is a monomorphism, we will identify each $\overline{P(x, y)} \in \mathbb{C}[\bar{x}, \bar{y}]$ with $\bar{\varphi}(\overline{P(x, y)}) = \overline{P(x, xz)} \in \mathbb{C}[\bar{x}, \bar{z}]$ (this prevents any confusion about the use of ⁻), and consequently $\mathbb{C}[\bar{x}, \bar{y}]$ with a subring of $\mathbb{C}[\bar{x}, \bar{z}]$.

Take $S \subset \mathbb{C}[\bar{x}, \bar{y}]$ to be the complement of the maximal ideal $\bar{\mathfrak{m}}_p = (\bar{x}, \bar{y})$ corresponding to p: the elements of S are the elements of $\mathbb{C}[\bar{x}, \bar{y}]$ which are non-zero at p. The set S is a multiplicative system of $\mathbb{C}[\bar{x}, \bar{y}]$; therefore it is also a multiplicative system of $\mathbb{C}[\bar{x}, \bar{z}]$ and there is an inclusion

$$\mathcal{O}_{C_1,p} = S^{-1}\mathbb{C}[\bar{x}, \bar{y}] \subset S^{-1}\mathbb{C}[\bar{x}, \bar{z}],$$

the equality due to 3.7.12. A quite important fact is:

Lemma 3.8.12 *The product of any element of $S^{-1}\mathbb{C}[\bar{x}, \bar{z}]$ by $\bar{x}^{e-1}$ belongs to $\mathcal{O}_{C_1,p}$.*

PROOF OF 3.8.12 Since $N(f)$ has a single side Γ with ends $(0, e)$ and $(e, 0)$, f may be written

$$f = y^e P_0 + y^{e-1} x P_1 + \cdots + y x^{e-1} P_{e-1} + x^e P_e$$

where $P_i \in \mathbb{C}[x, y]$ for $i = 0, \ldots, e$ (even $P_i \in \mathbb{C}[x]$ for $i > 0$) and $P_0(0, 0) \neq 0$. Then

$$\tilde{f} = z^e P_0 + z^{e-1} P_1 + \cdots + z P_{e-1} + P_e$$

and so in $\mathbb{C}[\bar{x}, \bar{z}]$ we have

$$0 = \bar{z}^e \bar{P}_0 + \hat{z}^{e-1} \bar{P}_1 + \cdots + \bar{z} \bar{P}_{e-1} + \bar{P}_e$$

from which

$$0 = \bar{z}^e + \bar{z}^{e-1} \xi_1 + \cdots + \bar{z}\xi_1 + \xi_e, \tag{3.25}$$

where the $\xi_i = \bar{P}_i/\bar{P}_0$, $i = 1, \ldots, e$, belong to $S^{-1}(\mathbb{C}[\bar{x}, \bar{y}])$. It is clear that any $\delta \in S^{-1}(\mathbb{C}[\bar{x}, \bar{z}])$ may be written in the form

$$\delta = \eta_r \bar{z}^r + \cdots + \eta_1 \bar{z} + \eta_0$$

with $\eta_i \in S^{-1}\mathbb{C}[\bar{x}, \bar{y}]$, $i = 1, \ldots, r$. If $r \geq e$, then the equality (3.25) may be used to turn the above equality into a similar one with degree $r - 1$ in $\bar{z}$ and so, inductively, to get an expression of δ as a linear combination of $\bar{z}^{e-1}, \ldots, \bar{z}, 1$ with coefficients in $S^{-1}\mathbb{C}[\bar{x}, \bar{y}] = \mathcal{O}_{C_1,p}$. In other words, we have seen that $S^{-1}(\mathbb{C}[\bar{x}, \bar{z}])$ is generated, as a $\mathcal{O}_{C_1,p}$-module, by $\bar{z}^{e-1}, \ldots, \bar{z}, 1$. Then, to get the claim it suffices to check that $\bar{x}^{e-1}\bar{z}^i = \bar{x}^{e-1-i}\bar{y}^i$ belongs to $\mathcal{O}_{C_1,p}$ for $i = 0, \ldots, e - 1$, and this is clear. ◇

By 3.7.11, the non-zero prime ideals of the affine ring $\mathcal{A}(\tilde{C}_1) = \mathbb{C}[\bar{x}, \bar{z}]$ are its maximal ideals, and they are the images $\bar{\mathfrak{m}}_q$ of the maximal ideals $\mathfrak{m}_q$ of $\mathbb{C}[x, z]$ corresponding to the points $q \in \tilde{C}_1$. If $q = (a, b)$ and $a \neq 0$, then $x - a \in \mathfrak{m}_q$ and its restriction belongs to S; it follows that $S^{-1}\bar{\mathfrak{m}}_q = (1)$. If, otherwise, $a = 0$, then q is one of the points $p_i \in \tilde{C}_1 \cap E$. In such a case $\bar{\mathfrak{m}}_q \cap S = \emptyset$: indeed, any $P(x, y) = P(x, xz)$ belonging to $\mathfrak{m}_q$ vanishes for $x = 0$ and $z = b$ and therefore has $P(0, 0) = 0$, which ensures that $\bar{P}$ does not belong to S. Applying 3.6.3, we see that the maximal ideals of $S^{-1}\mathbb{C}[\bar{x}, \bar{z}]$ are the ideals $S^{-1}\bar{\mathfrak{m}}_{p_i}$, $i = 1, \ldots, e$. Since, as seen above, $\mathbb{C}[\bar{x}, \bar{z}] - \bar{\mathfrak{m}}_{p_i} \supset S$, it is straightforward to check that, for $i = 1, \ldots, e$,

$$(S^{-1}\mathbb{C}[\bar{x}, \bar{z}])_{S^{-1}\bar{\mathfrak{m}}_{p_i}} = (\mathbb{C}[\bar{x}, \bar{z}])_{\bar{\mathfrak{m}}_{p_i}} = \mathcal{O}_{\tilde{C}_1, p_i},$$

the second equality due to 3.7.12(h'). As a consequence the localized rings of $S^{-1}\mathbb{C}[\bar{x},\bar{z}]$ in its maximal ideals are the local rings $\mathcal{O}_{\tilde{C}_1,p_i}$, $i = 1,\ldots,e$. Using 3.6.7 we get the last piece we need in order to achieve the proof of 3.8.8, namely:

Lemma 3.8.13 *We have*

$$S^{-1}\mathbb{C}[\bar{x},\bar{z}] = \bigcap_{i=1}^{e}\mathcal{O}_{\tilde{C}_1,p_i}.$$

Next we put all the pieces together in order to finish the proof of 3.8.8. For $i = 1,\ldots,e$, write $r_i = [C_2 \cdot \gamma_i]$ and $k_i = [C_3 \cdot \gamma_i]$. Then, by 3.8.10, $r_i = [\hat{C}_2 \cdot \tilde{C}_1]_{p_i}$ and $k_i = [\hat{C}_3 \cdot \tilde{C}_1]_{p_i}$. We know from 3.8.9 that $\tilde{C}_1$ and $E : x = 0$ are transverse at p_i; then 3.7.14 applies showing that the maximal ideal of $\mathcal{O}_{\tilde{C}_1,p_i}$ is generated by $\bar{x}$ and there are equalities

$$\bar{g} = u\bar{x}^r, \quad \bar{h} = v\bar{x}^k$$

where u and v are invertible elements of $\mathcal{O}_{\tilde{C}_1,p_i}$. Since by the hypothesis $k \geq r + e - 1$,

$$\bar{h}/\bar{x}^{e-1}\bar{g} = vu^{-1}\bar{x}^{k-r-e+1} \in \mathcal{O}_{\tilde{C}_1,p_i}.$$

The above being true for $i = 1,\ldots,e$, by 3.8.13, $\bar{h}/\bar{x}^{e-1}\bar{g} \in S^{-1}\mathbb{C}[\bar{x},\bar{z}]$, after which, by 3.8.12, $\bar{h}/\bar{g} \in \mathcal{O}_{C_1,p}$. This proves the claim because then $\bar{h}$ is a multiple of $\bar{g}$ in $\mathcal{O}_{C_1,p}$, say $\bar{g} = \bar{b}\bar{g}$, which, by 3.7.8(g), is equivalent to $h - bg$ being a multiple of f in $\mathcal{O}_{\mathbb{P}_2,p}$. This ends the proof of 3.8.8.   ◇

Putting together 3.8.6, 3.8.7 and 3.8.8 gives M. Noether's Fundamental Theorem:

Theorem 3.8.14 (M. Noether, 1873) *Assume that $C_1 : F = 0$, $C_2 : G = 0$ and $C_3 : H = 0$ are curves of $\mathbb{P}_2$, C_1 and C_2 sharing no irreducible component. If for each point $p \in C_1 \cap C_2$ either*

(a) *C_1 and C_2 share no tangent at p and $e_p(C_3) \geq e_p(C_1) + e_p(C_2) - 1$, or*

(b) *C_1 is irreducible, has an ordinary singularity at p and for each branch γ of C_1 at p*

$$[C_3 \cdot \gamma] \geq [C_2 \cdot \gamma] + e_p(C_1) - 1,$$

then there are homogeneous polynomials A, of degree $\deg H - \deg F$, and B, of degree $\deg H - \deg G$ such that $H = AF + BG$.

The reader may note that neither of the conditions (a), (b) is necessary, as neither of them is satisfied when taking $C_3 = C_2$ if $e_p(C_1) > 1$. The next corollary is a particular case of M. Noether's Fundamental Theorem, usually referred to as the *simple case*. Its proof is left to the reader.

Corollary 3.8.15 *Assume that $C_1 : F = 0$, $C_2 : G = 0$ and $C_3 : H = 0$ are curves of $\mathbb{P}_2$. If C_1 and C_2 intersect in exactly $\deg C_1 \deg C_2$ different points, then C_3 goes through all of them if and only if there are homogeneous polynomials A, of degree $\deg H - \deg F$, and B, of degree $\deg H - \deg G$ such that $H = AF + BG$.*

3.9   Exercises

3.1 Assume that $f \in \mathbb{C}\{x\}[y]$ has no factor x, $d = \deg_y f > 0$ and $\deg f(0,y) = 0$. Prove that, up to a non-zero constant factor, f is the product of d factors of the form $y - x^{-r}s$ where s is a convergent fractionary power series and $r > o_x s$. Use this to complete the factorization of 3.3.1, after which parameterizations of all parts of the curve ξ of 3.3.2 are evident. *Hint:* take $\bar{y} = 1/y$, apply the Puiseux theorem to $\bar{y}^d f(x, 1/\bar{y})$ and use Exercise 2.2.

3.2 Let $C : F = 0$ and $C' : G = 0$ be curves of $\mathbb{P}_2$. Take $f = F(1,x,y)$, $g = G(1,x,y)$ and let $R_y(f,g)$ be the resultant of f and g relative to y, with formal degrees equal to the effective degrees in y. Prove that $R_y(f,g)(a) = 0$ if and only if either a point $[1,a,b]$ belongs to $C \cap C'$ for some b, or the line $x_1 - ax_0$ is tangent to both C and C' at $[0,0,1]$.

3.3 Fixing $p \in \mathbb{P}_2$, for any irreducible curve C of $\mathbb{P}_2$, denote by $I(C)$ the ideal of C in $\mathcal{O}_{\mathbb{P}_2,p}$ (cf. Section 3.6). Let the decomposition of an arbitrary curve C of $\mathbb{P}_2$ into irreducible components be

$$C = \sum_{i=1}^{m} r_i C_i$$

and assume that the irreducible components of C through p are the C_i with $i \leq k$, $0 \leq k \leq m$. Prove that taking

$$I(C) = \prod_{i=1}^{k} I(C_i)^{r_i}$$

extends the definition of $I(C)$ to arbitrary curves in such a way that $I(C)$ is still generated by any equation of any affine part of C relative to an affine chart containing p. (As usual, an empty product of ideals is taken to be the whole ring.)

3.4 With the notations of Exercise 3.3, prove that the map

$$(C, C') \longmapsto \dim_{\mathbb{C}} \mathcal{O}_{\mathbb{P}_2,p}/(I(C) + I(C')) \in \mathbb{N} \cup \infty,$$

for any two curves C, C' of $\mathbb{P}_2$, satisfies the conditions of Exercise 2.12. Deduce that

$$[C \cdot C']_p = \dim_{\mathbb{C}} \mathcal{O}_{\mathbb{P}_2,p}/(I(C) + I(C'))$$

holds for any two curves C, C' of $\mathbb{P}_2$.

3.5 Let $\mathcal{P}$ be a pencil of curves of $\mathbb{P}_2$ and p a base point of $\mathcal{P}$. Prove that the following conditions are equivalent:

(i)   Any two $C, C' \in \mathcal{P}$ have $[C \cdot C']_p > 1$.

(ii)   There are $C, C' \in \mathcal{P}$, $C \neq C'$, that have $[C \cdot C']_p > 1$.

(iii)   There is a curve in $\mathcal{P}$ which has p as a singular point.

Prove also that if the above conditions fail, then mapping each $C \in \mathcal{P}$ to the tangent line to C at p is a projectivity between $\mathcal{P}$ and the pencil of lines of $\mathbb{P}_2$ through p.

3.6 Determine the base points of each of the three pencils of conics

$$\mathcal{P}_1 : \lambda(z_0^2 - z_1 z_2) + \mu(z_0^2 - z_1^2 + z_1 z_2) = 0,$$
$$\mathcal{P}_2 : \lambda(z_0^2 - z_1 z_2) + \mu z_0^2 = 0,$$
$$\mathcal{P}_3 : \lambda(z_0^2 - z_1 z_2) + \mu z_1^2 = 0,$$

determine those for which the conditions of Exercise 3.5 are satisfied and make explicit the projectivity of that exercise for the remaining ones.

3.7 Prove by a counterexample that, unlike claim (a) (see 3.4.4), claim (b) of 3.4.3 does not hold true if the branch γ is replaced with a curve and the intersections multiplicities are still taken at a fixed $p \in \mathbb{P}_2$.

3.8 Use that there is a pencil of conics through any four given points of $\mathbb{P}_2$ to prove that no irreducible quartic has four singular points, which of course also follows from 3.5.23.

3.9 Prove that no two different irreducible curves of degree $d \geq 6$ share $(d-1)(d-2)/2$ different singular points.

3.10 Let C be an irreducible curve of degree d of $\mathbb{P}_2$ having $(d-1)(d-2)/2$ singular points and call S the set of these points.

(1) Prove that all $q \in S$ are double points of C.

(2) Prove that the curves of degree $d - 2$ going through all points in S describe a linear system $\mathcal{L}$ of dimension at least $d - 2$.

(3) Take distinct points $q_1, \ldots, q_{d-2} \in C - S$. Prove that there is a curve $C' \in \mathcal{L}$ containing $q_1, \ldots, q_{d-2}$. Prove also that such a curve necessarily has $C \cap C' = S \cup \{q_1, \ldots, q_{d-2}\}$ and is smooth at every point of $C \cap C'$.

(4) Prove that the curve C' of (3) is unique. *Hint:* otherwise two distinct curves such as C' would span a pencil.

(5) Prove that $\dim \mathcal{L} = d - 2$.

3.11 Use 3.4.7 to prove Pascal's theorem (Pascal 1654):

   If the vertices of a hexagon all lie on an irreducible conic, then the intersections of its pairs of opposite sides are three aligned points.

3.12 Prove that for any $d > 1$ there is a smooth curve of $\mathbb{P}_2$ which has a flex of order $d - 2$. *Hint:* use Exercise 1.7.

3.13 In the conditions of Remark 3.5.10, prove that the affine part of the polar $\mathcal{P}_{q'}(C)$ relative to $q' = [c, a, b]$ has equation

$$a\frac{\partial f}{\partial x} + b\frac{\partial f}{\partial y} + c\left(dF - x\frac{\partial f}{\partial x} - y\frac{\partial f}{\partial y}\right) = 0,$$

$d = \deg C$.

3.14 Prove that if a curve C of $\mathbb{P}_2$ contains a line ℓ, $q \in \ell$ and $\mathcal{P}_q(C)$ is defined, then ℓ is an irreducible component of $\mathcal{P}_q(C)$.

3.15 Prove that if C is a reduced curve of $\mathbb{P}_2$ and $q \in \mathbb{P}_2$ is not a singular point of C, then, for all but finitely many lines L through q, C and L are transverse at all their common points. Deduce that, for any reduced curve C of $\mathbb{P}_2$, $\deg C = \max_L\{\sharp(C \cap L)\}$, where L runs over all the lines of $\mathbb{P}_2$ not contained in C and $\sharp$ stands for number of elements.

3.16 Prove that Plücker's formula 3.5.19 still holds if the point q is a simple point of C, provided the tangent line T_q to C at q is counted $[C \cdot T_q]_q$ times.

3.17 Extend Plücker's formula 3.5.19 to include the cases in which q belongs to a tangent line at a singular point of C (a node or an ordinary cusp), by assigning to each such tangent a suitable multiplicity in the count.

3.18 Plücker's second formula (Plücker, 1835). Let $C : F = 0$ be an irreducible curve of $\mathbb{P}_2$, of degree $d > 2$ and with homogeneous equation $F(x_0, x_1, x_2) = 0$. Take

$$H = \det\left(\frac{\partial^2 F}{\partial x_k \partial x_j}\right)_{k,j=0,1,2}.$$

(1) Prove that for any smooth point $p \in C$, if T_p is the tangent line to C at p and γ the only branch of C at p,

$$o_\gamma(H) = [C.T_p]_p - 2.$$

Deduce that $H \neq 0$ and that the curve $\mathbf{H} : H = 0$ (the *Hessian curve of* C) does not contain C. Note that the Hessian of C is the Jacobian curve of the net of polars of C (cf. Exercise 1.11).

(2) Prove that $p \in C \cap \mathbf{H}$ if and only if p is either a singular point or a flex of C, thus proving in particular that C has finitely-many flexes.

(3) Assume that C has no singular points other than δ nodes and κ ordinary cusps, and prove that the number of flexes of C is

$$\iota = 3d(d - 2) - 6\delta - 8\kappa,$$

where each flex p counts as many times as its order $[C.T_p]_p - 2$, T_p the tangent line to C at p. (The reader may see [11, 5.9] for details).

3.19 Reduced equations of smooth cubics. Assume that C is a smooth cubic of $\mathbb{P}_2$, q a flex of C (which does exist by Exercise 3.18) and T the tangent line to C at q.

(1) Prove that $[C \cdot T]_q = 3$ and $C \cap T = \{q\}$.

(2) Prove that the polar $\mathcal{P}_q(C)$ splits into two different lines, one of which is T, say $\mathcal{P}_q(C) = T + L$.

(3) Prove that $C \cap L$ is a set of three different points, none of which belongs to T. *Hint:* 3.5.11 may be useful.

(4) Prove that there is a reference of $\mathbb{P}_2$ relative to which C has equation

$$x_0 x_2^2 + x_1(x_1 - x_0)(x_1 - \alpha x_0) = 0, \quad \alpha \in \mathbb{C} - \{0, 1\}. \qquad (3.26)$$

(5) Prove that any curve given by an equation such as (3.26) above is a smooth cubic.

(6) Prove that, relative to other references of $\mathbb{P}_2$, C also has equations

$$y_0 y_2^2 + y_1(y_1 - y_0)(y_1 - \alpha' y_0) = 0$$

for $\alpha' = 1/\alpha, 1 - \alpha, 1/(1-\alpha), 1 - 1/\alpha, (\alpha-1)/\alpha$. *Hint:* use changes of coordinates of the form $x_0 = y_0$, $x_1 = ay_1 + by_0$, $x_2 = y_2$, $a \neq 0$.

3.20 Prove the following weak version of Noether's fundamental theorem (Lamé, 1818) directly from 3.4.6:

Assume that C_0 and C_1 are curves of degree d of $\mathbb{P}_2$ that intersect in different points p_i, $i = 1, \ldots d^2$. If an irreducible curve of degree d contains all these points, then it belongs to the pencil spanned by C_0 and C_1.

3.21 (Maclaurin, 1748) Assume that a line L intersects a smooth cubic C at three distinct points p_1, p_2, p_3. For $i = 1, 2, 3$, let T_i be the tangent to C at p_i and assume $T_i \cap C = \{p_i, q_i\}$, $q_i \neq p_i$. Prove that the points q_1, q_2, q_3 are aligned. *Hint:* apply Noether's fundamental theorem 3.8.14 to C, $2L$ and $T_1 + T_2 + T_3$, and consider the factor of the equation of $2L$.

3.22 Use intersection multiplicity to remove the hypothesis $q_i \neq p_i$ in Exercise 3.21, thus proving in particular that the line joining two different flexes of a smooth cubic C intersects C at a third flex.

3.23 Reprove 3.4.7 using Noether's fundamental theorem 3.8.14.

Chapter 4

The Intrinsic Geometry on a Curve

The study of an irreducible curve modulo the action of birational maps is called the *intrinsic geometry* on the curve. It may of course also be called *birationally invariant geometry*. The adjective *intrinsic* was also used by the classics to denote the notions and results belonging to the intrinsic geometry (that is, those invariant under birational maps). To avoid confusions with other uses of the term intrinsic we will not follow this practice, but use the terms *birationally invariant* instead. The first two sections of this chapter are devoted to introducing rational and birational maps and studying their action on curves; the remaining sections present the essentials of the intrinsic geometry on a curve.

4.1 Rational and Birational Maps

The *algebraic sets* of a projective space $\mathbb{P}_n$, with coordinates $x_0, \ldots, x_n$, are the sets the form

$$V(P_1, \ldots, P_r) = \\ \{p = [a_0, \ldots, a_n] \in \mathbb{P}_n \mid P_1(a_0, \ldots, a_n) = \cdots = P_r(a_0, \ldots, a_n) = 0\},$$

for given homogeneous polynomials $P_1, \ldots, P_r \in \mathbb{C}[x_0, \ldots, x_n]$, $r > 0$.

Similarly, the *algebraic sets* of an affine space $\mathbb{A}_n$, with coordinates $X_1, \ldots, X_n$, are the sets of the form

$$V(P_1, \ldots, P_r) = \\ \{p = (a_1, \ldots, a_n) \in \mathbb{A}_n \mid P_1(a_1, \ldots, a_n) = \cdots = P_r(a_1, \ldots, a_n) = 0\},$$

for given $P_1, \ldots, P_r \in \mathbb{C}[X_1, \ldots, X_n]$, $r > 0$.

The reader may easily check that in both cases the condition of being an algebraic set is independent of the choice of the coordinates.

© Springer Nature Switzerland AG 2019
E. Casas-Alvero, *Algebraic Curves, the Brill and Noether Way*, Universitext,
https://doi.org/10.1007/978-3-030-29016-0_4

Both $\mathbb{P}_n$ and $\mathbb{A}_n$ are algebraic sets (defined by the polynomial zero). The algebraic sets of $\mathbb{P}_n$ other than $\mathbb{P}_n$, as well as the algebraic sets of $\mathbb{A}_n$ other than $\mathbb{A}_n$, will be called *proper algebraic sets*. Also the empty set is an algebraic set, defined by any non-zero constant. The union and the intersection of two algebraic sets are algebraic sets too, because

$$V(P_1, \ldots, P_r) \cup V(Q_1, \ldots, Q_s) = V(P_i Q_j)_{\substack{i=1,\ldots,r \\ j=1,\ldots,s}} \tag{4.1}$$

and

$$V(P_1, \ldots, P_r) \cap V(Q_1, \ldots, Q_s) = V(P_1, \ldots, P_r, Q_1, \ldots, Q_s)$$

in both the projective and the affine cases. As a consequence, also in both cases, the finite unions and finite intersections of algebraic sets are algebraic sets.

Remark 4.1.1 Since a homogeneous polynomial (resp. polynomial) vanishes on the whole of $\mathbb{P}_n$ (resp. $\mathbb{A}_n$) if and only if it is zero (see the comment preceding 1.1.6), $V(P_1, \ldots, P_r)$ is a proper algebraic set unless $P_i = 0$, $i = 1, \ldots, r$. Then it follows from (4.1) that the union of any two (or finitely many) proper algebraic sets is a proper algebraic set.

Clearly, among the polynomials defining a proper and non-empty algebraic set of $\mathbb{P}_n$ or $\mathbb{A}_n$ there is a non-constant one. It follows that any proper algebraic set is contained in the set of points of a hypersurface. This, together with 1.1.6, make obvious the following lemma:

Lemma 4.1.2 *If V is a proper algebraic set of $\mathbb{P}_n$ and G a homogeneous polynomial that vanishes on all the points of $\mathbb{P}_n - V$, then $G = 0$.*

Assume we have fixed two projective spaces $\mathbb{P}_n$, of dimension n with projective coordinates $x_0, \ldots, x_n$, and $\mathbb{P}_m$, of dimension m with projective coordinates $y_0, \ldots, y_m$. Fix an ordered set $\mathcal{P} = (P_0, \ldots, P_m)$, where $P_0, \ldots, P_m \in \mathbb{C}[x_0, \ldots, x_n]$ are homogeneous polynomials, all of the same degree d and one at least non-zero. Then, $V(\mathcal{P}) = V(P_0, \ldots, P_m)$ is a proper algebraic set and we consider the map

$$\Phi_{\mathcal{P}} : \mathbb{P}_n - V(\mathcal{P}) \longrightarrow \mathbb{P}_m$$
$$p = [a_0, \ldots, a_n] \longmapsto [P_0(a_0, \ldots, a_n), \ldots, P_m(a_0, \ldots, a_n)].$$

It is clearly well defined, because, on one hand, taking other coordinates of p, namely $p = [\lambda a_0, \ldots, \lambda a_n]$, $\lambda \in \mathbb{C} - \{0\}$, does not change $\Phi_{\mathcal{P}}(p)$, but just multiplies its coordinates by λ^d. On the other hand, $p \notin V(P_0, \ldots, P_m)$ ensures that at least one of the coordinates of $\Phi_{\mathcal{P}}(p)$ is non-zero, as due.

Since we have seen above that it does not depend on the choice of the coordinates $a_0, \ldots, a_n$ of p, in the sequel we will write

$$[P_0(a_0, \ldots, a_n), \ldots, P_m(a_0, \ldots, a_n)] = [P_0(p), \ldots, P_m(p)].$$

Replacing the polynomials $P_0, \ldots, P_m$ with their products $PP_0, \ldots, PP_m$, by a non-zero homogeneous factor P, gives rise to a similar map which is just the restriction of $\Phi_{\mathcal{P}}$ to the smaller – still non-empty – set

$$\mathbb{P}_n - V(PP_0, \ldots, PP_m) = \mathbb{P}_n - V(P_0, \ldots, P_m) \cup V(P).$$

Similarly, cancelling a homogeneous factor common to $P_0, \ldots, P_m$ gives rise to a map which is an extension of $\Phi_{\mathcal{P}}$.

We are not interested in distinguishing $\Phi_{\mathcal{P}}$ from the above restrictions and extensions. Therefore we will proceed as for rational functions in section 1.4: assume we have maps $\Phi_{\mathcal{P}}$ and $\Phi_{\mathcal{Q}}$, each defined by an ordered set of homogeneous polynomials of the same degree, not all zero, $\mathcal{P} = (P_0, \ldots, P_m)$ and $\mathcal{Q} = (Q_0, \ldots, Q_m)$; then $\Phi_{\mathcal{P}}$ and $\Phi_{\mathcal{Q}}$ are taken as equivalent if and only if they agree on the complement of a proper algebraic set V, $V \supset V(\mathcal{P}) \cup V(\mathcal{Q})$. Checking that the above is an equivalence relation is straightforward and left to the reader, who may use 4.1.1 for the transitive property. The equivalence classes for this equivalence relation will be called *rational maps* from $\mathbb{P}_n$ to $\mathbb{P}_m$, written $\Phi : \mathbb{P}_n \to \mathbb{P}_m$.

The polynomials $P_0, \ldots P_m$ defining the representative $\Phi_{\mathcal{P}}$, $\mathcal{P} = (P_0, \ldots P_m)$, of a rational map Φ are said to define – or determine – Φ. The equalities

$$y_0 = P_0(x_0, \ldots, x_n)$$

$$\vdots$$

$$y_m = P_m(x_0, \ldots, x_n)$$

are called *equations* of $\Phi_{\mathcal{P}}$, and also of Φ.

Proposition 4.1.3 *Let* $\mathcal{P} = (P_0, \ldots, P_m)$ *and* $\mathcal{Q} = (Q_0, \ldots Q_m)$ *be ordered sets of homogeneous polynomials in* $x_0, \ldots, x_n$, *each set being composed of polynomials of the same degree, one at least non-zero. The following conditions are equivalent:*

(i) *The maps* $\Phi_{\mathcal{P}}$ *and* $\Phi_{\mathcal{Q}}$ *are equivalent.*

(ii) $\Phi_{\mathcal{P}}$ *and* $\Phi_{\mathcal{Q}}$ *agree on all the points of* $\mathbb{P}_n - V(\mathcal{P}) \cup V(\mathcal{Q})$.

(iii) $P_i Q_j - P_j Q_i = 0$ *for all* $i, j = 0, \ldots, m$.

(iv) *If* $P_r \neq 0$ *for some* $r = 0, \ldots, m$ *then* $Q_r \neq 0$ *and* $P_j / P_r = Q_j / Q_r$ *in* $\mathbb{C}(x_0, \ldots, x_m)$ *for any* $j = 0, \ldots, m$.

PROOF: We will prove (ii) $\Rightarrow$ (i) $\Rightarrow$ (iii) $\Rightarrow$ (iv) $\Rightarrow$ (ii). That (ii) $\Rightarrow$ (i) is clear. Assume (i); then, for some proper algebraic set V, each of the polynomials $P_i Q_j - P_j Q_i$, $i, j = 0, \ldots, m$, vanishes on $\mathbb{P}_n - V$ and therefore, by 4.1.2, is the polynomial zero. This proves (iii). Assume now (iii) and that $P_r \neq 0$: if $Q_r = 0$, then since there is an s with $Q_s \neq 0$, the equality $P_r Q_s - P_s Q_r = 0$ gives a contradiction, hence $Q_r \neq 0$. Using this, the second claim of (iv) is clear from the equalities $P_r Q_j - P_j Q_r = 0$, $j = 0, \ldots, m$. To close, assume (iv) and take any $p = [a_0, \ldots, a_n] \notin V(\mathcal{P}) \cup V(\mathcal{Q})$; then we have $P_r(p) \neq 0$ for some r, which guarantees $P_r \neq 0$, and hence that the equalities of (iv) hold. These equalities giving

$$Q_j(a_0, \ldots, a_n) = \frac{Q_r(a_0, \ldots, a_n)}{P_r(a_0, \ldots, a_n)} P_j(a_0, \ldots, a_n)$$

for $j = 0, \ldots, m$, it is clear that $\Phi_{\mathcal{P}}(p)$ and $\Phi_{\mathcal{Q}}(p)$ are both defined and agree, thus proving (ii). $\diamond$

It is worth highlighting part of 4.1.2 in the form of the next corollary:

Corollary 4.1.4 *Different representatives $\Phi_{\mathcal{P}}$ and $\Phi_{\mathcal{Q}}$ of a rational map Φ have $\Phi_{\mathcal{P}}(p) = \Phi_{\mathcal{Q}}(p)$ for all the points p at which both are defined.*

From 4.1.4 it is clear that the different representatives $\Phi_{\mathcal{P}}$ of a rational map Φ patch together to give the map $p \mapsto \Phi_{\mathcal{P}}(p)$ if a representative $\Phi_{\mathcal{P}}$ of Φ is defined at p, defined in the set of points of $\mathbb{P}_n$ at which at least one representative of Φ is defined. This map obviously determines and is in turn determined by Φ, and therefore we will make no distinction between it and Φ: we say that Φ is *defined* at a point p if and only if one of its representatives is defined or *regular* at p; the *image* $\Phi(p)$ of p under Φ is then defined as being the image of p under any representative of Φ defined at p.

When Φ is defined at a point p, p is equivalently said to be a *regular point* of Φ. The points at which Φ is not defined are called *points of indetermination* of Φ. In the sequel, when writing $\Phi(p)$ we will implicitly assume that p is a regular point of Φ.

The identity map Id $: \mathbb{P}_n \to \mathbb{P}_n$ arises as a rational map, given by the set of polynomials $(x_0, \ldots, x_n)$. Similarly the constant map $\mathbb{P}_n \to \mathbb{P}_m$ with image $[a_0, \ldots, a_n]$ is the rational map given by the set of degree-zero polynomials $(a_0, \ldots, a_m)$.

Actually any rational map agrees, as a map, with one of its representatives. This is due to the factoriality of the ring $\mathbb{C}[x_0, \ldots, x_n]$ and fails to be true in more general situations:

Proposition 4.1.5 *Let Φ be a rational map from $\mathbb{P}_n$ to $\mathbb{P}_m$ and $\Phi_{\mathcal{P}}$, $\mathcal{P} = (P_0, \ldots, P_m)$, a representative of Φ. If $D = \gcd(P_0, \ldots, P_m)$, $P_i' = P_i/D$, $i = 0, \ldots, m$ and $\mathcal{P}' = (P_0', \ldots, P_m')$, then $\Phi_{\mathcal{P}'}$ is a representative of Φ and any other representative of Φ is a restriction of $\Phi_{\mathcal{P}'}$.*

PROOF: From the definitions it is clear that $V(\mathcal{P}') \supset V(\mathcal{P})$ and also that $\Phi_{\mathcal{P}}$ and $\Phi_{\mathcal{P}'}$ agree on $\mathbb{P}_n - V(\mathcal{P})$, which shows that $\Phi_{\mathcal{P}'}$ represents Φ. If $\Phi_{\mathcal{Q}}$, $\mathcal{Q} = (Q_0, \ldots, Q_m)$ is any representative of Φ, then, by 4.1.3, $P_i' Q_j - P_j' Q_i = 0$, $i, j = 0, \ldots, m$. Fix any i: since there is a non-zero P_j', these equalities show that if $P_i' = 0$, then also $Q_i = 0$. If, otherwise, $P_i' \neq 0$ and for an irreducible polynomial F, F^r divides P_i', there is a j for which F does not divide P_j' and therefore F^r divides Q_i. This proves that each non-zero P_i' divides the corresponding Q_i, say $Q_i = G_i P_i'$. Using again the equalities

$P_i'Q_j - P_j'Q_i = 0$, $i, j = 0, \ldots, m$, shows that the G_i are the same for all i, say $G_i = G$ for all i for which $P_i' \neq 0$. This gives equalities $Q_i = GP_i'$, $i = 0, \ldots, m$, and hence the claim. $\diamond$

The notations being as in 4.1.5, clearly $\Phi_{\mathcal{P}'}$ of is the only representative of Φ which is an extension of any representative of Φ: it is called the *maximal representative* of Φ. From now on, the reader may safely identify each rational map and its maximal representative. Nevertheless, this does not completely eliminate the use of non-maximal representatives, as sometimes rational maps are defined using them; the composition of two rational maps, defined below, is an example.

Remark 4.1.6 We have seen in the proof of 4.1.5 that the polynomials defining any representative of Φ have the form $Q_i = GP_i'$, $i = 0, \ldots, m$, where the P_i' define the maximal representative of Φ and G is a non-zero polynomial, necessarily homogenous by 1.1.1.

The points at which the maximal representative of a rational map Φ is defined are its regular points, while those at which the maximal representative is not defined are the indetermination points of Φ.

Remark 4.1.7 The polynomials defining the maximal representative of a rational map $\Phi : \mathbb{P}_2 \to \mathbb{P}_m$ being coprime, the set of indetermination points of such a Φ is finite, by 3.2.3.

The image of the maximal representative of a rational map Φ – which obviously equals the union of the images of all representatives of Φ – is called the *image* of Φ, denoted in the sequel by $\mathrm{Im}(\Phi)$.

Here is a relevant example, it will be used in the next section:

Example 4.1.8 Once a triangle T, with vertices p_0, p_1, p_2, has been fixed in $\mathbb{P}_2$, let x_0, x_1, x_2 be coordinates relative to a reference of $\mathbb{P}_2$ whose vertices are the vertices of T. Then the equations

$$\bar{x}_0 = x_1 x_2$$
$$\bar{x}_1 = x_2 x_0$$
$$\bar{x}_2 = x_0 x_1$$

define (the maximal representative of) a rational map $\mathfrak{Q} : \mathbb{P}_2 \to \mathbb{P}_2$ which is called a *standard* (or *ordinary*) *quadratic transformation* with *fundamental triangle* T. Denote by ℓ_i the side of T opposite p_i, $i = 0, 1, 2$. The reader may easily check that

(1) The points of indetermination of $\mathfrak{Q}$ are the vertices p_0, p_1, p_2 of the fundamental triangle T. They are often called the *fundamental points* of $\mathfrak{Q}$.

(2) For $i = 0, 1, 2$, $\mathfrak{Q}$ maps all non-fundamental points on the side ℓ_i to the opposite vertex p_i.

(3) The restriction of $\mathfrak{Q}$ to $\dot{T} = \mathbb{P}_2 - \ell_0 \cup \ell_1 \cup \ell_2$ is the bijection

$$\dot{T} \longrightarrow \dot{T}$$
$$[a,b,c] \mapsto [\frac{1}{a}, \frac{1}{b}, \frac{1}{c}].$$

(4) The image of $\mathfrak{Q}$ is the set

$$(\mathbb{P}_2 - \ell_0 \cup \ell_1 \cup \ell_2) \cup \{p_0, p_1, p_2\}.$$

A rational map Φ is determined by its restriction to the complement of any proper algebraic set containing all the indetermination points of Φ:

Proposition 4.1.9 *If rational maps $\Phi, \Psi : \mathbb{P}_n \to \mathbb{P}_m$ satisfy $\Phi(p) = \Psi(p)$ for $p \in \mathbb{P}_n - W$, where W is a proper algebraic subset $W \subset \mathbb{P}_n$ (containing all the indetermination points of Π or Ψ), then $\Phi = \Psi$.*

PROOF: By 4.1.3, any representative $\Phi_{\mathcal{P}}$ of Φ satisfies $\Phi_{\mathcal{P}}(p) = \Phi(p)$ for any $p \in \mathbb{P}_n - W \cup V(\mathcal{P})$, and also any representative $\Psi_{\mathcal{Q}}$ of Ψ satisfies $\Psi_{\mathcal{Q}}(p) = \Psi(p)$ for any $p \in \mathbb{P}_n - W \cup V(\mathcal{Q})$. By the hypothesis, these representatives do agree on $\mathbb{P}_n - W \cup V(\mathcal{P}) \cup V(\mathcal{Q})$ and therefore define the same rational map. ◇

Let $\Phi_{\mathcal{P}}$ be a representative of a rational map $\Phi : \mathbb{P}_n \to \mathbb{P}_m$. If $\mathcal{P} = (P_0, \ldots, P_m)$, by the definition, $P_i \neq 0$ for some i; by 4.1.3 (or 4.1.6) this condition is independent of the representative, and it is obviously equivalent to $\mathrm{Im}(\Phi)$ not being included in the hyperplane $y_i = 0$ of $\mathbb{P}_m$. In the sequel we will fix our attention on the case in which $P_0 \neq 0$, the other cases being dealt with similarly. Let $\mathbb{A}_m$ be the 0-th affine chart of the target space $\mathbb{P}_m$ and take on it the affine coordinates $Y_i = y_i/y_0$, $i = 1, \ldots, m$. If φ_i are the rational functions $\varphi_i = P_i/P_0$, $i = 1, \ldots, m$, then the composite map

$$\mathbb{P}_n - V(F_0) \longrightarrow \mathbb{A}_n \hookrightarrow \mathbb{P}_m$$
$$p \longmapsto (\varphi_1(p), \ldots, \varphi_m(p))$$

is the restriction of $\Phi_{\mathcal{P}}$ to $\mathbb{P}_n - V(F_0)$ and therefore, by 4.1.9 above, determines Φ as the only rational map extending it. This map is given by the equations

$$Y_1 = \varphi_1(p)$$
$$\vdots$$
$$Y_m = \varphi_m(p).$$

The map itself, and also its equations above, are called an *affine representation* of Φ (relative to the 0-th affine chart of the target). Often these equations are presented using affine representations of the φ_i, that is, using affine coordinates of p after restricting it to vary in an affine chart $\mathbb{A}_n$ of $\mathbb{P}_n$.

This in fact restricts further $\Phi_{\mathcal{P}}$, namely to $(\mathbb{P}_n - V(F_0)) \cap \mathbb{A}_n$; since the latter is still the complement of a proper algebraic set, this restriction of $\Phi_{\mathcal{P}}$ still determines Φ.

The ordered set of polynomials $\mathcal{P}$ defining a representative $\Phi_{\mathcal{P}}$ of the above rational map Φ and the ordered set $(\varphi_1, \dots, \varphi_m)$, of the rational functions appearing in the above affine representation of Φ, are equivalent data and either may be used to define Φ. To effectively recover Φ from its affine representation it suffices to write the φ_i as quotients of homogeneous polynomials of the same degree in the coordinates $x_0, \dots, x_n$ of $\mathbb{P}_n$ with the same denominator, say $\varphi_i = Q_i/Q$, $i = 1, \dots, m$; then $\Phi_{\mathcal{Q}}$, $\mathcal{Q} = (Q, Q_1, \dots, Q_m)$, may be taken as a representative of Φ. A similar rule applies if the affine representation is relative to another affine chart of the target.

Consider for instance the map

$$\mathbb{A}_2 - V(x) \longrightarrow \mathbb{A}_2'$$
$$(x, y) \mapsto (x, y/x)$$

already used in the proof of 3.8.8. By writing

$$x = \frac{x_1}{x_0} = \frac{x_1^2}{x_0 x_1} \quad \text{and} \quad \frac{y}{x} = \frac{x_2}{x_1} = \frac{x_0 x_2}{x_0 x_1},$$

it appears to be an affine representation of the rational map given by the polynomials $x_0 x_1, x_1^2, x_0 x_2$.

Remark 4.1.10 If $m = 1$, the affine representations may be understood as relative to the choice of an absolute coordinate on $\mathbb{P}_1$. Such a coordinate being fixed, an affine representation of a rational map Φ consists of a single rational function φ, which, as explained above, determines Φ. By contrast, Φ does not determine φ, as the latter depends also on the choice of the absolute coordinate: a rational function ψ is an affine representation of Φ relative to another choice of the absolute coordinate if and only if $\psi = (a\varphi + b)/(c\varphi + d)$ with $a, b, c, d \in \mathbb{C}$ and $ad - bc \neq 0$ (see 1.2.2). Thus, rational maps with target $\mathbb{P}_1$ and rational functions have to be retained as essentially different objects; addition, product and notions such as zero or pole, which are well defined for rational functions, do not translate from an affine representation of Φ to Φ itself due to the non-intrinsic character of their relationship.

Remark 4.1.11 If the polynomials $P_0, \dots, P_m$ defining the maximal representative of Φ – or, equivalently by 4.1.6, those defining an arbitrary representative – are linearly dependent, say $\sum_{i=0}^{m} \lambda_i P_i = 0$ with some $\lambda_i \neq 0$, then the image of Φ is contained in a hyperplane of the target, namely in $\sum_{i=0}^{m} \lambda_i y_i = 0$. The converse is also true: if the image of Φ is contained in $\sum_{i=0}^{m} \lambda_i y_i = 0$ then the polynomial $\sum_{i=0}^{m} \lambda_i P_i$ is zero because it vanishes on the complement of a proper algebraic set.

Assume to be given rational maps

$$\Phi : \mathbb{P}_n \longrightarrow \mathbb{P}_m \quad \text{and} \quad \Psi : \mathbb{P}_m \longrightarrow \mathbb{P}_r$$

with maximal representatives $\Phi_{\mathcal{P}}$, $\mathcal{P} = (P_0, \ldots, P_m)$ and $\Psi_{\mathcal{Q}}$, $\mathcal{Q} = (Q_0, \ldots, Q_r)$. If $\mathrm{Im}(\Phi) = \mathrm{Im}(\Phi_{\mathcal{P}}) \subset V(\mathcal{Q})$, then the composition $\Psi \circ \Phi$, of Φ and Ψ, is taken as not defined. Otherwise at least one of the polynomials

$$H_i = Q_i(P_0, \ldots, P_m) \in \mathbb{C}[x_0, \ldots, x_n],$$

obtaining by replacing in each Q_i the variable y_j with P_j, $j = 0, \ldots, m$, is non-zero. Since they obviously are homogeneous of the same degree, they define a representative

$$\Theta_{\mathcal{H}} : \mathbb{P}_n - V(\mathcal{H}) \longrightarrow \mathbb{P}_r,$$

$\mathcal{H} = (H_0, \ldots, H_r)$, of a rational map $\Theta : \mathbb{P}_n \to \mathbb{P}_r$ which is called the *composition* of Φ and Ψ, written $\Theta = \Psi \circ \Phi$.

As is clear, $\Theta_{\mathcal{H}}(p) = \Psi_{\mathcal{Q}}(\Phi_P(p))$ for any $p \in \mathbb{P}_n - V(H_0, \ldots, H_r)$. $\Theta_{\mathcal{H}}$ may not be the maximal representative of Θ, see Example 4.1.12 below.

The reader may easily check that the composition defined above satisfies associativity, and also that $\Phi \circ \mathrm{Id}_{\mathbb{P}_n} = \mathrm{Id}_{\mathbb{P}_m} \circ \Phi = \Phi$ for any rational map $\Phi : \mathbb{P}_n \to \mathbb{P}_m$. It is also direct to check that if, in the definition of $\Psi \circ \Phi$ above, the maximal representatives of Φ and Ψ are replaced with other representatives $\Phi_{\mathcal{P}'}$ and $\Psi_{\mathcal{Q}'}$, subject to the condition $\mathrm{Im}(\Phi_{\mathcal{P}'}) \not\subset V(\mathcal{Q}')$, then another representative of $\Psi \circ \Phi$ is obtained.

Let $\Phi : \mathbb{P}_n \to \mathbb{P}_m$ be a rational map. If there exists another rational map $\Psi : \mathbb{P}_m \to \mathbb{P}_n$ such that

$$\Phi \circ \Psi = \mathrm{Id}_{\mathbb{P}_n} \quad \text{and} \quad \Psi \circ \Phi = \mathrm{Id}_{\mathbb{P}_m}, \tag{4.2}$$

then the rational map Φ is said to be a *birational map*. The rational map Ψ is then uniquely determined by the conditions (4.2) and is called the *inverse* of Φ (if Ψ' is another inverse, just compute $\Psi' \circ \Phi \circ \Psi$ in two ways). The inverse rational map Ψ is of course birational too. The easiest examples of birational maps are the projectivities. Next is a more interesting example:

Example 4.1.12 Consider the standard quadratic transformation $\mathfrak{Q}$ of 4.1.8. Performing the substitution of polynomials as prescribed to get $\mathfrak{Q}^2 = \mathfrak{Q} \circ \mathfrak{Q}$ gives rise to the polynomials

$$x_0^2 x_1 x_2, x_0 x_1^2 x_2, x_0 x_1 x_2^2,$$

which obviously define a non-maximal representative of the identity. Therefore $\mathfrak{Q}$ is a birational map which equals its own inverse.

Actually, the existence of a birational map $\Phi : \mathbb{P}_n \to \mathbb{P}_m$ implies $m = n$. We will not prove this fact here, as we will make no use of it in the sequel and giving a proof of it would take us too long.

Assume now that $C : F = 0$ is an irreducible curve of $\mathbb{P}_2$, in which coordinates x_0, x_1, x_2 have been fixed. We are interested in restricting to (the set of points of) C rational maps $\Phi : \mathbb{P}_2 \to \mathbb{P}_m$. Clearly, restricting to C a representative $\Phi_{\mathcal{P}}$ of Φ makes sense if and only if $\Phi_{\mathcal{P}}$ is defined at some point of C. Next is an important remark regarding these maps:

Remark 4.1.13 If $\Phi_\mathcal{P}$, $\mathcal{P} = (P_0, \ldots, P_m)$ is defined at some point of C, then, by 3.6.14(a), $P_i \notin (F)$ for some i. Then $\Phi_\mathcal{P}$ – and therefore also Φ – is defined at all but finitely many points of C, namely at least at all those $p \in C$ for which $P_i(p) \neq 0$; these are all but finitely many points of C due to 3.6.14(a).

In the sequel we will restrict ourselves to considering rational maps $\mathbb{P}_2 \to \mathbb{P}_m$ that are defined at some point of C. As noted above, each of them is then defined at all but finitely many points of C. As already with rational functions, the restrictions of different rational maps to C may patch together to give a map defined in a larger subset of C: we are not interested in distinguishing between the different restrictions that patch together and the extended map they give rise to. To this end, we introduce, once again, an equivalence relation: rational maps $\Phi, \Psi : \mathbb{P}_2 \to \mathbb{P}_m$, both defined at some point of C, are said to be *equivalent on C* if and only if they are defined and agree at all but finitely many points of C. This is indeed an equivalence relation, as is directly checked; the corresponding equivalence classes are called *rational maps from C to $\mathbb{P}_m$*.

Proposition 4.1.14 *If representatives $\Phi_\mathcal{P}$, $\mathcal{P} = (P_0, \ldots, P_m)$, and $\Psi_\mathcal{Q}$, $\mathcal{Q} = (Q_0, \ldots, Q_m)$, of rational maps $\Phi, \Psi : \mathbb{P}_2 \to \mathbb{P}_m$, are each defined at some point of an irreducible curve C of $\mathbb{P}_2$, then the following conditions are equivalent:*

(i)   *Φ and Ψ are equivalent on C.*

(ii)   *Φ and Ψ agree at all points of C at which they both are defined.*

(iii)   *If, for some i, P_i is non-zero at some point of C, then also Q_i is non-zero at some point of C and the rational functions on $\mathbb{P}_2$ P_j/P_i and Q_j/Q_i restrict to the same rational function on C, for $j = 0, \ldots, m$, $j \neq i$.*

(iv)   *$P_i Q_j - P_j Q_i \in (F)$ for all $i, j = 1, \ldots, m$, $i \neq j$.*

PROOF: As noted in 4.1.13, by the hypothesis, both Φ and Ψ are defined at all but finitely many points of C, from which (ii) $\Rightarrow$ (i) is clear.

Assume that (i) is satisfied. Then, since $\Phi(p) = \Phi_\mathcal{P}(p)$ and $\Psi(p) = \Psi_\mathcal{Q}(p)$ for all points p at which both $\Phi_\mathcal{P}$ and $\Psi_\mathcal{Q}$ are defined, also $\Phi_\mathcal{P}$ and $\Psi_\mathcal{Q}$ are defined and agree at all but finitely points of C. As a consequence, the polynomials $P_i Q_j - P_j Q_i$, $i, j = 1, \ldots, m$, $i \neq j$, are zero at all but finitely many points of C and (iv) follows from 3.6.14(a).

Assume (iv) and that P_i is non-zero at some point of C, or, equivalently, that $P_i \notin (F)$. By the hypothesis on $\Phi_\mathcal{Q}$, there is a j for which $Q_j \notin (F)$. If $i \neq j$ and $Q_i \in (F)$, $P_i Q_j - P_j Q_i \in (F)$ yields $P_i Q_j \in (F)$ and so, by the irreducibility of F, either $P_i \in (F)$ or $Q_j \in (F)$, a contradiction in any case. Thus $Q_i \notin (F)$. Both P_j/P_i and Q_j/Q_i may thus be restricted to C and their restrictions agree due to (iv) and 3.7.3. This proves (iii).

To close, assume (iii). We will check first that (iii) also holds for any other representatives $\Phi'_{\mathcal{P}'}$, $\mathcal{P}' = (P'_0, \ldots, P'_m)$ and $\Psi'_{\mathcal{Q}'}$, $\mathcal{Q}' = (Q'_0, \ldots, Q'_m)$ of Φ and Ψ, provided each is defined at a point of C. Indeed, by 4.1.19 we have equalities

$$P'_i P_j = P'_j P_i \quad \text{and} \quad Q'_i Q_j = Q'_j Q_i \qquad (4.3)$$

for $i, j = 1, \ldots, m$. Assume that P'_i is non-zero at some point of C. Then $P'_i \notin (F)$, and, $\Phi_{\mathcal{P}}$ being defined at some point of C, there is a j for which $P_j \notin (F)$; F being irreducible, it follows that $P_i P'_j = P'_i P_j \notin (F)$; then $P_i \notin (F)$, after which it is non-zero at some point of C. By (iii), Q_i is non-zero at some point of C and then, repeating the preceding argument, Q'_i is non-zero at some point of C. Thus, for any $j \neq i$, both the rational functions P'_j/P'_i and Q'_j/Q'_i may be restricted to C. Since, by the equalities (4.3), $P'_j/P'_i = P_j/P_i$ and $Q'_j/Q'_i = Q_j/Q_i$ as rational functions of $\mathbb{P}_n$, the restrictions of P'_j/P'_i and Q'_j/Q'_i to C agree.

Now, assume both Φ and Ψ to be defined at a point $p \in C$. By the above, we may assume that representatives of them, $\Phi_{\mathcal{P}}$, $\mathcal{P} = (P_0, \ldots, P_m)$, and $\Psi_{\mathcal{Q}}$, $\mathcal{Q} = (Q_0, \ldots, Q_m)$ are both defined at p and satisfy (iii). There is thus an i for which P_i is not zero at p and then (iii) ensures that the restrictions of P_j/P_i and Q_j/Q_i to C are defined and agree for $j = 1, \ldots, m$, $j \neq i$. By 3.7.7(a) applied to $P_j/P_i - Q_j/Q_i$, $P_i Q_j - P_j Q_i$ is zero on C, still for $j = 1, \ldots, m$, $j \neq i$, and hence

$$\Phi(p) = [P_0(p), \ldots, P_m(p)] = [Q_0(p), \ldots, Q_m(p)] = \Psi(p),$$

as claimed in (ii).   ◇

By 4.1.14(ii), the restrictions of the different representatives of a rational map $\mathfrak{f}$ on C patch together to define a map of a subset of C with finite complement. This map clearly determines and is determined by $\mathfrak{f}$ and will be identified with $\mathfrak{f}$ from now on: we will say that $\mathfrak{f}$ is *defined* at $p \in C$ – or that p is a *regular point* of $\mathfrak{f}$ – if and only if one of the representatives of $\mathfrak{f}$ is defined at p; then the *image* $\mathfrak{f}(p)$ of p under $\mathfrak{f}$ is defined as being the image of p under any representative of $\mathfrak{f}$ defined at p. The set of all the images of regular points of $\mathfrak{f}$ is called the *image of* $\mathfrak{f}$, denoted $\mathrm{Im}(\mathfrak{f})$. The points at which $\mathfrak{f}$ is not defined are called *points of indetermination* of $\mathfrak{f}$. As for rational functions, the rational map $\mathfrak{f}$ is often referred to as the restriction to C of any of its representatives.

Obviously any constant map from an irreducible curve C of $\mathbb{P}_2$ to $\mathbb{P}_m$ is a rational map, as it is the restriction of a constant map $\mathbb{P}_2 \to \mathbb{P}_m$.

We will not make use of rational maps defined on reducible curves. Therefore, in the sequel, when considering a rational map $\mathfrak{f}$ defined on a curve C, we will assume, sometimes implicitly, that C is irreducible. Also, when writing $\mathfrak{f}(p)$ we will implicitly assume that $\mathfrak{f}$ is defined at p.

Remark 4.1.15 If two rational maps $\mathfrak{f}, \mathfrak{f}'$, from an irreducible curve C to $\mathbb{P}_m$, are defined and agree on all but finitely many points $p \in C$, then, by the definition of the equivalence on C, $\mathfrak{f} = \mathfrak{f}'$.

If homogeneous polynomials $P_0, \ldots, P_n$ define a representative of a rational map $\mathfrak{f} : C \to \mathbb{P}_m$, we will say that $\mathfrak{f}$ is *defined* by $P_0, \ldots, P_n$, or by the rule

$$\mathfrak{f}(p) = [P_0(p), \ldots, P_n(p)]$$

which clearly holds for all but finitely many $p \in C$. If furthermore P_0 is not zero at some point of C, then the rational functions $\varphi_1, \ldots, \varphi_m$ on C represented by $P_1/P_0, \ldots, P_n/P_0$ still determine $\mathfrak{f}$. Indeed, if $y_0, \ldots, y_m$ are the homogeneous coordinates on $\mathbb{P}_m$, then, excluding the finitely many points $p \in C$ for which $P_0(p) = 0$, for all but finitely many $p \in C$ the values $\varphi_i(p)$ are the affine coordinates $Y_i = y_i/y_0$ of $\mathfrak{f}(p)$, namely

$$\mathfrak{f}(p) = (\varphi_0(p), \ldots, \varphi_n(p)).$$

As for the rational maps on $\mathbb{P}_n$, this equality is called an *affine representation* of $\mathfrak{f}$.

When $m = 1$, Remark 4.1.10 still applies to the present case: once an absolute coordinate has been chosen in $\mathbb{P}_1$, a rational map $\mathfrak{f} : C \to \mathbb{P}_1$ is determined by the single rational function φ_1 appearing in its affine representation, but φ_1 is not determined by $\mathfrak{f}$. Therefore also φ_1 and $\mathfrak{f}$ have to be retained as different objects.

The next proposition is a consequence of the principality of the local rings of the smooth points of curves.

Proposition 4.1.16 *If p is a smooth point of an irreducible curve C of $\mathbb{P}_2$, then any rational map $\mathfrak{f} : C \to \mathbb{P}_m$ is defined at p.*

PROOF: Take projective coordinates x_0, x_1, x_2 on $\mathbb{P}_2$ such that $p = [1, 0, 0]$ and the line $x_1 = 0$ is not tangent to C at p. In the affine chart $x_0 \neq 0$ we take affine coordinates $x = x_1/x_0$, $y = x_2/x_0$. Assume that $\mathfrak{f}$ is the restriction to C of a rational map given by homogeneous polynomials of degree d, $F_i \in \mathbb{C}[x_0, x_1, x_2]$, $i = 0, \ldots, m$. Consider the rational functions $f_i = F_i/x_0^d \in \mathcal{O}_{\mathbb{P}_2, p}$, still for $i = 0, \ldots, m$, and denote by a bar $\bar{}$ the classes in $\mathcal{O}_{C, p}$ of the elements of $\mathcal{O}_{\mathbb{P}_2, p}$, that is, their restrictions to C. Since $\bar{x}$ is a generator of the maximal ideal $\mathfrak{M}_{C, p}$ of $\mathcal{O}_{C, p}$ (3.7.14), we may write

$$\bar{f}_i = \bar{x}^{r_i} \bar{g}'_i,$$

where $r_i \geq 0$, $g'_i \in \mathcal{O}_{\mathbb{P}_2, p}$ and $g'_i(p) \neq 0$, for $i = 0, \ldots, m$. By taking $r = \min\{r_1, \ldots, r_m\}$, and $g_i = x^{r_i - r} g'_i$, we have

$$\bar{f}_i = \bar{x}^r g_i, \tag{4.4}$$

$r \geq 0$, still $g_i \in \mathcal{O}_{\mathbb{P}_2, p}$ for $i = 0, \ldots, m$ and $g_i(p) \neq 0$ for at least one value of the index i.

Write $g_i = G_i/G$, $i = 0, \ldots, m$, the G_i and G being homogeneous polynomials of the same degree in x_0, x_1, x_2, $G(p) \neq 0$: then $G_i(p) \neq 0$ for at least one i. By the equalities (4.4), for each $i = 0, \ldots, m$, the rational functions

$$\frac{F_i}{x_0^d} \quad \text{and} \quad \frac{x_1^r G_i}{x_0^r G}$$

agree on all but finitely many points of C. As a consequence, the rational maps $\mathbb{P}_2 \to \mathbb{P}_m$ defined by $(F_0, \dots, F_m)$ and $(G_0, \dots, G_m)$ are equivalent on C and therefore both represent $\mathfrak{f}$. Since we know that $G_i(p) \neq 0$ for some i, the latter is defined at p and therefore so is $\mathfrak{f}$, as claimed. $\diamond$

The claim of 4.1.16 cannot be extended to arbitrary points; here is an example:

Example 4.1.17 An easy computation shows that the cubic curve $C : x_0 x_1^2 - x_0 x_2^2 + x_2^3 = 0$, of a projective plane $\mathbb{P}_2$, is irreducible and has a node at $O = [1, 0, 0]$. Using affine coordinates $x = x_1/x_0$, $y = x_2/x_0$ on the affine chart $\mathbb{A}_2 : x_0 \neq 0$, representatives γ_1, γ_2 of the branches of C at O have Puiseux series

$$s = x + x^2 s_1 \quad \text{and} \quad s' = -x + x^2 s_1'$$

for some $s_1, s_1' \in \mathbb{C}\{x\}$.

Take $\mathbb{P}_1$ to be a projective line with fixed coordinates and $\Phi : \mathbb{P}_2 \to \mathbb{P}_1$ the rational map defined by the ordered pair of polynomials (x_1, x_2).

For any small enough neighbourhood U of O in $\mathbb{A}_2$, the restriction of Φ to $\gamma_1 \cap U - \{O\}$ is $(x, y) \mapsto [x, x + x^2 s_1(x)]$, whose unique continuous extension to $\gamma_1 \cap U$ is $(x, y) \mapsto [1, 1 + x s_1(x)]$ and maps O to $[1, 1]$. Similarly, up to replacing U with a smaller neighbourhood, the restriction of Φ to $\gamma_2 \cap U - \{O\}$ has as unique continuous extension to $\gamma_2 \cap U$ the map $(x, y) \mapsto [1, -1 + x s_1'(x)]$ which maps O to $[1, -1]$. Since $[-1, 1] \neq [1, 1]$, the restriction of Φ to $C \cap U - \{O\}$ cannot be extended to a continuous map defined on the whole of $C \cap U$.

Let $\mathfrak{f} : C \to \mathbb{P}_1$ be the rational map represented by Φ. Any other representative Ψ of $\mathfrak{f}$ agrees with Φ on all but finitely many points of C, and so, in particular, up to a suitable shrinking of U above, on all points of $C \cap U - \{O\}$. If O where a regular point of Ψ, the restriction of Ψ to $C \cap U$ would be a continuous extension of the restriction of Φ to $C \cap U - \{O\}$, against the above. Therefore O is a point of indetermination of all the representatives of $\mathfrak{f}$, hence a point of indetermination of $\mathfrak{f}$.

The following is a quite important claim, stronger than 4.1.15:

Proposition 4.1.18 *If two rational maps* $\mathfrak{f}, \mathfrak{f}' : C \to \mathbb{P}_m$ *are defined and agree at all points of an infinite subset* $T \subset C$, *then* $\mathfrak{f} = \mathfrak{f}'$.

PROOF: Assume $\mathfrak{f}$ and $\mathfrak{f}'$ to be represented, respectively, by rational maps of $\mathbb{P}_2$, Φ and Ψ, and these in turn by maps $\Phi_{\mathcal{P}}$ and $\Psi_{\mathcal{Q}}$ each defined at some point of C, $\mathcal{P} = (P_0, \dots, P_m)$, $\mathcal{Q} = (Q_0, \dots, Q_m)$. The polynomials $P_i Q_j - P_j Q_i$, $i, j = 0, \dots, m$, vanish at all points $p \in T$, because either one of the representatives $\Phi_{\mathcal{P}}, \Psi_{\mathcal{Q}}$ is undefined at p, or they both have the same value at p due to the hypothesis. Therefore, the polynomials $P_i Q_j - P_j Q_i$, $i, j = 0, \dots, m$, vanish at all points $p \in C$ due to 3.6.14(a). Then $\Phi_{\mathcal{P}}(p) = \Psi_{\mathcal{Q}}(p)$ for all $p \in C$ at which both $\Phi_{\mathcal{P}}$ and $\Psi_{\mathcal{Q}}$ are defined, and so $\Phi(p) = \Psi(p)$ for all but finitely many points of C. The rational maps Φ and Ψ are thus equivalent on C and hence $\mathfrak{f} = \mathfrak{f}'$ as claimed. $\diamond$

Corollary 4.1.19 *If C is an irreducible plane curve and $\mathfrak{f} : C \to \mathbb{P}_m$ a non-constant rational map, then:*

(a) *For any $q \in \mathrm{Im}(\mathfrak{f})$, $\mathfrak{f}^{-1}(q)$ is a finite set.*

(b) $\mathrm{Im}(\mathfrak{f})$ *is an infinite set.*

PROOF: The rational map $\mathfrak{f}$ and the constant map with image $q \in \mathrm{Im}(\mathfrak{f})$ are defined and agree at all points of $\mathfrak{f}^{-1}(q)$; if $\mathfrak{f}^{-1}(q)$ is an infinite set, then they agree by 4.1.18. For the second claim, note that the set of points at which $\mathfrak{f}$ is defined may be written

$$\bigcup_{q \in \mathrm{Im}(\mathfrak{f})} \mathfrak{f}^{-1}(q),$$

and we already know it to be an infinite set. Then $\mathrm{Im}(\mathfrak{f})$ cannot be a finite set due to claim (a). ◇

Let C' be an irreducible curve of a projective plane $\mathbb{P}'_2$. The rational maps $\mathfrak{f} : C \to \mathbb{P}'_2$ with $\mathrm{Im}(\mathfrak{f}) \subset C'$, taken as maps $\mathfrak{f} : C \to C'$ defined at all but finitely many points of C, will be called *rational maps* from C to C'. For any two plane irreducible curves C, C', the constant maps $C \to C'$ and, if $C' = C$, the identity map of C, obviously are rational maps $C \to C'$.

In fact, any rational map $\mathfrak{f} : C \to \mathbb{P}'_2$ has its image contained in an irreducible curve C', and may therefore be taken as a rational map between irreducible curves. Since we will make no use of this fact for a while, we delay its proof to Section 4.11, see 4.11.17.

By identifying any abstract $\mathbb{P}_1$ to the line $\ell : x_2 = 0$ of a projective plane $\mathbb{P}_2$, we will not distinguish between an arbitrary rational map

$$C \longrightarrow \mathbb{P}_1$$
$$p \longmapsto [P_0(p), P_1(p)]$$

and the rational map between curves

$$C \longrightarrow \ell$$
$$p \longmapsto [P_0(p), P_1(p), 0].$$

Lemma 4.1.20 *Assume that $\mathfrak{f} : C \to \mathbb{P}'_2$, C an irreducible curve of $\mathbb{P}_2$, is a rational map. Assume also that a representative of $\mathfrak{f}$ is given by homogeneous polynomials P_0, P_1, P_2, from which at least one is non-zero at some point of C. If C' is an irreducible curve of $\mathbb{P}'_2$, in order to have $\mathrm{Im}(\mathfrak{f}) \subset C'$ it suffices to have $[P_0(p), P_1(p), P_2(p)] \in C'$ for all $p \in C$ for which $[P_0(p), P_1(p), P_2(p)]$ is a well-defined point, namely $(P_0(p), P_1(p), P_2(p)) \neq (0, 0, 0)$.*

PROOF: Let the curve C' be $C' : G = 0$. Assume $q \in \mathrm{Im}(\mathfrak{f})$; then $q = [Q_0(p'), Q_1(p'), Q_2(p')]$ for some $p' \in C$ and polynomials Q_0, Q_1, Q_2, not all zero at p' and defining a representative of $\mathfrak{f}$. Then, for all but finitely many points $p \in C$, both $[P_0(p), P_1(p), P_2(p)]$ and $[Q_0(p), Q_1(p), Q_2(p)]$ are well-defined points and agree. By the hypothesis, $[Q_0(p), Q_1(p), Q_2(p)]$ is also a well-defined point that belongs to C' for all but finitely $p \in C$ and hence, for all these points, $G(Q_0(p), Q_1(p), Q_2(p)) = 0$. By 3.6.14, $G(Q_0(p), Q_1(p), Q_2(p)) = 0$ for all $p \in C$, in particular $G(Q_0(p'), Q_1(p'), Q_2(p')) = 0$ and $q \in C'$ as wanted.   ◇

Let $\mathfrak{f} : C \to C'$ and $\mathfrak{g} : C' \to C''$ be rational maps between irreducible curves of projective planes $\mathbb{P}_2$, $\mathbb{P}'_2$ and $\mathbb{P}''_2$, respectively. Assume first that $\mathfrak{f}$ is the constant map with image $p \in C'$. If $\mathfrak{g}$ is defined at p, then the *composition* $\mathfrak{g} \circ \mathfrak{f}$ is defined as being the constant map $C \to C''$ with image $\mathfrak{g}(p)$; if $\mathfrak{g}$ is not defined at p, then the composition is left undefined. If $\mathfrak{f}$ is not constant, assume that it is represented by a rational map Φ of $\mathbb{P}_2$. Then Φ is defined on all points of C but those belonging to a finite subset T of C. Since the image of $\mathfrak{f}$ is infinite (by 4.1.19), so is $\Phi(C - T)$ and therefore any representative Ψ of $\mathfrak{g}$ may be composed with Φ, by 4.1.7. The rational map $C \to C''$ represented by $\Psi \circ \Phi$ is called the *composition* of $\mathfrak{f}$ with $\mathfrak{g}$, denoted $\mathfrak{g} \circ \mathfrak{f}$. Next we check that it does not depend on the choice of the representatives Φ and Ψ. Assume Φ' and Ψ' to also be representatives of $\mathfrak{f}$ and $\mathfrak{g}$. Then there is a finite subset $K' \subset C'$ such that Ψ and Ψ' are defined and agree at all points of $C' - K'$, and similarly, a finite $K \subset C$ such that Φ and Φ' are defined and agree at all points of $C - K$. If L is the set of the points $p \in C - K$ for which $\Phi(p) = \Phi'(p) \in K'$, then L is finite by 4.1.19 and clearly $\Psi(\Phi(p))$ and $\Psi'(\Phi'(p))$ are defined and agree for $p \in C - K \cup L$.

Remark 4.1.21 If the rational map $\mathfrak{f} : C \to C'$ is not constant, then the composition $\mathfrak{g} \circ \mathfrak{f}$ is defined for any rational map $\mathfrak{g} : C' \to C''$.

A useful description of the composite map is provided next:

Proposition 4.1.22 *Let $\mathfrak{f} : C \to C'$ and $\mathfrak{g} : C' \to C''$ be rational maps between irreducible plane curves. Then:*

(a) *If $\mathfrak{g} \circ \mathfrak{f}$ is not defined, then for no $p \in C$ is $\mathfrak{g}$ defined at $\mathfrak{f}(p)$. Otherwise,*

(b) *if $\mathfrak{g}$ is defined at $\mathfrak{f}(p)$, then $(\mathfrak{g} \circ \mathfrak{f})$ is defined at p and $(\mathfrak{g} \circ \mathfrak{f})(p) = \mathfrak{g}(\mathfrak{f}(p))$.*

(c) *$\mathfrak{g} \circ \mathfrak{f}$ is the only rational map from C to C'' satisfying (b).*

PROOF: If $\mathfrak{f}$ is constant, then all claims are clear. Assume otherwise. Then (b) is clear after choosing representatives Φ and Ψ, of $\mathfrak{f}$ and $\mathfrak{g}$, defined at p and $\mathfrak{f}(p)$ respectively. Claim (c) follows from 4.1.18, because the hypothesis of (b) is satisfied by all but finitely many $p \in C$ due to 4.1.19.   ◇

The next proposition directly follows from 4.1.22:

Proposition 4.1.23

(a) *For any rational maps between irreducible plane curves*

$$\mathfrak{f} : C \longrightarrow C', \quad \mathfrak{g} : C' \longrightarrow C'', \quad \mathfrak{h} : C'' \longrightarrow C'''$$

we have

$$\mathfrak{h} \circ (\mathfrak{g} \circ \mathfrak{f}) = \mathfrak{h} \circ (\mathfrak{g} \circ \mathfrak{f})$$

in the sense that if one of the sides is defined, so is the other and then they agree.

(b) *For any rational map between irreducible plane curves* $\mathfrak{f} : C \to C'$.

$$\mathrm{Id}_{C'} \circ \mathfrak{f} = \mathfrak{f} \circ \mathrm{Id}_C = \mathfrak{f}.$$

In the sequel we will mainly deal with non-constant rational maps; Remark 4.1.21 and Lemma 4.1.24 below will be frequently used, often with no explicit reference.

Lemma 4.1.24 *If both the rational maps* $\mathfrak{f} : C \to C'$ *and* $\mathfrak{g} : C' \to C''$ *are non-constant, then so is* $\mathfrak{g} \circ \mathfrak{f}$.

PROOF: By 4.1.19 applied to both $\mathfrak{f}$ and $\mathfrak{g}$, the image of $\mathfrak{f}$ contains infinitely many points, and only finitely many of them can have the same image by $\mathfrak{g}$. ◇

A rational map $\mathfrak{f} : C \to C'$ is said to be *birational* if and only if there is another rational map $\mathfrak{g} : C' \to C$ which may be composed with $\mathfrak{f}$ in both senses yielding $\mathfrak{f} \circ \mathfrak{g} = \mathrm{Id}_C$ and $\mathfrak{g} \circ \mathfrak{f} = \mathrm{Id}_{C'}$, or, equivalently by 4.1.22, $\mathfrak{f}(\mathfrak{g}(q)) = q$ for all but finitely many $q \in C'$ and $\mathfrak{g}(\mathfrak{f}(p)) = p$ for all but finitely many $p \in C$. Clearly, such a $\mathfrak{g}$ is then uniquely determined by $\mathfrak{f}$; it is called the *inverse* of $\mathfrak{p}$, denoted $\mathfrak{p}^{-1}$.

Remark 4.1.25 As a direct consequence of the definition, no birational map is constant; therefore, by 4.1.22, the composition of birational maps

$$\mathfrak{f} : C \longrightarrow C', \quad \mathfrak{g} : C' \longrightarrow C''$$

is always defined and obviously the composite map $\mathfrak{g} \circ \mathfrak{f}$ is birational too, its inverse being $\mathfrak{f}^{-1} \circ \mathfrak{g}^{-1}$. The inverse $\mathfrak{f}^{-1}$ of any birational map $\mathfrak{f} : C \longrightarrow C'$ is birational too, as its inverse is $\mathfrak{f}$. The identical map of any irreducible curve is of course birational.

Proposition 4.1.26 *If* $\mathfrak{f} : C \to C'$ *is birational, then there are finite subsets* $T \subset C$ *and* $T' \subset C'$ *such that* $\mathfrak{f}$ *and* $\mathfrak{f}^{-1}$ *induce reciprocal bijections between* $C - T$ *and* $C' - T'$.

PROOF: By the hypothesis, there are finite subsets $T_1 \subset C$ and $T_1' \subset C'$ such that

$$\mathfrak{f}^{-1}(\mathfrak{f}(p)) = p \quad \text{for} \quad p \in C - T_1$$

and

$$\mathfrak{f}(\mathfrak{f}^{-1}(q)) = q \quad \text{for} \quad q \in C' - T_1'.$$

Take

$$T_2 = \{p \in C \mid \mathfrak{f}(p) \in T_1'\}$$

and

$$T_2' = \{q \in C' \mid \mathfrak{f}^{-1}(q) \in T_1\}$$

and then

$$T = T_1 \cup T_2 \quad \text{and} \quad T' = T_1' \cup T_2',$$

after which the claim follows by a direct check which is left to the reader. $\diamond$

The rest of this section is devoted to studying the action of rational maps on rational functions. Assume that $\mathfrak{f} : C \to C'$ is a non-constant rational map between irreducible plane curves C of $\mathbb{P}_2$ and C' of $\mathbb{P}_2'$. Since rational functions on C' are in fact rational maps with a supplementary structure on the target, they may be composed with $\mathfrak{f}$ to give rational functions on C. Namely, assume that φ is a rational function on C'; if y_0, y_1, y_2 are coordinates on $\mathbb{P}_2'$ assume that φ is represented by $h = H_1/H_0$, where $H_1, H_0 \in \mathbb{C}[y_0, y_1, y_2]$, are homogeneous of the same degree and $H_0(q) \neq 0$ for all but finitely many points $q \in C'$. If x_0, x_1, x_2 are coordinates on $\mathbb{P}_2$, assume $\mathfrak{f}$ to be represented by a rational map between the projective planes given by polynomials (P_0, P_1, P_2). Then, for all but finitely many points $p \in C$, $[P_0(p), P_1(p), P_2(p)]$ is a point of C'. Furthermore, $\mathfrak{f}$ being not constant, by 4.1.19, still $[P_0(p), P_1(p), P_2(p)] \in C'$ and $H_0(P_0(p), P_1(p), P_2(p)) \neq 0$ for all but finitely many points of C. It follows that $h(P_0, P_1, P_2) \in \mathbb{C}(x_0, x_1, x_2)$ defines a rational function on $\mathbb{P}_2$ which is defined on all but finitely many points of C: the rational function on C it represents is called the *pull-back* (or *inverse image*) of φ by $\mathfrak{f}$, usually denoted $\mathfrak{f}^*(\varphi)$, and also $\varphi \circ \mathfrak{f}$.

It is clear from its definition above that $\mathfrak{f}^*(\varphi)$ satisfies $\mathfrak{f}^*(\varphi)(p) = \varphi(\mathfrak{f}(p))$ for all but finitely many points $p \in C$. By 3.7.6, $\mathfrak{f}$ and φ being given, this property uniquely determines $\mathfrak{f}^*(\varphi)$, which in particular proves that $\mathfrak{f}^*(\varphi)$ does not depend of the many choices made in its definition. Summarizing we have proved:

Proposition 4.1.27 *If* $\mathfrak{f} : C \to C'$ *is a non-constant rational map between irreducible plane curves C and C', and $\varphi \in \mathbb{C}(C')$, then there is one and only one rational function $\mathfrak{f}^*(\varphi) \in \mathbb{C}(C')$ that satisfies $\mathfrak{f}^*(\varphi)(p) = \varphi(\mathfrak{f}(p))$ for all but finitely many points $p \in C$.*

Remark 4.1.28 In the conditions of 4.1.27, if $\mathfrak{f}$ is defined at p and φ is regular at $\mathfrak{f}(p)$, then $\mathfrak{f}^*(\varphi)$ is regular at p and $\mathfrak{f}^*(\varphi)(p) = \varphi(\mathfrak{f}(p))$, because in the definition of $\mathfrak{f}^*(\varphi)$ we can use representatives of $\mathfrak{f}$ and φ that are defined at p and $\mathfrak{f}(p)$, respectively.

Example 4.1.29 As the reader may check, if $\mathfrak{f} : C \to \mathbb{P}_1$ is a non-constant rational map and $z \in \mathbb{C}(\mathbb{P}_1)$ is an absolute coordinate on $\mathbb{P}_1$, then $\mathfrak{f}(p) = (\mathfrak{f}^*(z)(p))$ is an affine representation of $\mathfrak{f}$, and any affine representation of $\mathfrak{f}$ arises in this way.

The next proposition states the main properties of pulling back rational functions:

Proposition 4.1.30 *If* $\mathfrak{f} : C \to C'$ *and* $\mathfrak{g} : C' \to C''$ *are non-constant rational maps between irreducible plane curves, then:*

(a) *For any* $\varphi \in \mathbb{C}(C'')$,

$$(\mathfrak{g} \circ \mathfrak{f})^*(\varphi) = \mathfrak{f}^*(\mathfrak{g}^*(\varphi)).$$

(b) *For any* $\varphi \in \mathbb{C}(C)$,
$$(\mathrm{Id}_C)^*(\varphi) = \varphi.$$

(c) *For any* $\varphi, \psi \in \mathbb{C}(C')$,

$$\mathfrak{f}^*(\varphi + \psi) = \mathfrak{f}^*(\varphi) + \mathfrak{f}^*(\psi) \quad and \quad \mathfrak{f}^*(\varphi\psi) = \mathfrak{f}^*(\varphi)\mathfrak{f}^*(\psi).$$

(d) *If* κ_a *and* κ'_a *are, respectively, the constant maps from* C *and* C' *with value* $a \in \mathbb{C}$, *then*
$$\mathfrak{f}^*(\kappa'_a) = \kappa_a.$$

PROOF: Direct from 4.1.27, as by 3.7.3 it suffices to check that both sides of each equality take the same value at all but finitely many points of C.   ◇

Remark 4.1.31 The field of rational functions $\mathbb{C}(C)$ of any irreducible curve C is obviously an extension of $\mathbb{C}$ by identifying the complex numbers with the constant functions. Claims (c) and (d) of 4.1.30 ensure that mapping each $\varphi \in \mathbb{C}(C')$ to $\mathfrak{f}^*(\varphi)$ is a non-zero (and therefore injective) homomorphism of fields $\mathfrak{f}^* : \mathbb{C}(C') \to \mathbb{C}(C)$ which leaves invariant the elements of $\mathbb{C}$, hence a $\mathbb{C}$-algebra homomorphism; it is called the *pull-back of rational functions.*

The next proposition shows an important relationship between the irreducible plane curves and their fields of rational functions.

Theorem 4.1.32 *If* C *and* C' *are irreducible plane curves, mapping* $\mathfrak{f}$ *to* $\mathfrak{f}^*$ *is a bijection between the set* $\mathrm{Rat}(C, C')$ *of the non-constant rational maps from* C *to* C' *and the set* $\mathrm{Hom}_{\mathbb{C}}(\mathbb{C}(C'), \mathbb{C}(C))$ *of the* $\mathbb{C}$-*algebra homomorphisms from* $\mathbb{C}(C')$ *to* $\mathbb{C}(C)$.

PROOF: If $\mathbb{P}_2$ is the projective plane C belongs to, take on it projective coordinates x_0, x_1, x_2. On the plane $\mathbb{P}'_2$ of C' take projective coordinates y_0, y_1, y_2 such that C' is not $y_0 = 0$, and affine coordinates $z = y_1/y_0$ and $t = y_2/y_0$ on the affine chart $y_0 \neq 0$. Denote by $\bar{z}$ and $\bar{t}$ the rational functions on C' represented by z and t.

For the injectivity, assume we have rational maps $\mathfrak{f}, \mathfrak{g} \in \mathrm{Rat}(C, C')$. For all but finitely many points of C, $(\mathfrak{f}^*(z))(p) = \bar{z}(\mathfrak{f}(p))$ and $(\mathfrak{f}^*(t))(p) = \bar{t}(\mathfrak{f}(p))$ are the first and second affine coordinates of $\mathfrak{f}(p)$, and similarly for $\mathfrak{g}$. Therefore, if $\mathfrak{f}^* = \mathfrak{g}^*$, then, for all but finitely points $p \in C$, $\mathfrak{f}(p)$ and $\mathfrak{g}(p)$ have equal coordinates and therefore agree. This gives $\mathfrak{f} = \mathfrak{g}$, by 4.1.18.

For the exhaustivity, assume given a $\mathbb{C}$-algebra homomorphism $\omega : \mathbb{C}(C') \to \mathbb{C}(C)$; ω being a non-zero homomorphism of fields, it is injective. Take representatives of $\omega(z)$ and $\omega(t)$ written in the form P_1/P_0 and P_2/P_0, with $P_0, P_1, P_2 \in \mathbb{C}[x_0, x_1, x_2]$ homogeneous of the same degree and $P_0(p) \neq 0$ for all but finitely many $p \in C$. The polynomials P_0, P_1, P_2 define a rational map $\Phi : \mathbb{P}_2 \to \mathbb{P}'_2$ that represents a rational map $\mathfrak{f} : C \to \mathbb{P}'_2$. We will first check that $\mathfrak{f}$ is a rational map $\mathfrak{f} : C \to C'$. To this end assume that

$$G = \sum_{i,j=0}^{d} a_{i,j} y_0^{d-i-j} y_1^i y_2^j$$

is an equation of C'. Then the rational function

$$\frac{G}{y_0^d} = \sum_{i,j=0}^{d} a_{i,j} z^i t^j$$

takes value zero at all the points of C' in the affine chart $y_0 \neq 0$, which are all but finitely many points of C'. Its restriction to C' is thus the function zero, and so we have the equality

$$0 = \sum_{i,j=0}^{d} a_{i,j} \bar{z}^i \bar{t}^j.$$

Since ω is a $\mathbb{C}$-algebra homomorphism, it follows that

$$0 = \sum_{i,j=0}^{d} a_{i,j} \omega(\bar{z})^i \omega(\bar{t})^j. \tag{4.5}$$

The rational function on $\mathbb{P}_2$

$$\sum_{i,j=0}^{d} a_{i,j} \left(\frac{P_1}{P_0} \right)^i \left(\frac{P_2}{P_0} \right)^j$$

is a representative of the right-hand side of (4.5) and therefore takes value zero at all but finitely many points of C. As a consequence, by 3.6.14, the polynomial

$$G(P_0, P_1, P_2) = \sum_{i,j=0}^{d} a_{i,j} P_0^{d-i-j} P_1^i P_2^j$$

is zero on all points of C and, by 4.1.20, $\mathrm{Im}(\mathfrak{f}) \subset C'$ as wanted.

Once we know that f is a rational map $f : C \to C'$, note that, by the definition of the polynomials P_i, for all but finitely many points p of C, $f(p)$ has affine coordinates

$$\bar{z}(f(p)) = \omega(\bar{z})(p) \quad \text{and} \quad \bar{t}(f(p)) = \omega(\bar{t})(p). \tag{4.6}$$

Then it is clear that f is not constant, as otherwise both $\omega(\bar{z})$ and $\omega(\bar{t})$ would be constant, and then so would be $\bar{z}$ and $\bar{t}$ by the injectivity of ω, against the fact that the affine part of C' has infinitely many points. Thus f^* is defined and the equalities (4.6) ensure that for all but finitely many $p \in C$,

$$(f^*(\bar{z})(p) = \bar{z}(f(p)) = \omega(\bar{z})(p)$$

and

$$(f^*(\bar{t})(p) = \bar{t}(f(p)) = \omega(\bar{t})(p).$$

Therefore

$$f^*(\bar{z}) = \omega(\bar{z}) \quad \text{and} \quad f^*(\bar{t}) = \omega(\bar{t}).$$

Since $\bar{z}$ and $\bar{t}$ generate $\mathbb{C}(C') = \mathbb{C}(\bar{z}, \bar{t})$ (3.7.8) as an extension of $\mathbb{C}$, and both f^* and ω leave the elements of $\mathbb{C}$ invariant, it follows that $f^* = \omega$ and the proof is complete. ◇

Corollary 4.1.33 *A non-constant rational map* $f : C \to C'$ *between irreducible plane curves is birational if and only if* f^* *is an isomorphism of $\mathbb{C}$-algebras.*

PROOF: The *only if* part is clear, as if f is birational then $(f^{-1})^*$ is the inverse of f^* due to the claims (a) and (b) of 4.1.23. For the *if* part, assume that f^* has inverse ω. Then, by 4.1.32 $\omega = g^*$ for a rational map $g : C' \to C$ and so

$$(g \circ f)^* = f^* \circ g^* = \mathrm{Id}_{\mathbb{C}(C')} = \mathrm{Id}_{C'}^*$$

and

$$(f \circ g)^* = g^* \circ f^* = \mathrm{Id}_{\mathbb{C}(C)} = \mathrm{Id}_C^*.$$

Since the map $\mathfrak{h} \mapsto \mathfrak{h}^*$ has been seen to be injective in 4.1.32, it follows that

$$g \circ f = \mathrm{Id}_{C'} \quad \text{and} \quad f \circ g = \mathrm{Id}_C,$$

which prove that f is birational. ◇

4.2 Reducing Singularities

The goal of this section is to prove that a sequence of standard quadratic transformations (see Example 4.1.8) may be used to transform any irreducible plane curve into one having only ordinary singular points.

As explained in Examples 4.1.8 and 4.1.12, once the vertices p_0, p_1, p_2 of a triangle T of $\mathbb{P}_2$ have been taken as the vertices of a projective reference, a standard quadratic transformation with fundamental triangle T is the birational map $\mathfrak{Q} : \mathbb{P}_2 \to \mathbb{P}_2$ defined by the equations

$$\bar{x}_0 = x_1 x_2$$
$$\bar{x}_1 = x_2 x_0$$
$$\bar{x}_2 = x_0 x_1,$$

the point $[\bar{x}_0, \bar{x}_1, \bar{x}_2]$ being, when defined, the image of $[x_0, x_1, x_2]$. As already noted in 4.1.8 and 4.1.12:

- $\mathfrak{Q}$ equals its own inverse,

- the points of indetermination of $\mathfrak{Q}$ are the vertices of T,

- $\mathfrak{Q}$ maps the points of each side of T other than the vertices to the opposite vertex, and

- $\mathfrak{Q}$ restricts to a bijection from the set of points belonging to no side of T onto itself.

Let $C : F = 0$ be a curve of $\mathbb{P}_2$ of degree d. The reader may easily check that the polynomial $\bar{F}(x_0, x_1, x_2) = F(x_1 x_2, x_2 x_0, x_0 x_1)$ is not zero and has degree $2d$. It thus defines a curve $\bar{C}$ of $\mathbb{P}_2$, of degree $2d$, which is called the *total transform* of C by $\mathfrak{Q}$.

Remark 4.2.1 Directly from its definition, q belongs to $\bar{C}$ if and only if either q is a point of indetermination of $\mathfrak{Q}$, or q is a regular point of $\mathfrak{Q}$ and $\mathfrak{Q}(q) \in C$.

Denote by ℓ_i the side of the fundamental triangle T opposite p_i. The next lemma describes the total transform $\bar{C}$.

Lemma 4.2.2 *If C contains no side of T, then its total transform $\bar{C}$ decomposes as*

$$\bar{C} = e_{p_0}(C)\ell_0 + e_{p_1}(C)\ell_1 + e_{p_2}(C)\ell_2 + \tilde{C},$$

where $\tilde{C}$ contains no side of T and has degree $2d - e_{p_0}(C) - e_{p_1}(C) - e_{p_2}(C) > 0$.

PROOF: Write $e_i = e_{p_i}(C)$, $i = 0, 1, 2$. The same arguments applying to the other vertices, let us fix our attention on p_0. If the equation of C is written in the form

$$F = x_0^{d-e} f_e(x_1, x_2) + x_0^{d-e-1} f_{e+1}(x_1, x_2) + \cdots + f_d(x_1, x_2)$$

where each f_j, $j = e, \dots, d$, is a homogeneous polynomial of degree j and $f_e \neq 0$, then an equation of the 0-th affine part of C is

$$f_e(x, y) + f_{e+1}(x, y) + \cdots + f_d(x, y)$$

and therefore, by 1.3.1 and 1.3.11, $e = e_0$ and $f_e(x_1, x_2) = 0$ is the tangent cone to C at p_0.

By a direct computation, the equation $\bar{F}$ of $\bar{C}$ has the form

$$\bar{F} = x_0^e \big(x_1^{d-e} x_2^{d-e} f_e(x_2, x_1) $$
$$+ x_0 x_1^{d-e-1} x_2^{d-e-1} f_{e+1}(x_2, x_1) + \cdots + x_0^{d-e} f_d(x_2, x_1) \big),$$

in which a factor x_0^e is evident. Replacing x_0 with 0 in the other factor gives $x_1^{d-e} x_2^{d-e} f_e(x_2, x_1) \neq 0$; therefore the multiplicity of x_0 as an irreducible factor of $\bar{F}$ is $e = e_0$, as claimed.

Using similar arguments with the other vertices, we get a factorization

$$\bar{F} = x_0^{e_0} x_1^{e_1} x_2^{e_2} \tilde{F},$$

where $\tilde{F}$ has no factor x_i, $i = 0, 1, 2$, and, obviously, has degree $2d - e_0 - e_1 - e_2$. Assume $2d - e_0 - e_1 - e_2 = 0$: this forces $e_i < d$ for some i; all cases being similar, assume $i = 0$. Then $e_1 + e_2 > d$, which, by the definition of multiplicity and 3.3.9, forces the line ℓ_0 and C to share an irreducible component. The line ℓ_0 being irreducible, it is contained in C against the hypothesis. Therefore $\deg \tilde{F} > 0$ and the proof ends by taking $\tilde{C}$ to be $\tilde{C} : \tilde{F} = 0$.   ◇

The curve $\tilde{C}$ of 4.2.2 is called the *strict transform* of C by $\mathfrak{Q}$; as seen, it contains no side of the fundamental triangle.

Remark 4.2.3 If $C = C_1 + C_2$ contains no side of T, then the same holds for C_1 and C_2, and a direct computation from the above definitions proves that $\tilde{C} = \tilde{C}_1 + \tilde{C}_2$.

Lemma 4.2.4 *The strict transform of $\tilde{C}$ is C.*

PROOF: Use all notations as in the proof of 4.2.2. On one hand performing again the substitution of variables into $\bar{F} = F(x_1 x_2, x_2 x_0, x_0 x_1)$ gives

$$\bar{\bar{F}} = F(x_0^2 x_1 x_2, x_0 x_1^2 x_2, x_0 x_1 x_2^2) = x_0^d x_1^d x_2^d F(x_0, x_1, x_2) \qquad (4.7)$$

where $d = \deg C$.

On the other, performing the same substitution on $\bar{F}$ written as

$$\bar{F} = x_0^{e_0} x_1^{e_1} x_2^{e_2} \tilde{F}$$

gives

$$\bar{\bar{F}} = x_0^{e_1 + e_2} x_1^{e_2 + e_0} x_2^{e_2 + e_0} \tilde{F}(x_1 x_2, x_2 x_0, x_0 x_1), \qquad (4.8)$$

where $\tilde{F}(x_1 x_2, x_2 x_0, x_0 x_1)$ is an equation of the total transform of $\tilde{C}$. Since this equation in turn may be written as the product of some powers of the variables by the equation G of the strict transform of $\tilde{C}$, (4.8) takes the form

$$\bar{\bar{F}} = x_0^{n_0} x_1^{n_1} x_2^{n_2} G \qquad (4.9)$$

for some non-negative integers n_0, n_1, n_2. By comparing (4.9) and (4.7) we get $F = G$ because neither has a factor x_i, $i = 0, 1, 2$. This completes the proof.   ◇

Corollary 4.2.5 *Still assuming that the curve C contains no side of the fundamental triangle, C is irreducible if and only if its strict transform $\tilde{C}$ is irreducible.*

PROOF: The *if* part has been seen in 4.2.3. For the converse just use 4.2.4.
◇

Standard quadratic transformations are useful mainly due to their effect on a singular point of a curve when it is taken as a vertex of the fundamental triangle: this effect is described next, see also Figure 4.1:

Proposition 4.2.6 *Let $\mathfrak{Q}$ be a standard quadratic transformation of $\mathbb{P}_2$. Fix any vertex p_0 of its fundamental triangle T, call p_1, p_2 the other vertices and let ℓ_i be the side of T opposite p_i, $i = 0, 1, 2$. Then there is a bijection τ between the pencil of lines through p_0 and the line ℓ_0 which satisfies the following conditions:*

(a) $\tau(\ell_1) = p_2$, $\tau(\ell_2) = p_1$.

(b) *If C is any curve containing no side of T, call H_C the set of lines tangent to C at p_0 other than ℓ_1, ℓ_2. Then τ maps H_C onto $\tilde{C} \cap \ell - \{p_1, p_2\}$ and the multiplicity of each $\ell \in L$ as a tangent to C at p_0 equals $[\tilde{C} \cdot \ell_0]_{\tau(\ell)}$.*

PROOF: Take the notations as in the proof of 4.2.2, in particular an equation F of C has the form

$$F = x_0^{d-e} f_e(x_1, x_2) + x_0^{d-e-1} f_{e+1}(x_1, x_2) + \cdots + f_d(x_1, x_2),$$

with f_j homogeneous of degree j, $j = e, \ldots, d$, and $f_e \neq 0$. According to the computations made in the proof of 4.2.2, an equation of $\tilde{C}$ is

$$\tilde{F} = x_1^{-e_1} x_2^{-e_2} \big(x_1^{d-e} x_2^{d-e} f_e(x_2, x_1)$$
$$+ x_0 x_1^{d-e-1} x_2^{d-e-1} f_{e+1}(x_2, x_1) + \cdots + x_0^{d-e} f_d(x_2, x_1) \big).$$

Therefore

$$x_1^{d-e-e_1} x_2^{d-e-e_2} f_e(x_2, x_1) \tag{4.10}$$

is an equation of $\tilde{C} \cap \ell_0$.

As recalled in the proof of 4.2.2, $f_e(x_1, x_2)$ is an equation of the tangent cone to C at p_0; assume that it factorizes into pairwise coprime factors

$$f_e(x_1, x_2) = x_1^{r_1} x_1^{r_2} \prod_{j=1}^{s} (a_j x_1 + b_j x_2)^{\mu_j}$$

with $r_1, r_2 \geq 0$ and $a_j, b_j \neq 0$, $j = 1, \ldots, r$. Then, on one hand, the tangents to C at p_0 other than $\ell_1 : x_1 = 0$ and $\ell_2 : x_2 = 0$ are the lines

$$t_j : a_j x_1 + b_j x_2 = 0, \quad j = 1, \ldots, s,$$

the multiplicity of each t_j as a tangent being μ_j. On the other hand, from (4.10), the points of $\tilde{C} \cap \ell_0 - \{p_0, p_1\}$ are

$$q_j = [0, a_j, -b_j], \quad j = 1, \ldots, s,$$

and

$$[\tilde{C} \cdot \ell_0]_{q_j} = \mu_j.$$

After the above, just define τ as mapping the line $ax_1 + bx_2 = 0$ to the point $[0, a, -b]$ and the claim is satisfied. ◇

Next we give a few remarks; in all of them the notations are as in 4.2.6.

Remark 4.2.7 The bijection τ of 4.2.6 is a projectivity, although we will make no use of this fact in the sequel.

Remark 4.2.8 Conditions (a) and (b) in 4.2.6 clearly determine τ.

Remark 4.2.9 By 4.2.6, $\tilde{C} \cap \ell - \{p_1, p_2\} \neq \emptyset$ yields $p_0 \in C$.

Remark 4.2.10 It follows also from 4.2.6 that if C has an ordinary singularity at p_0, then $[\tilde{C} \cdot \ell_0]_q = 1$ for all $q \in \tilde{C} \cap \ell_0 - \{p_1, p_2\}$. In particular, all points in $\tilde{C} \cap \ell_0 - \{p_1, p_2\}$ are non-singular points of $\tilde{C}$.

Remark 4.2.11 By 4.2.4, the roles of C and $\tilde{C}$ in the claim of 4.2.6 may be swapped over. Therefore, also the tangents to $\tilde{C}$ at each vertex p_i of T, other than the sides of T through p_i, correspond one to one with the intersections of C and the opposite side ℓ_i, other than the vertices of T on ℓ_i, and equalities of multiplicities similar to those of 4.2.6 hold.

As one could expect,

Proposition 4.2.12 *For any irreducible curve C, other than a side of the fundamental triangle of $\mathfrak{Q}$, $\mathfrak{Q}$ restricts to a birational map from C to $\tilde{C}$.*

PROOF: We will check that the image under $\mathfrak{Q}$ of any point $p \in C$ other than a vertex of T belongs to $\tilde{C}$. If $p \in \ell_i$, then $\mathfrak{Q}(p) = p_i$ and $p_i \in \tilde{C}$ by 4.2.9 applied to $\tilde{C}$ according to 4.2.11. If p belongs to no side of T, then $q = \mathfrak{Q}(p)$ belongs to no side of T either. Therefore $\mathfrak{Q}(q) = p$, which by 4.2.1 ensures that q belongs to $\bar{C}$, and hence also to $\tilde{C}$ because it belongs to no side of T. ◇

The last result we need about standard quadratic transformations is that they preserve both the multiplicity and the configuration of tangents of a curve at all the points belonging to no side of the fundamental triangle:

Proposition 4.2.13 *If the curve C contains no side of the fundamental triangle T and $p \in C$ belongs to no side of T, then there is a one to one correspondence preserving multiplicities between the set of lines tangent to C at p and the set of lines tangent to $\tilde{C}$ at $\mathfrak{Q}(p)$. In particular, $e_p(C) = e_{\mathfrak{Q}(p)}(\tilde{C})$.*

PROOF: Since the strict and total transforms of C differ by irreducible components that miss p, it will suffice to prove that the total transform of C fulfills the conditions of the claim. As above, take projective coordinates x_0, x_1, x_2 such that the equations of $\mathfrak{Q}$ are $\bar{x}_i = x_j x_k$, $\{i, j, k\} = \{0, 1, 2\}$, the vertices of the reference then being the vertices of T. Take $p = [1, a, b]$, $a \neq 0$, $b \neq 0$. On the affine chart $x_0 \neq 0$ take the affine coordinates

$$x = \frac{x_1}{x_0} - a, \quad y = \frac{x_2}{x_0} - b$$

whose origin is p. Write the equation of the affine part of C in the form

$$f_e(x, y) + f_{e+1}(x, y) + \cdots + f_d(x, y),$$

where $d = \deg C$, each f_i is homogeneous of degree i and $f_e \neq 0$. Then $e = e_p(C)$ and a homogeneous equation of C may be recovered by replacing x and y with $x_1/x_0 - a$ and $x_2/x_0 - b$, respectively, and multiplying by x_0^d, which gives

$$F = x_0^{d-e} f_e(x_1 - ax_0, x_2 - bx_0)$$
$$+ x_0^{d-e-1} f_{e+1}(x_1 - ax_0, x_2 - bx_0) + \cdots + f_d(x_1 - ax_0, x_2 - bx_0).$$

An equation of the total transform is thus

$$\bar{F} = x_1^{d-e} x_2^{d-e} f_e(x_0 x_2 - ax_1 x_2, x_0 x_1 - bx_1 x_2)$$
$$+ x_1^{d-e-1} x_2^{d-e-1} f_{e+1}(x_0 x_2 - ax_1 x_2, x_0 x_1 - bx_1 x_2) + \cdots$$
$$+ f_d(x_0 x_2 - ax_1 x_2, x_0 x_1 - bx_1 x_2).$$

Now we take on the affine chart new affine coordinates x', y' with origin

$$p' = \mathfrak{Q}(p) = [ab, b, a] = [1, 1/a, 1/b],$$

namely

$$x' = \frac{1}{a} - \frac{x_1}{x_0}, \quad y' = \frac{1}{b} - \frac{x_2}{x_0}.$$

Then

$$\frac{x_1}{x_0} = \frac{1}{a} - x', \quad \frac{x_2}{x_0} = \frac{1}{b} - y',$$

and so replacing in $\bar{F}$ the homogeneous coordinates x_0, x_1, x_2 with $1, \frac{1}{a} - x', \frac{1}{b} - y'$ gives the following equation of the affine part of $\bar{C}$:

$$\left(\frac{1}{a} - x'\right)^{d-e} \left(\frac{1}{b} - y'\right)^{d-e} f_e\left(\frac{a}{b}x' - ax'y', \frac{b}{a}y' - bx'y'\right) +$$
$$\left(\frac{1}{a} - x'\right)^{d-e-1} \left(\frac{1}{b} - y'\right)^{d-e-1} f_{e+1}\left(\frac{a}{b}x' - ax'y', \frac{b}{a}y' - bx'y'\right) + \cdots$$
$$+ \left(\frac{a}{b}x' - ax'y', \frac{b}{a}y' - bx'y'\right).$$

From it, only the first summand has a non-zero homogeneous part of degree e, which is

$$\frac{1}{a^{d-e}} \frac{1}{b^{d-e}} f_e\left(\frac{a}{b}x', \frac{b}{a}y'\right).$$

It may be taken as an equation of the affine part of the tangent cone to $\bar{C}$ at p'; since there is an obvious multiplicity preserving one to one correspondence between its linear factors and those of $f_e(x, y)$, the claim follows.   ◇

Now we reach the main goal of this section:

Theorem 4.2.14 *If C is an irreducible curve of $\mathbb{P}_2$, then there is a finite sequence of standard quadratic transformations of $\mathbb{P}_2$ such that the iterated strict transform of C by the transformations of the sequence has no singularities other than ordinary singular points.*

PROOF: As before, take $d = \deg C$. If C has no singular point other than ordinary singularities, the empty sequence (or $\mathfrak{Q}, \mathfrak{Q}$, for any standard quadratic transformation $\mathfrak{Q}$) does the job. Otherwise, let p be a non-ordinary singular point of C and take $e = e_p(C)$. We select a triangle T in the following way:

(1) Take $p = p_0$ as the first vertex of T.

(2) As allowed by 3.5.13, take the sides ℓ_1, ℓ_2 of T through p to be two different lines through p, each different from any tangent to C at p and missing all intersections of C and the polar $\mathcal{P}_p(C)$. By 3.5.11 the intersection of each of these sides and C consists of p and $d - e$ different points, all different from p, at which the intersection multiplicity of C and the side is one.

(3) Choose the vertex p_1 to be any point of ℓ_2 not belonging to C.

(4) Choose the third side ℓ_0 to be any line through p_1 missing all points in $C \cap \ell_1$ or $C \cap \mathcal{P}_{p_1}(C)$. Again by 3.5.11 $C \cap \ell_0$ consists of d different points, the intersection multiplicity of C and ℓ_0 at each of them being equal to one.

Let $\mathfrak{Q}$ be a standard quadratic transformation with fundamental triangle T. The strict transform $\tilde{C}$ of C by $\mathfrak{Q}$ has degree $2d - e$. Furthermore, by the choices (2) and (4) above and 4.2.11, $\tilde{C}$ has d simple tangents at p and $d - e$ simple tangents at each of the vertices p_1 and p_2. In particular, the multiplicities of $\tilde{C}$ at p and p_1 are at least d and $d - e$. Since $d + d - e = 2d - e = \deg \tilde{C}$ and $\ell_2 \not\subset \tilde{C}$, by 3.3.9 these multiplicities cannot be higher. Therefore we have $e_p(\tilde{C}) = d$, $e_{p_1}(\tilde{C}) = d - e$, which shows that $\tilde{C}$ has exactly d different tangents at p, all simple, exactly $d - e$ different tangents at p_1, all simple too, and no point other than p and p_1 on ℓ_2.

A similar argument applying to p, p_2 and ℓ_1, we conclude that the only points of $\tilde{C}$ on ℓ_1 or ℓ_2 are p, p_1, p_2 and these are ordinary singularities of multiplicities d, $d - e$ and $d - e$, respectively. In addition $\tilde{C}$ has points $q_1, \ldots, q_s \in \ell_0 - \{p_1, p_2\}$, each corresponding to one of the tangents to C

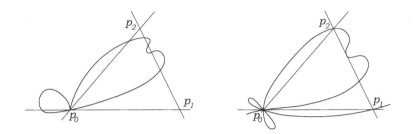

Figure 4.1: On the left, a quartic C, with a two-branched triple point p_0, and the fundamental triangle of a standard quadratic transformation $\mathfrak{Q}$ in the conditions of the proof of 4.2.14. On the right, the strict transform of C by $\mathfrak{Q}$. (Hand drawings, not plots of actual curves).

at p. By 4.1.8 and 4.2.13, $\mathfrak{Q}$ restricts to a multiplicity-preserving bijection between the sets of points of C and $\tilde{C}$ that belong to no side of T. Therefore, $\tilde{C}$ has no singular points other than:

(1) The fundamental points p, p_1, p_2, which are ordinary singularities of $\tilde{C}$, of multiplicities d, $d - e$ and $d - e$, respectively.

(2) The non-fundamental points $q_1, \ldots, q_s$ of C on ℓ_0. We will write $e_i = e_{q_i}(\tilde{C})$, $i = 1, \ldots, s$.

(3) The images $q' = \mathfrak{Q}(q)$ of the singular points q of C other than p. For each such q, $e_q(C) = e_{q'}(\tilde{C})$ and furthermore q' is an ordinary singularity of $\tilde{C}$ if and only if q is an ordinary singularity of C (by 4.2.13).

Now we will compare the deficiencies of C and $\tilde{C}$. According to the definition in 3.5.24

$$\sigma(C) = \frac{(d-1)(d-2)}{2} - \frac{e(e-1)}{2} - R$$

where $R = \sum e_q(e_q - 1)/2$, the summation running over the singular points of C other than p. By 4.2.13, $R = \sum e_{\mathfrak{Q}(q)}(e_{\mathfrak{Q}(q)} - 1)/2$, the summation running as above. Therefore, according to the description of the singular points of $\tilde{C}$ already given,

$$\sigma(\tilde{C}) = \frac{(2d - e - 1)(2d - e - 2)}{2} - \frac{d(d-1)}{2}$$
$$- 2\frac{(d-e)(d-e-1)}{2} - \sum_{i=1}^{s} \frac{e_i(e_i-1)}{2} - R.$$

After computation we have

$$\sigma(C) - \sigma(\tilde{C}) = \sum_{i=1}^{s} \frac{e_i(e_i-1)}{2}.$$

It follows that either $\sigma(C) > \sigma(\tilde{C})$, or $\sigma(C) = \sigma(\tilde{C})$ and $e_i = 1$, $i = 1, \ldots, s$. In the latter case, the number of non-ordinary singular points of $\tilde{C}$ is one less than the number of non-ordinary singular points of C.

Now, if $\tilde{C}$ has no non-ordinary singularities, we are done. Otherwise we repeat the procedure from $\tilde{C}$ and a non-ordinary singular point of it, and so on. At each step either the deficiency drops, or the deficiency stays constant and the number of non-ordinary singular points drops; since the deficiency has been seen to be always non-negative in 3.5.24, after finitely many steps the iterated strict transform of C has no non-ordinary singular points, as wanted.   ◇

A direct consequence of 4.2.14 and 4.2.12 is:

Corollary 4.2.15 *For any irreducible curve C of $\mathbb{P}_2$ there is an irreducible curve C' of $\mathbb{P}_2$, all of whose singular points are ordinary, and a birational map $\mathfrak{T} : C \to C'$.*

4.3 Birational Equivalence and Intrinsic Geometry

Two irreducible plane curves C and C' are said to be *birationally equivalent* if and only if there is a birational map $\mathfrak{f} : C \to C'$. Birational equivalence is indeed an equivalence relation due to 4.1.25.

Remark 4.3.1 Corollary 4.2.15 just says that any irreducible curve of $\mathbb{P}_2$ is birationally equivalent to an irreducible curve of $\mathbb{P}_2$ with only ordinary singularities.

As said at the beginning of this chapter, the intrinsic geometry on algebraic curves studies the properties of irreducible curves that are invariant under birational equivalence. The objects associated to an irreducible curve C which – up to isomorphism – depend only on the birational equivalence class of C, are said to be *birational invariants*. The field of rational functions $\mathbb{C}(C)$ of C, as a $\mathbb{C}$-algebra, is a birational invariant due to 4.1.33.

Remark 4.3.2 Proposition 4.1.32 and Corollary 4.1.33 ensure that two irreducible plane curves are birationally equivalent if and only if their fields of rational functions are isomorphic as $\mathbb{C}$-algebras. Therefore, the intrinsic geometry on an irreducible curve C depends only on the $\mathbb{C}$-algebra $\mathbb{C}(C)$; in fact it can be developed making exclusive use of $\mathbb{C}(C)$, with no use of C itself or other geometric objects such as points, curves, etc. For such a purely algebraic approach, the reader is referred to [5].

Although it is also related to projective problems (see for instance Exercise 4.17), the intrinsic geometry on algebraic curves did appear, by the middle of the 19th century, as a geometric counterpart to the study of algebraic functions in a single variable, already undertaken by analysts like Abel, Jacobi and Weierstrass. According to the by then quite usual definition, an *algebraic function* $u(x)$, of the complex variable x, is a multi-valued function whose values for $x = \alpha \in \mathbb{C}$ are the solutions of the equation in y, $P(\alpha, y) = 0$, where P is a fixed polynomial, $P \in \mathbb{C}[x, y]$, $\deg_y P > 0$. The rational functions arise for $\deg_y P = 1$ and the easiest non-rational algebraic function is the square root, for which $P = y^2 - x$. It is usual to take P irreducible – the function u then being called *irreducible* too – as the general case is reduced to the irreducible one by splitting the algebraic function into the ones defined by the irreducible factors of P.

To turn the algebraic functions into single-valued functions, Riemann proposed replacing them by closely related functions of a variable point on an irreducible algebraic curve. To be precise, consider the above algebraic function u, assumed irreducible, and, with it, all functions rational in u and the free variable x, namely the $R(x, u(x))$, $R \in \mathbb{C}(x, y)$. Take C to be the projective closure of the affine curve $P(x, y) = 0$ and denote by a bar ‾ the restrictions to C. Riemann proposed to replace u with $\bar{y}$, and consequently each $R(x, u(x))$, $R \in \mathbb{C}(x, y)$ with $R(\bar{x}, \bar{y})$. For instance $x/\sqrt{x^3 - 1}$, as a function of the free variable x, would be replaced with $\bar{x}/\bar{y}$, as a function of the variables $\bar{x}, \bar{y}$ related by the equality $\bar{y}^2 = \bar{x}^3 - 1$, or, which is the same, as a function of the point $(\bar{x}, \bar{y})$ varying on the curve $y^2 - x^3 + 1 = 0$. This replaces the family of algebraic functions $\{R(x, u(x)) \mid R \in \mathbb{C}(x, y)\}$ with just $\mathbb{C}(C)$. The original multi-valued functions are of course recovered from their replacements by considering the values of the latter as functions of the first coordinate of $p \in C$, and not as functions of p itself. Riemann's proposal was successful to the point that today an algebraic function in a single variable is usually understood to be a rational function on an algebraic curve.

Next we will start building up the geometric setting needed for Brill and Noether's approach to the intrinsic geometry on curves. Our first job is to show that restricting to branches eliminates all the indeterminations of the rational maps and the rational functions at the points of a curve.

Proposition 4.3.3 *Let C be an irreducible curve of $\mathbb{P}_2$, γ a branch of C with origin O and $\mathfrak{f} : C \to \mathbb{P}_m$ a rational map. There is an open neighbourhood U of 0 in $\mathbb{C}$ and a uniformizing map $\varphi : U \to C$ such that the composition of $\varphi_{|U-\{0\}}$ with $\mathfrak{f}$ is a well-defined map*

$$\mathfrak{f} \circ \varphi_{|U-\{0\}} : U - \{0\} \longrightarrow \mathbb{P}_m$$

which has a unique extension to an analytic map

$$\mathfrak{f}_U : U \longrightarrow \mathbb{P}_m.$$

Furthermore the germ $\mathfrak{f}_\gamma$ of $\mathfrak{f}_U$ at 0 is uniquely determined by $\mathfrak{f}$ and γ up to isomorphism on the source.

PROOF: Take on $\mathbb{P}_2$ projective coordinates x_0, x_1, x_2 for which $O = [1, 0, 0]$ and the affine coordinates $x = x_1/x_0$, $y = x_2/x_0$ on the 0-th affine chart. Let $\varphi' : U' \to C$ be a uniformizing map of γ given by $\varphi(t) = (u(t), v(t))$

for $t \in U'$, where u and v are integral power series convergent in U'. After taking homogeneous coordinates $y_0, \ldots, y_m$ on $\mathbb{P}_m$, there are homogeneous polynomials $P_i \in \mathbb{C}[x_0, x_1, x_2]$ such that $\mathfrak{f}(p) = [P_0(1, x, y), \ldots, P_m(1, x, y)]$ for all $p = (x, y) \in C \cap \mathbb{A}_2$ other than finitely many points $q_1, \ldots, q_r$. Then there is a smaller open neighbourhood $U \subset U'$ of 0 such that no point in $U - \{0\}$ has image q_i, $i = 1, \ldots, r$ (use 2.4.10). Take $\varphi = \varphi'_{|U}$; then, clearly, $\mathfrak{f} \circ \varphi_{|U-\{0\}}$ is well defined and given by the rule

$$t \longmapsto [P_0(1, u(t), v(t)), \ldots, P_m(1, u(t), v(t))].$$

If each of the series above is written

$$P_i(1, u(t), v(t)) = t^{r_i} w_i(t), \quad w_i(0) \neq 0,$$

and, up to renumbering the coordinates on $\mathbb{P}_m$, we assume $r_0 \leq r_i$ for $i > 0$, then we take affine coordinates $Y_i = y_i/y_0$ on the 0-th affine chart of $\mathbb{P}_n$. Using them, we have

$$\mathfrak{f} \circ \varphi(t) = \left(\frac{t^{r_1} w_1(t)}{t^{r_0} w_0(t)}, \ldots, \frac{t^{r_1} w_1(t)}{t^{r_0} w_0(t)} \right).$$

for $t \in U - \{0\}$, and just cancelling the factor t^{r_0} provides the wanted extension to U, namely

$$\mathfrak{f}_U(t) = \left(\frac{t^{r_1 - r_0} w_1(t)}{w_0(t)}, \ldots, \frac{t^{r_1 - r_0} w_1(t)}{w_0(t)} \right).$$

Such an extension is unique by continuity.

For the second claim, assume that $\bar{\varphi} : \bar{U} \to C$ is a second uniformizing map for which $\mathfrak{f} \circ \bar{\varphi}_{|\bar{U}-\{0\}}$ is defined on $\bar{U} - \{0\}$ and extends to an analytic map $\mathfrak{f}_{\bar{U}}$ defined on $\bar{U}$. Then, by 2.4.7(b), up to reduction of $\bar{U}$, there is a representative $\sigma : W \to \bar{U}$ of an analytic isomorphism such that φ and $\bar{\varphi} \circ \sigma$ have the same germ at 0; then they restrict to the same map $\psi : V \to C$ for a certain open neighbourhood V of 0 in $\mathbb{C}$, $V \subset U \cap W$. It follows that both $\mathfrak{f} \circ \varphi_{|U-\{0\}}$ and $\mathfrak{f} \circ \bar{\varphi}_{|\bar{U}-\{0\}} \circ \sigma_{|W-\{0\}}$ restrict to $\mathfrak{f} \circ \psi_{|V-\{0\}}$: by continuity, $\mathfrak{f}_U$ and $\mathfrak{f}_{\bar{U}} \circ \sigma$ agree on V and thus give rise to the same germ at 0. ◇

Remark 4.3.4 Proceeding as in the proof of 4.3.3, with no need of renumbering the coordinates on $\mathbb{P}_n$, for any $t \in U$ we have

$$\mathfrak{f}_U(t) = \left[\frac{P_0(1, u(t), v(t))}{t^r}, \ldots, \frac{P_m(1, u(t), v(t))}{t^r} \right]$$

where $r = \min\{o_t P_i(1, u(t), v(t))\}_{i=0,\ldots,m}$.

Remark 4.3.5 If the rational map $\mathfrak{f}$ of 4.3.3 is defined at the origin O of γ, then, for a suitable uniformizing map φ, $\mathfrak{f} \circ \varphi$ is analytic and therefore may be taken as $\mathfrak{f}_U$. By the uniqueness of $\mathfrak{f}_\gamma$, $\mathfrak{f}_\gamma(0) = \mathfrak{f}_U(0) = \mathfrak{f}(O)$.

A result similar to Proposition 4.3.3 holds for rational functions as well; it may be presented as a consequence of 4.3.3, but giving a direct proof is easier. In the sequel $\mathbb{C} \cup \{\infty\}$ denotes the Riemann sphere.

Proposition 4.3.6 *Let C be an irreducible curve of $\mathbb{P}_2$, γ a branch of C with origin $O \in C$ and $f \in \mathbb{C}(C)$ a rational function. There is an open neighbourhood U of 0 in $\mathbb{C}$ and a uniformizing map $\varphi : U \to C$ such that the composition of $\varphi_{|U-\{0\}}$ with f is a well-defined map*

$$ f \circ \varphi_{|U-\{0\}} : U - \{0\} \longrightarrow \mathbb{C} $$

which has a unique extension to a meromorphic function

$$ f_U : U \longrightarrow \mathbb{C} \cup \{\infty\}. $$

Furthermore, the germ f_γ of f_U at 0 is uniquely determined by f and γ up to isomorphism on the source.

PROOF: As before, take projective coordinates x_0, x_1, x_2 on $\mathbb{P}_2$ such that the origin of γ is $[1, 0, 0]$, the affine chart $x_0 \neq 0$ and coordinates $x = x_1/x_0$, $y = x_2/x_0$ on it. Assume that f is the restriction of the rational function P_1/P_0 of $\mathbb{P}_2$, where $P_0, P_1 \in \mathbb{C}[x_0, x_1, x_2]$ are homogeneous and have the same degree, and $P_1(p) \neq 0$ for all but finitely many $p \in C$. Assume also that a uniformizing map of γ is

$$ \varphi : U \longrightarrow C $$
$$ t \longmapsto (u(t), v(t)) $$

where U is an open neighbourhood of 0 in $\mathbb{C}$ and $u, v \in \mathbb{C}\{t\}$ are convergent in U and have $u(0) = v(0) = 0$. Up to replacing U with a smaller neighbourhood, it is not restrictive to assume that $P_1(1, u(t), v(t)) \neq 0$ for $t \in U - \{0\}$ (use 2.4.10). Then,

$$ f(\varphi(t)) = \frac{P_1(1, u(t), v(t))}{P_0(1, u(t), v(t))} $$

for all $t \in U - \{0\}$ and so, not only is $f \circ \varphi_{|U-\{0\}}$ a well-defined map, but also it obviously extends to the wanted meromorphic function f_U by just cancelling all factors t shared by $P_1(1, u(t), v(t))$ and $P_0(1, u(t), v(t))$. The uniqueness of the extension is clear by continuity. The second claim follows from an argument similar to the one used in the proof of 4.3.3, the details of which are left to the reader.   ◇

Remark 4.3.7 Notations and hypothesis being as in the proof of 4.3.6, if $r = \min\{o_t P_1(1, u(t), v(t)), o_t P_0(1, u(t), v(t))\}$, then

$$ f_U(t) = \frac{t^{-r} P_1(1, u(t), v(t))}{t^{-r} P_0(1, u(t), v(t))} \tag{4.11} $$

for $t \in U$. Therefore, if the rational function f of 4.3.6 is defined at the origin O of γ, then P_1/P_0 may be chosen with $P_0(1, 0, 0) \neq 0$, after which

$r = 0$ and f_U and f_γ are, respectively, the restriction to U and the germ at 0 of $f \circ \varphi$. In particular, in such a case, as for rational maps in 4.3.5, $f_\gamma(0) = f_U(0) = f(O)$.

Remark 4.3.8 If $f, g \in \mathbb{C}(C)$, using the equality (4.11) for both f and g, or just the uniqueness claimed in 4.3.6, we have

$$(f + g)_\gamma = f_\gamma + g_\gamma \quad \text{and} \quad (fg)_\gamma = f_\gamma g_\gamma.$$

Let C be, as above, an irreducible curve of $\mathbb{P}_2$. The set of all branches of C will be called the *Riemann surface of C*, denoted $X(C)$ in the sequel. The word *surface* is used because $X(C)$ may be given a topology with which it becomes a topological variety of real dimension two. The elements of $X(C)$ will be referred to as *points* of the Riemann surface as well as branches of C. Clearly, taking $\chi_C(\gamma)$ equal to the origin of γ, for all branches γ of C, defines an exhaustive map

$$\chi_C : X(C) \longrightarrow C$$

through which the smooth points of C are in one to one correspondence with the branches with origin at them, due to 2.4.5. In particular, if C is smooth, then χ_C is a bijection; in such a case it is usual to identify C and $X(C)$ through χ_C, making no distinction between each point of a smooth curve C and the only branch with origin at it.

Back to the case of a possibly singular C, the Riemann surface $X(C)$ may be viewed, heuristically, as a modified copy of C in which the smooth points of C remain unaltered, while each singular point O of C has been turned into finitely many different points of $X(C)$, one for each branch of C with origin at O. The aim is to separate the coincident origins of different branches into different points, and this is achieved, formally, by taking as points the branches themselves.

Let f be a rational function on C. If γ is a branch of C – that is, a point of $X(C)$ – we define the *value* of f at γ as being $X(f)(\gamma) = f_\gamma(0) \in \mathbb{C} \cup \{\infty\}$, f_γ the meromorphic germ of 4.3.6. The fact that f_γ is determined up to isomorphism on the source guarantees that the definition does not depend on any choice. The rational function f thus lifts to a well-defined map

$$X(f) : X(C) \longrightarrow \mathbb{C} \cup \{\infty\}$$
$$\gamma \longmapsto X(f)(\gamma)$$

which, by 4.3.7, satisfies $X(f)(\gamma) = f(\chi_C(\gamma))$ for all branches γ at whose origin $\chi_C(\gamma)$ the function f is defined. We have seen – in Example 4.1.17 – that in general the indeterminacy of a rational function at certain points of C cannot be removed preserving continuity: the map $X(f)$ may be understood as the result of removing the indeterminacy of f by separating the origins of the different branches of C into different points (of $X(C)$). In the sequel we will write just $f(\gamma)$ instead of $X(f)(\gamma)$ if no confusion may result. We have:

Lemma 4.3.9 *For any $f, g \in \mathbb{C}(C)$ and any $\gamma \in X(C)$,*

$$(f + g)(\gamma) = f(\gamma) + g(\gamma) \quad \text{and} \quad (fg)(\gamma) = f(\gamma)g(\gamma).$$

PROOF: Direct from 4.3.8.   ◇

If the germ of meromorphic function f_γ is not zero, then it may be represented by a Laurent series $t^r u(t)$ where $r \in \mathbb{Z}$, u is a convergent power series and $u(0) \neq 0$. Again, the fact that f_γ is determined up to isomorphism on the source ensures that r is determined by f and γ. We define the *order of f along γ* – or *at γ* if γ is thought of as a point of $X(C)$ – as $o_\gamma f = r$. If $r > 0$, γ is said to be a *zero* of f of *order* – or *multiplicity* – r. When $r < 0$ or, equivalently, $f(\gamma) = \infty$, γ is said to be a *pole* of f of *order* – or *multiplicity* – r. As usual we take $o_\gamma 0 = \infty$ for any branch γ.

If a rational function Φ of $\mathbb{P}_2$ can be restricted to C, then its *order along γ* is taken to be the order along γ of its restriction; by 4.3.7, this extends the definition given in Section 2.6 for Φ regular at the origin of γ.

Proposition 4.3.10 *For any $f, g \in \mathbb{C}(C)$ and any $\gamma \in X(C)$, using the customary rules with ∞,*

$$o_\gamma(f + g) \geq \min(o_\gamma f, o_\gamma g),$$

the equality holding if $o_\gamma f \neq o_\gamma g$, and

$$o_\gamma(fg) = o_\gamma f + o_\gamma g.$$

PROOF: Direct from the definitions and 4.3.8.   ◇

The preceding definitions are the main reason for considering the Riemann surface $X(C)$ instead of the projective curve C: all the rational functions on C can be evaluated at the points of $X(C)$, while, in general, they cannot at all the points of C. Avoiding much of our formalism, the classical geometers took the points of $X(C)$ as the true points of the curve C regarding its intrinsic geometry; the fact that two of these points become coincident when $X(C)$ is mapped into $\mathbb{P}_2$ via χ_C was regarded by them just as, in their own words, a *projective accident*.

Next we will lift non-constant rational maps between irreducible curves to well-defined maps between their Riemann surfaces using 4.3.3. Assume C and C' to be irreducible curves, of projective planes $\mathbb{P}_2$ and $\mathbb{P}'_2$, and $\mathfrak{f}: C \to C'$ a non-constant rational map between them. If γ is a branch of C, then, for a certain $O' \in C'$, $\mathfrak{f}_\gamma$ is a germ of analytic map

$$\mathfrak{f}_\gamma : (\mathbb{C}, 0) \longrightarrow (C', O')$$

which, by 2.4.7, factorizes through the uniformizing germ of a uniquely determined branch γ' of C' with origin O'. We set $X(\mathfrak{f})(\gamma) = \gamma'$, thus defining a map

$$X(\mathfrak{f}) : X(C) \longrightarrow X(C').$$

Remark 4.3.11 If $\mathfrak{f}$ is defined at the origin O of γ, then, by 4.3.5, $X(\mathfrak{f})(\gamma)$ has origin $O' = \mathfrak{f}_\gamma(0) = \mathfrak{f}(O)$. The diagram

$$
\begin{array}{ccc}
X(C) & \xrightarrow{X(\mathfrak{f})} & X(C') \\
\chi_C \downarrow & & \downarrow \chi_{C'} \\
C & \xrightarrow{\mathfrak{f}} & C'
\end{array}
$$

is thus commutative, in the sense that when $\mathfrak{f}(\chi_C(\gamma))$ is defined, then it equals $\chi_{C'}(X(\mathfrak{f})(\gamma))$.

Example 4.3.12 Fix homogeneous coordinates x_0, x_1, x_2 on P_2 and let $\mathfrak{Q}$ be the standard quadratic transformation

$$\mathfrak{Q}: P_2 \longrightarrow P_2$$
$$[x_0, x_1, x_2] \longmapsto [x_1 x_2, x_0 x_2, x_0 x_1].$$

Let C be an irreducible curve through $q = [1,0,0]$, not a side of the fundamental triangle, and γ a branch of C with origin at q. Then γ has a uniformizing map

$$t \longmapsto [1, t^e u(t), t^e v(t)]$$

where $u, v \in \mathbb{C}\{t\}$ and either $u(0) \neq 0$ or $v(0) \neq 0$. Then $e = e_q(\gamma)$ and the tangent to γ is the line $\ell : v(0)x_1 - u(0)x_2 = 0$ (as it has the highest intersection multiplicity with γ). If $\bar{\mathfrak{Q}}$ is the restriction of $\mathfrak{Q}$ to C, $\bar{\mathfrak{Q}}_\gamma$ is given by

$$t \longmapsto [t^e u(t)v(t), v(t), u(t)]$$

and hence the origin of $\bar{\mathfrak{Q}}_\gamma$ is the point $q' = [0, v(0), u(0)]$. Still naming τ the bijection of 4.2.6, the reader may easily check that $q' = \tau(\ell)$ (see the end of the proof of 4.2.6).

We have:

Proposition 4.3.13 *For any irreducible projective plane curves C, C' and C'':*

(a) $X(\mathrm{Id}_C) = \mathrm{Id}_{X(C)}$.

(b) *If $\mathfrak{f}: C \to C'$ and $\mathfrak{g}: C' \to C''$ are non-constant rational maps, then so is $\mathfrak{g} \circ \mathfrak{f}$ and*

$$X(\mathfrak{g} \circ \mathfrak{f}) = X(\mathfrak{g}) \circ X(\mathfrak{f}).$$

(c) *If $\mathfrak{f}: C \to C'$ is a birational map, then $X(\mathfrak{f})$ is a bijection and*

$$X(\mathfrak{f})^{-1} = X(\mathfrak{f}^{-1}).$$

PROOF: Claim (a) is clear, while claim (c) is straightforward from claims (a) and (b). Next we prove claim (b). First note that $\mathfrak{g} \circ \mathfrak{f}$ is non-constant due to 4.1.24. Take any $\gamma \in X(C)$ and write $\gamma' = X(\mathfrak{f})(\gamma)$. By the definition of $\gamma'' = X(\mathfrak{g})(\gamma')$, there is a uniformizing map $\varphi' : U' \to C'$ of γ' such that the extension $\mathfrak{g}_{U'}$ of $\mathfrak{g} \circ \varphi'|_{U'-\{0\}}$ factorizes through a uniformizing map $\varphi'' : U'' \to C''$ of γ''. For a uniformizing map $\varphi : U \to C$ of γ, the extension $\mathfrak{f}_U$ of $\mathfrak{f} \circ \varphi|_{U-\{0\}}$ factorizes in turn through a uniformizing map of γ' which, up to a reduction of U, may be assumed to be φ'. There is thus a commutative diagram

$$
\begin{array}{ccccc}
C & \xrightarrow{\;\mathfrak{f}\;} & C' & \xrightarrow{\;\mathfrak{g}\;} & C'' \\
\varphi\uparrow & \nearrow\mathfrak{f}_U & \varphi'\uparrow & \nearrow\mathfrak{g}_{U'} & \uparrow\varphi'' \\
U & \xrightarrow{\;\psi\;} & U' & \xrightarrow{\;\psi'\;} & U''
\end{array}
$$

where ψ and ψ' are analytic maps. Now, it is clear from the diagram that $\mathfrak{g}_{U'} \circ \psi$ is the extension of $(\mathfrak{g} \circ \mathfrak{f}) \circ \varphi|_{U-\{0\}}$, and also that it factorizes through the uniformizing map φ'' of γ''. It follows that $X(\mathfrak{g} \circ \mathfrak{f})(\gamma) = \gamma''$, as wanted. ◇

In the sequel, if no confusion may result, we will write just $\mathfrak{f}(\gamma)$ for $X(\mathfrak{f})(\gamma)$. Next is a sort of local refinement of 4.2.15. It is an example of what is usually called a *local uniformization* result:

Proposition 4.3.14 *Let C be an irreducible curve of $\mathbb{P}_2$ and γ a branch of C. Then there is a birational map $\mathfrak{f} : C \to \bar{C}$ such that all singular points of $\bar{C}$ are ordinary and the origin of $\mathfrak{f}(\gamma)$ is a non-singular point of $\bar{C}$.*

PROOF: Take the birational map $\mathfrak{T} : C \to C'$ of 4.2.15. If the origin q of $\mathfrak{T}(\gamma)$ is a non-singular point of C', then the claim is proved. Otherwise q is an ordinary singularity of C'. We take a standard quadratic transformation $\mathfrak{Q}$ whose fundamental triangle T has its first vertex at q and satisfies the conditions (2) to (4) posed in the proof of 4.2.14: we will prove that it suffices to take $\mathfrak{f} = \bar{\mathfrak{Q}} \circ \mathfrak{T}$, where $\bar{\mathfrak{Q}}$ is the birational map between C' and its strict transform $\widetilde{C}'$ induced by $\mathfrak{Q}$. Indeed, on one hand, as seen in the proof of 4.2.14, $\mathfrak{Q}$ does not introduce new non-ordinary singularities and therefore, in this case, $\widetilde{C}'$ has all its singular points ordinary. On the other hand, we have seen in 4.3.12 that the origin q' of $\bar{\mathfrak{Q}}(\mathfrak{T}(\gamma))$ is one of the points of the strict transform $\widetilde{C}'$ on the side of T opposite q, not a vertex. Since the singularity of C' at Q is ordinary, this point is non-singular, by 4.2.10. ◇

Before continuing, we need to recall an elementary notion relative to germs of analytic functions. Assume that $\psi : (\mathbb{C}, 0) \to (\mathbb{C}, 0)$ is a germ of analytic map. If ψ is represented by a power series $u = \sum_{i \geq r} a_i t^i$, $r > 0$, $a_r \neq 0$, then the integer r is called the *ramification index* of ψ. As the reader may easily see, if $\tilde{\psi}$ is any representative of ψ, the ramification index r equals the number of elements of the inverse image $\tilde{\psi}^{-1}(t)$ that approach 0 when t approaches 0. From this fact, and also from a direct computation, it follows that the ramification index of ψ equals the ramification index of any composition $\rho \circ \psi \circ \delta$, where ρ and δ are analytic isomorphisms.

Back to considering a non-constant rational map $\mathfrak{f} : C \to C'$ and $\gamma \in X(C)$, we have seen that $\mathfrak{f}_\gamma$ factorizes through a uniformizing germ φ' of $\mathfrak{f}(\gamma)$ in the form

$$\mathfrak{f}_\gamma = \varphi' \circ \psi,$$

where $\psi : (\mathbb{C}, 0) \to (\mathbb{C}, 0)$ is a uniquely determined germ of analytic map (2.4.7). Such a ψ may be understood as describing the local behaviour of the restriction of $\mathfrak{f}$ to a representative of γ because there is a commutative diagram

$$\begin{array}{ccc} U & \xrightarrow{\tilde{\varphi}} & C \\ \tilde{\psi} \downarrow & & \downarrow \mathfrak{f} \\ U' & \xrightarrow{\tilde{\varphi}'} & C' \end{array}$$

where $\tilde{\psi}$ is a representative of ψ and $\tilde{\varphi}$ and $\tilde{\varphi}'$ are injective representatives of φ and φ' respectively. The germ ψ is determined up to composition with analytic isomorphisms on both sides, due to the determination of the uniformizing germs up to analytic isomorphisms on their sources, but this, as noticed above, does not affect its ramification index: then the *ramification index* of $\mathfrak{f}$ at γ is defined as being the ramification index of ψ at γ.

Proposition 4.3.15 *If $\mathfrak{f} : C \to C'$ is a non-constant rational map between irreducible projective plane curves, then for any $g \in \mathbb{C}(C')$ and any $\gamma \in X(C)$ we have*

$$\mathfrak{f}^*(g)(\gamma) = g(\mathfrak{f}(\gamma))$$

and

$$o_\gamma \mathfrak{f}^*(g) = r o_{\mathfrak{f}(\gamma)} g,$$

where r is the ramification index of $\mathfrak{f}$ at γ.

PROOF: Arguing as in the proof of 4.3.13, there is a commutative diagram

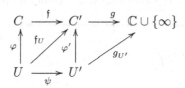

in which φ is a uniformizing map of γ, $\mathfrak{f}_U$ the extension of $\mathfrak{f} \circ \varphi_{|U-0}$, which factorizes through the uniformizing map φ' of $\mathfrak{f}(\gamma)$, and $g_{U'}$ the meromorphic extension of $g \circ \varphi'_{|U'-0}$. Regarding the first claim, it is clear that $\mathfrak{f}^*(g) \circ \varphi = g \circ \mathfrak{f} \circ \varphi_{|U-0}$ has meromorphic extension $g_{U'} \circ \psi$ and hence $\mathfrak{f}^*(g)(\gamma) = g_{U'}(\psi(0)) = g_{U'}(0) = g(\mathfrak{f}(\gamma))$ as claimed.

For the second claim, if r is the ramification index of $\mathfrak{f}$ at γ and $\ell = o_\gamma g$, for t and $\bar{t}$ close enough to 0 we have $\psi(t) = t^r u(t)$ and $g_{U'}(\bar{t}) = \bar{t}^\ell v(t)$ with u and v convergent series, $u(0) \neq 0$ and $v(0) \neq 0$. By replacing $\bar{t}$ with $t^r u(t)$ the claim follows, because, as seen above, $g_{U'} \circ \psi$ is the meromorphic extension of $\mathfrak{f}^*(g) \circ \varphi_{|U-0}$. ◇

Corollary 4.3.16 *For any irreducible projective plane curves C, C' and C'':*

(a) *All ramification indices of Id_C equal one.*

(b) *If $\mathfrak{f} : C \to C'$ and $\mathfrak{g} : C' \to C''$ are non-constant rational maps with ramification indices r at $\gamma \in X(C)$ and r' at $\mathfrak{f}(\gamma)$ respectively, then the ramification index of $\mathfrak{g} \circ \mathfrak{f}$ at γ is rr'.*

PROOF: The first claim is obvious, and follows also from the second one using $\mathrm{Id}_C^2 = \mathrm{Id}_C$.

For the second claim take $h \in \mathbb{C}(C'')$ with $o_{\mathfrak{g}(\mathfrak{f}(\gamma))} h > 0$ (any of the coordinate functions will do if affine coordinates with origin the origin of $\mathfrak{g}(\mathfrak{f}(\gamma))$ are taken). Then, if r'' is the ramification index of $\mathfrak{g} \circ \mathfrak{f}$ at γ, using 4.3.15 three times,

$$r'' o_{\mathfrak{g}(\mathfrak{f}(\gamma))} h = o_\gamma (\mathfrak{g} \circ \mathfrak{f})^*(h) = o_\gamma (\mathfrak{f}^*(\mathfrak{g}^*(h))) = r o_{\mathfrak{f}(\gamma)} (\mathfrak{g}^*(h)) = r' r o_{\mathfrak{g}(\mathfrak{f}(\gamma))} h,$$

thus proving the claim. ◇

Corollary 4.3.17 *If $\mathfrak{f} : C \to C'$ is a birational map between irreducible plane curves, then the ramification index of $\mathfrak{f}$ at any $\gamma \in X(C)$ is one, and so, for any $g \in \mathbb{C}(C')$ and any $\gamma \in X(C)$,*

$$o_\gamma \mathfrak{f}^*(g) = o_{\mathfrak{f}(\gamma)} g.$$

PROOF: Direct from the existence of $\mathfrak{f}^{-1}$, 4.3.15 and 4.3.16. ◇

Still let C denote an arbitrary irreducible curve of $\mathbb{P}_2$. The claims belonging to the intrinsic geometry on C will make use, exclusively, of the Riemann surface $X(C)$ of C, the field $\mathbb{C}(C)$ of rational functions on C, the values and orders of these functions at the points of $X(C)$ and further objects defined using them. We have seen above that a birational map

$$\mathfrak{f} : C \longrightarrow C'$$

induces a bijection (4.3.13)

$$X(C) \longrightarrow X(C')$$
$$\gamma \longmapsto \mathfrak{f}(\gamma)$$

and an isomorphism of $\mathbb{C}$-algebras (4.1.33)

$$\mathfrak{f}^* : \mathbb{C}(C') \longrightarrow \mathbb{C}(C)$$

which preserve values and orders, namely

$$(\mathfrak{f}^*(g))(\gamma) = g(\mathfrak{f}(\gamma)) \quad \text{and} \quad o_{\mathfrak{f}(\gamma)} g = o_\gamma \mathfrak{f}_*(g)$$

for any $\gamma \in X(C)$ and any $g \in \mathbb{C}(C')$ (4.3.15, 4.3.17). As a consequence, if a claim in terms of points of the Riemann surface, rational functions and their values and orders, holds for a curve C, then it holds for any C' birationally equivalent to C as well, and therefore belongs to the birationally invariant geometry of C. However, in the sequel, the proofs of these claims will often be achieved by fixing a suitable representative of the birational equivalence class of C, usually referred to as a *projective model*, and using properties of it that are not birationally invariant. In old books, these properties, which are dependent on the model as a curve of $\mathbb{P}_2$, are sometimes called *projective*, as opposite to birationally invariant.

4.4 Divisors and Linear Series

Throughout this section, C denotes an irreducible curve of $\mathbb{P}_2$. From now on the points of $X(C)$ – the branches of C – will be often represented by letters such as p rather than by Greek letters as before.

We start by setting an important birationally invariant definition:

A *divisor* on an irreducible curve C of $\mathbb{P}_2$ is a formal linear combination of points of $X(C)$ with integer coefficients, in other words, a formal expression

$$D = \sum_{p \in X(C)} a_p p$$

where $a_p \in \mathbb{Z}$ for all p and $a_p = 0$ for all but finitely many p. The points p for which $a_p \neq 0$ are called the *points of* – or *belonging to* – D, and a_p the *multiplicity* of p in D. We will often write the above D just as $D = \sum_p a_p p$ if no explicit reference to $X(C)$ is needed. As usual with formal linear combinations, we will make no distinction between any $p \in X(C)$ and the divisor $1p + \sum_{q \neq p} 0q$. The integer $\deg(D) = \sum_p a_p$ is called the *degree* of D.

By its own definition, the set of all divisors on C is a commutative group $\mathrm{Div}(C)$ with the addition

$$\sum_p a_p p + \sum_p b_p p = \sum_p (a_p + b_p)p.$$

The zero of $\mathrm{Div}(C)$ is obviously the divisor with all coefficients equal to zero, denoted just 0 in the sequel. As is clear, $\deg(D) + \deg(D') = \deg(D + D')$ and $\deg(-D) = -\deg(D)$ for any $D, D' \in \mathrm{Div}(C)$.

Divisors on C are partially ordered by the rule

$$\sum_p a_p p \geq \sum_p b_p p \Leftrightarrow a_p \geq b_p \text{ for all } p,$$

and, as usual, $D > D'$ means $D \geq D'$ and $D \neq D'$.

Obvious properties of compatibility of this ordering, addition and degree are:

Remark 4.4.1 For any divisors D, D', T, T' on C.

(a) $D \geq D'$ and $T \geq T'$ yield $D + T \geq D' + T'$ and $-D \leq -D'$.

(b) Assume $D \geq D'$: then $\deg D \geq \deg D'$ and the equality of degrees holds if and only if $D = D'$

Divisors D with $D \geq 0$ will be called *effective divisors* or *groups of points*, or just *groups* if no confusion may result. When D and D' are effective divisors on C and $D \leq D'$ it is often said that D is *contained* in D', or that D' *contains* D.

As the reader may easily check, any divisor D has a unique expression $D = D_+ - D_-$ where D_+ and D_- share no point and are both effective. They will be referred to in the sequel as the *positive* and the *negative part* of D.

Next we will introduce two examples of a divisor which will be intensively used in the sequel. Assume that C' is a curve of $\mathbb{P}_2$ which does not contain C. Then $C' \cap C$ is a finite set (3.2.6) and the set of branches of C which have positive intersection multiplicity with C' is finite too (2.6.2(b)). Therefore

$$\sum_{p \in X(C)} [C' \cdot p] p$$

(note the double role of p as a branch of C and as a point of $X(C)$) is an effective divisor on C called the *divisor cut out by C' on C*, or the *section of C by C'*. It will be denoted by $C' \cdot C$.

Remark 4.4.2 If p belongs to the divisor $C' \cdot C$, then its origin $\chi_C(p)$ belongs to $C' \cap C$, by 2.6.7(c).

Remark 4.4.3 It follows from 2.6.2(a) that $(C_1 + C_2) \cdot C = C_1 \cdot C + C_2 \cdot C$ for any two curves C_1, C_2 not containing C.

Remark 4.4.4 Bézout's theorem 3.3.5 gives

$$\deg(C' \cdot C) = \deg C' \deg C$$

for any curve C' not containing C.

If D is an effective divisor on C we will say that a curve C' *goes through* D if and only if either $C' \supset C$ or $C' \not\supset C$ and $C' \cdot C \geq D$. Different curves may of course cut the same divisor on C. This characterization will be useful in the sequel:

Lemma 4.4.5 *Assume that $C' : G_1 = 0$ and $C'' : G_2 = 0$ are curves of $\mathbb{P}_2$ of the same degree, neither containing C. Then there is a homogeneous polynomial $G = \lambda_1 G_1 + \lambda_2 G_2$, $(\lambda_1, \lambda_2) \in \mathbb{C}^2 - \{(0,0)\}$, which is zero on C if and only if $C_1 \cdot C = C_2 \cdot C$.*

PROOF: For the direct part, if $G = 0$, then $C_1 = C_2$ and the claim is obvious. Thus assume $G \neq 0$. Then we may consider the curve $L : G = 0$, which contains C. Due to the hypothesis on C_1 and C_2, $\lambda_1 \neq 0$, $\lambda_2 \neq 0$. Thus, after replacing G_i with $\lambda_i G_i$, $i = 1, 2$, we may assume $G = G_1 + G_2$. If γ is any branch of C, then γ is a branch of L too (2.4.3), and therefore the inequality of 2.6.2(d) is

$$\infty = [L \cdot \gamma] \geq \min\{[C_1 \cdot \gamma], [C_2 \cdot \gamma]\}.$$

On the other hand γ is not a branch of C_1 or C_2 because neither of them contains C (2.5.7); the right-hand member of the inequality above is thus finite, the inequality is strict and, by 2.6.2(d), $[C_1 \cdot \gamma] = [C_2 \cdot \gamma]$. This being true for any branch γ of C, the claim follows.

For the converse, the case in which the equations G_1 and G_2 are proportional is clear. Assume otherwise and consider the pencil of curves

$$\{C_\alpha : \alpha_1 G_1 + \alpha_2 G_2 = 0 \mid (\alpha_1, \alpha_2) \in \mathbb{C}^2 - \{(0,0)\}\}.$$

Pick any $p \in X(C)$ not belonging to $C_1 \cdot C = C_2 \cdot C$ and take $E : \lambda_1 G_1 + \lambda_2 G_2 = 0$ to be a curve of the pencil going through the origin of p (by 1.5.3). Then p belongs to $E \cdot C$, after which $E \neq C_i$, $i = 1, 2$, and λ_1, λ_2 are both non-zero. If $E \supset C$ we are done. If not, using $\lambda_1 G_1$ and $\lambda_2 G_2$ as equations of C_1, C_2, we have $E \cdot C \geq C_1 \cdot C = C_2 \cdot C$ due to 2.6.2(d). Further, the inequality is strict because $[E \cdot p] > 0$ (2.6.2(b)) and p does not belong to $C_1 \cdot C$. The curves C_1 and E being of the same degree, $\deg(E \cdot C) = \deg(C_1 \cdot C)$ by 4.4.4, which contradicts 4.4.1(b). $\diamond$

The proposition below introduces the second example of a divisor and relates it to the former one; as usual, x_0, x_1, x_2 denote homogeneous coordinates on $\mathbb{P}_2$.

Proposition 4.4.6 *If $f \in \mathbb{C}(C)$ is not zero, then*

(a) *$(f) = \sum_{p \in X(C)} (o_p f)p$ is a divisor on C.*

(b) *Assume that f is the restriction of a rational function F/G of $\mathbb{P}_2$, where $F, G \in \mathbb{C}[x_0, x_1, x_2]$ are homogeneous, have the same degree, and none is zero at all points of C. If C_1 and C_2 are the curves $C_1 : F = 0$, $C_2 : G = 0$, then*

$$(f) = C_1 \cdot C - C_2 \cdot C$$

PROOF: Fix any $p \in X(C)$. In the situation described in claim (b), let $H \in \mathbb{C}[x_0, x_1, x_2]$ be a homogeneous polynomial which is not zero at the origin of p and has the same degree as F and G. Then, according to the definition of intersection multiplicity with a branch in Section 2.6,

$$o_p(F/H) = [C_1 \cdot p] \quad \text{and} \quad o_p(G/H) = [C_2 \cdot p].$$

Since $F/G = (F/H)(G/H)^{-1}$, it follows that

$$o_p f = [C_1 \cdot p] - [C_2 \cdot p]. \tag{4.12}$$

This proves that $o_p f = 0$ but for the $p \in X(C)$ whose origin belongs to $C \cap C_1$ or to $C \cap C_2$. In particular, $o_p f = 0$ for all but finitely many $p \in X(C)$ and so (f) is, indeed, a divisor. This proved, claim (b) directly follows by adding up the equalities (4.12) for all $p \in X(C)$.   ◇

Divisors of the form (f) for $f \in \mathbb{C}(C) - \{0\}$ are called *principal divisors*; more precisely, (f) is called the (principal) *divisor of* – or *associated to* – f. It is usual to write $(f)_0$ and $(f)_\infty$ for the positive and the negative part of (f), respectively, namely

$$(f)_0 = \sum_{p \text{ a zero of } f} (o_p f)p \quad \text{and} \quad (f)_\infty = - \sum_{p \text{ a pole of } f} (o_p f)p,$$

and call them the *divisor of zeros* and the *divisor of poles* of f. Both are effective divisors, share no points and

$$(f) = (f)_0 - (f)_\infty.$$

The definitions and 4.3.10 give:

Lemma 4.4.7 *For any* $f, g \in \mathbb{C}(C) - \{0\}$,

$$(fg) = (f) + (g) \quad \text{and} \quad (f^{-1}) = -(f),$$

while for any $a \in \mathbb{C} - \{0\}$, $(a) = 0$. *In particular, the principal divisors compose a subgroup of* $\mathrm{Div}(C)$.

An important direct consequence of 4.4.6(b) and Bézout's theorem is:

Corollary 4.4.8 *Any principal divisor has degree zero.*

PROOF: Just note that C_1 and C_2 in 4.4.6 have the same degree and then use 4.4.20.   ◇

This in turn gives rise to a couple of quite relevant results:

Corollary 4.4.9 *If* $f \in \mathbb{C}(C) - \{0\}$ *has no zeros in* $X(C)$, *then* f *is constant. The same holds if* f *has no poles.*

PROOF: By 4.4.8, if f has no zeros, then it has no poles either. Pick any $p \in X(C)$; if the function $f - f(p)$ is not zero, then it has an obvious zero at p, and still no pole, by 4.3.10. Since this contradicts 4.4.8, $f - f(p)$ is the zero function and therefore f is constant. If f has no poles, then, by 4.4.8, it has no zeros either and the above applies.   ◇

Corollary 4.4.10 *If* $f, g \in \mathbb{C}(C) - \{0\}$ *and* $(f) = (g)$, *then there is* $a \in \mathbb{C} - \{0\}$ *such that* $f = ag$.

PROOF: According to the hypothesis, by 4.4.7, $(f/g) = 0$; then 4.4.9 applies showing that f/g is constant.   ◇

Corollary 4.4.10 asserts that a non-zero rational function f is determined by its principal divisor (f) – that is, by its zeros and poles weighted with multiplicities – up to a constant factor. This is a crucial fact: it allows us to deal with their principal divisors rather than with the rational functions themselves, thus opening the door to the reformulation of the theory of algebraic functions in one variable into the geometry of linear series on algebraic curves. Linear series are introduced next.

Assume that D is a divisor on C and F a finite-dimensional linear subspace of $\mathbb{C}(C)$ such that $D + (f) \geq 0$ for all $f \in F - \{0\}$. Then the set of effective divisors

$$\mathcal{L} = \{D + (f) \mid f \in F - \{0\}\}$$

is called a *linear series* on C. The empty set appears as a linear series, defined by any divisor D and the subspace $F = \{0\}$. If D is an effective divisor and $F = \mathbb{C}$, then the linear series $\mathcal{L}$ defined by D and F has a single divisor, namely $\mathcal{L} = \{D\}$. Next is a more interesting example:

Example 4.4.11 Consider any non-constant $g \in \mathbb{C}(C)$ and take $D = (g)_\infty$ and $F = \langle 1, g \rangle$. Then the corresponding linear series $\mathcal{L}_g$ is composed of the group of poles of g, $(g)_\infty = (g)_\infty + (1)$ and the groups $(g)_\infty + (-a + g)$, for any $a \in \mathbb{C}$. Since $(-a + g)_\infty = (g)_\infty$ due to 4.3.10, $(g)_\infty + (-a + g) = (-a + g)_0$. Therefore, each group of the linear series other than $(g)_\infty$ is the group of the points of $X(C)$ at which g takes a certain fixed value $a \in \mathbb{C}$, each point taken with multiplicity equal to its multiplicity as a zero of $f - a$, and conversely. Using this characterization, the reader may easily check that if $g' = (ag + b)/(cg + d)$, for $a, b, c, d \in \mathbb{C}$, $ad - bc \neq 0$, then $\mathcal{L}_g = \mathcal{L}_{g'}$. We will refer to $\mathcal{L}_g$ as the *linear series associated to* g.

All groups of points in a non-empty linear series $\mathcal{L}$, defined as above by D and F, have degree equal to the degree of D, by 4.4.8: such a common degree is called the *degree* of $\mathcal{L}$. The degree of the empty linear series is left undefined; therefore, in the sequel, when mentioning the degree of a linear series $\mathcal{L}$ we will implicitly assume $\mathcal{L} \neq \emptyset$.

The word *contained* and the symbol $\subset$ will have the usual meaning when applied to pairs of linear series, namely $\mathcal{L} \subset \mathcal{L}'$ if and only if any divisor in $\mathcal{L}$ belongs to $\mathcal{L}'$. When any divisor in $\mathcal{L}$ is contained in a divisor belonging to $\mathcal{L}'$, it is often said that $\mathcal{L}$ is *partially contained* in $\mathcal{L}'$. If this is the case and $\mathcal{L}$ is non-empty, then so is $\mathcal{L}'$ and $\deg \mathcal{L} \leq \deg \mathcal{L}'$; furthermore, equality holds if and only if $\mathcal{L} \subset \mathcal{L}'$.

Different pairs D, F may define the same linear series; the next lemma characterizes them:

Lemma 4.4.12 *If D, F and D', G are pairs divisor-subspace defining the same non-empty linear series $\mathcal{L}$, then there is a $g \in \mathbb{C}(C) - \{0\}$ such that $D' = D - (g)$ and $G = gF$. Conversely, for any $g \in \mathbb{C}(C) - \{0\}$, $D' = D - (g)$ and $G = gF$ define the same linear series as D, F.*

PROOF: If both D, F and D', G define $\mathcal{L}$, pick any $D'' = D + (g_1) = D' + (g_2) \in \mathcal{L}$, $g_1 \in F$, $g_2 \in G$. By taking $g = g_2/g_1$ we have $D' = D - (g)$. In addition, for any $f \in F - \{0\}$, $D + (f)$ may be written

$$D + (f) = D' + (f') = D + (f') - (g) = D + (f'/g)$$

for some $f' \in G - \{0\}$. It follows that $(f) = (f'/g)$ and therefore (4.4.10) $f = af'/g$ for some $a \in \mathbb{C}$. Thus $gf = af' \in G$ for all $f \in F - \{0\}$, which yields $gF \subset G$. The symmetric argument proves $g^{-1}G \subset F$ and therefore completes the proof of the direct claim. For the converse, just note that for any $f \in F - \{0\}$,

$$D + (f) = D - (g) + (g) + (f) = D' + (gf).$$

$\diamond$

Next we will see that any linear series has a projective space structure inherited from any of the subspaces defining it. To cover all cases, we will formally take the empty set as the projective space of dimension -1 associated to the subspace $\{0\}$. This obvious case covered, assume $\mathcal{L}$ non-empty and consider the map

$$\pi_F : F - \{0\} \longrightarrow \mathcal{L}$$
$$f \longmapsto D + (f).$$

It is exhaustive by the definition of $\mathcal{L}$, while non-zero $f, f' \in F$ have the same image if and only if $(f) = (f')$, which, by 4.4.10, is equivalent to being $f' = af$ for some $a \in \mathbb{C}$. This gives to $\mathcal{L}$ the structure of a projective space of dimension $\dim F - 1$. Further, this structure does not depend on the choice of D and F, because if D' and G also define $\mathcal{L}$, taking g as in 4.4.12, the diagram

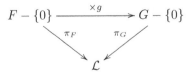

is commutative, as the reader may easily check. In particular, the dimension $\dim F - 1$ of $\mathcal{L}$ as a projective space depends only on $\mathcal{L}$: it is called the *dimension* of $\mathcal{L}$, denoted $\dim \mathcal{L}$.

Remark 4.4.13 If $f_0, \ldots, f_d$ is a basis of F, then $a_0, \ldots, a_d \in \mathbb{C}$ may be taken as projective coordinates of $D + (a_0 f_0 + \cdots + a_d f_d)$ in the linear series defined by D and F.

Remark 4.4.14 Assume that a non-empty linear series $\mathcal{L}$ is defined as above by D and F. Then for any $D' \in \mathcal{L}$ we have $D' = D + (f)$, with $f \in F$. By 4.4.10, D' and $G = f^{-1}F$ also define $\mathcal{L}$. Therefore, any non-empty linear series $\mathcal{L}$ may be defined by any $D' \in \mathcal{L}$, which in particular is an effective divisor, and a subspace containing 1.

Lemma 4.4.15 *If $\mathcal{L}'$ is linear series on an irreducible curve C and $\mathcal{L}$ is a subset of $\mathcal{L}'$, then $\mathcal{L}$ is a linear series on C if and only if $\mathcal{L}$ is a linear variety of $\mathcal{L}'$.*

PROOF: If $\mathcal{L}'$ is defined by a divisor D and a subspace F', then, by the definition of the projective structure on $\mathcal{L}'$, any linear variety $\mathcal{L}$ of $\mathcal{L}'$ is of the form $\mathcal{L} = \pi(F - \{0\}) = \{D + (f) \mid f \in F - \{0\}\}$ for a subspace F of F', and therefore $\mathcal{L}$ is a linear series, namely the one defined by D and F.

For the converse, the case $\mathcal{L} = \emptyset$ being obvious, assume otherwise and take $D \in \mathcal{L}$. Since $\mathcal{L} \subset \mathcal{L}'$, we may assume, by 4.4.14, $\mathcal{L}$ to be defined by D and a subspace F, and $\mathcal{L}'$ by the same D and a subspace F'. Next we will prove that $F \subset F'$, which obviously implies the claim. For, take any $f \in F$, $f \neq 0$. Then $D + (f)$ belongs to $\mathcal{L}$, and hence also to $\mathcal{L}'$. Thus $D + (f) = D + (f')$ for some $f' \in F'$, for which $(f) = (f')$. Then 4.4.10 gives $f = af'$ for some $a \in \mathbb{C}$, and so $f \in F'$, as wanted.   ◇

A direct consequence of 4.4.15 is

Corollary 4.4.16 *If $\mathcal{L}$ and $\mathcal{L}'$ are linear series on an irreducible curve C, then*

(a) *$\mathcal{L} \subset \mathcal{L}'$ implies $\dim \mathcal{L} \leq \dim \mathcal{L}'$.*

(b) *$\mathcal{L} \subset \mathcal{L}'$ and $\dim \mathcal{L} = \dim \mathcal{L}'$ imply $\mathcal{L} = \mathcal{L}'$.*

Again, let $\mathcal{L}$ be a non-empty linear series, defined by a divisor D and a subspace F. For each $p \in X(C)$ take u_p to be the minimal multiplicity of p in the divisors $D' \in \mathcal{L}$. Then $\mathrm{fp}(\mathcal{L}) = \sum_p u_p p$ clearly is an effective divisor (possibly 0) contained in all groups of $\mathcal{L}$, and is in fact the maximal divisor contained in all groups of $\mathcal{L}$: it will be called the *fixed part* of $\mathcal{L}$ and its points, the *fixed points* of $\mathcal{L}$. A point $p \in X(C)$ is thus a fixed point of a non-empty linear series $\mathcal{L}$ if and only if $D - p \geq 0$ for all $D \in \mathcal{L}$.

If $\mathcal{L}$ is a linear series, defined as above by D and F, and T an effective divisor, then T may be added to all divisors of $\mathcal{L}$: the resulting groups $D' + T$, $D' \in \mathcal{L}$ are obviously effective and describe the linear series $\mathcal{L} + T$ defined by $D + T$ and F; clearly such a linear series has dimension $\dim \mathcal{L}$ and group of fixed points $\mathrm{fp}(\mathcal{L}) + T$.

Subtracting an effective divisor T from the groups of $\mathcal{L}$ that contain it is a more interesting operation. To this end consider

$$F_T = \{f \in F \mid f = 0 \text{ or, otherwise, } (f) \geq -D + T\} :$$

by 4.3.10, it is a subspace of F. Then the set

$$\mathcal{L} - T = \{D' - T \mid D' \in \mathcal{L}, D' \geq T\}$$

is clearly the linear series defined by $D - T$ and F_T. The series $\mathcal{L} - T$ is called the *residual linear series of $\mathcal{L}$ with respect to T*. Its elements are usually referred to as *residual groups* or *residual divisors*. By its definition, $\dim(\mathcal{L} - T) \leq \dim \mathcal{L}$. The residual series $\mathcal{L} - T$ may of course be empty; otherwise its degree is $\deg \mathcal{L} - \deg T$. The reader may note that the residual linear series $\mathcal{L} - T$ arises by first selecting the groups of $\mathcal{L}$ that contain T and then subtracting T from each of them.

The next lemma may be easily proved by the reader using just the definition of residual series:

Lemma 4.4.17 *If T and T' are effective divisors, for any linear series $\mathcal{L}$,*

$$\mathcal{L} - (T + T') = (\mathcal{L} - T) - T'.$$

Remark 4.4.18 The reader may note that, in the hypothesis of 4.4.17, an equality of the type $(\mathcal{L} - T) + T' = \mathcal{L} + (T' - T)$ makes no sense in general, because $T' - T$ need not be effective. Even in the case when $T' - T \geq 0$, the equality may fail: for instance $(\mathcal{L} - T) + T \neq \mathcal{L}$ unless T is contained in the fixed part of $\mathcal{L}$.

Remark 4.4.19 The above definition of residual linear series applies in particular to the case in which $T \leq \mathrm{fp}(\mathcal{L})$. Then, $F_T = F$ and therefore $\mathcal{L}$ and the residual linear series $\mathcal{L} - T$ – the result of removing part of the fixed points from all groups of points in $\mathcal{L}$ – have the same dimension. In particular, the differences $D' - \mathrm{fp}(\mathcal{L})$, for $D' \in \mathcal{L}$, describe a linear series which has the same dimension as $\mathcal{L}$ and no fixed point: it is called the *variable part* of $\mathcal{L}$.

The dimension of a residual series is not so obvious in the general case. We start by considering the case in which T is a single point:

Proposition 4.4.20 *If $\mathcal{L}$ is a positive-dimensional linear series on C and $p \in X(C)$, then the residual series $\mathcal{L} - p$ has degree $\deg \mathcal{L} - 1$. If p is a fixed point of $\mathcal{L}$, then $\dim(\mathcal{L} - p) = \dim \mathcal{L}$; otherwise, $\dim(\mathcal{L} - p) = \dim \mathcal{L} - 1$.*

PROOF: Only the claim about the dimension needs a proof, and the case in which p is a fixed point of $\mathcal{L}$ has been dealt with in 4.4.19.

If p is not a fixed point of $\mathcal{L}$, then there is a divisor $D \in \mathcal{L}$ that does not contain p. By 4.4.14, such a D and a subspace F containing 1, say with basis $1, f_1, \ldots, f_r$, may be used to define $\mathcal{L}$. An arbitrary divisor $D' \in \mathcal{L}$ is thus of the form

$$D' = D + (\lambda_0 + \lambda_1 f_1 + \cdots + \lambda_r f_r),$$

for $\lambda_0, \ldots, \lambda_r \in \mathbb{C}$, not all zero. Proceeding as in the definition of residual series, D' contains p if and only if

$$\lambda_0 + \lambda_1 f_1(p) + \cdots + \lambda_r f_r(p) = 0$$

(we are using 4.3.9). As a consequence the divisors $D' \in \mathcal{L}$ that contain p are those of the form $D + (f)$, for f in the subspace

$$F_p = \{\lambda_0 + \sum_{i=1}^{r} \lambda_i f_i \mid \lambda_0 + \sum_{i=1}^{r} \lambda_i f(p) = 0\},$$

which obviously has codimension one in F. The residual series $\mathcal{L} - p$ being defined by $D - p$ and F_p, it has dimension $r - 1$, as claimed.   ◇

An important consequence of 4.4.20 is:

Corollary 4.4.21 *If a non-empty linear series $\mathcal{L}$ has degree n and dimension r, then $n \geq r$.*

PROOF: The claim being obvious if $r = 0$, proceed by induction on r. If $\mathcal{L}$ has degree n and dimension $r > 0$, choose any $p \in X(C)$ which is not a fixed point of $\mathcal{L}$. Then, by 4.4.20, the residual series of $\mathcal{L}$ with respect to p is a linear series of degree $n - 1$ and dimension $r - 1$ to which the induction hypothesis applies.   ◇

Corollary 4.4.22 *If $\mathcal{L}$ is a linear series on C of dimension $r > 0$ and T a non-zero effective divisor on C of degree $k \leq r$, then*

$$\dim(\mathcal{L} - T) \geq r - k.$$

Furthermore, for each $k \leq r$ there is a choice of an effective divisor T of degree k for which equality holds.

PROOF: For the inequality, the case $k = 1$ is 4.4.20 and the case $k > 1$ follows by an easy induction on k, picking a point p belonging to T and using the equality

$$\mathcal{L} - T = (\mathcal{L} - p) - T', \quad T' = T - p,$$

which is a particular case of 4.4.17.

For the equality, if $k = 1$ there is equality if $T = p$ is not a fixed point of $\mathcal{L}$, by 4.4.20; again the general case follows by induction.   ◇

Remark 4.4.23 Notations and hypothesis being as above, consider the linear series defined by D and F_T, whose members are the groups of $\mathcal{L}$ that contain T, namely

$$\mathcal{L}_T = \{D' \mid D' \in \mathcal{L}, D' \geq T\} = (\mathcal{L} - T) + T.$$

Being defined by the same subspace, $\dim \mathcal{L}_T = \dim(\mathcal{L} - T)$. Hence, by 4.4.22, also $\dim \mathcal{L}_T \geq \dim \mathcal{L} - \deg T$ and equality holds for a suitable choice of the effective divisor T, provided $\deg T \leq \dim \mathcal{L}$. In particular, for a certain choice of an effective divisor T of degree $\deg T = \dim \mathcal{L}$, we have $\dim \mathcal{L}_T = 0$. Then $\mathcal{L}_T$ contains a single divisor D' which is the only divisor in $\mathcal{L}$ that contains T. The dimension of $\mathcal{L}$ thus appears as the minimal number of points that need to be fixed in order to determine a group of $\mathcal{L}$, an interpretation of the dimension of a linear series quite usual in old texts. In them 4.4.21 is often proved by just saying that a group of points in a linear series cannot contain fewer points than the number of points that need to be fixed for determining it. In our terms, if T has $\deg T = \dim \mathcal{L}$, then $\dim \mathcal{L}_T \geq 0$ by 4.4.22, and picking any $D \in \mathcal{L}_T$, $\deg \mathcal{L} = \deg D \geq \deg T = \dim \mathcal{L}$.

The notions of divisor, linear series and related ones introduced up to now in this section, are all defined in terms of points of the Riemann surface, rational functions, values and orders, and therefore are all preserved by birational maps. The next three propositions state this. In all of them $\mathfrak{f} : C \to C'$ is a birational map between irreducible curves C and C' of projective planes $\mathbb{P}_2, \mathbb{P}'_2$.

Proposition 4.4.24 *The bijection*

$$X(\mathfrak{f}) : X(C) \longrightarrow X(C')$$

extends to a group isomorphism, still denoted $X(\mathfrak{f})$,

$$X(\mathfrak{f}) : \mathrm{Div}(C) \longrightarrow \mathrm{Div}(C')$$

$$\sum_{p \in X(C)} a_p p \longmapsto \sum_{p \in X(C)} a_p X(\mathfrak{f})(p)$$

which preserves ordering and degree and so, in particular, maps effective divisors to effective divisors.

PROOF: Direct from the definitions.   ◇

The reader may note that the properties of 4.3.13 still hold for the above extension of $X(f)$ to divisors. As for points of $X(C)$, in the sequel we will write $X(\mathfrak{f})(D) = \mathfrak{f}(D)$ for any $D \in \mathrm{Div}(C)$ unless some confusion may occur.

Proposition 4.4.25 *For any $g \in \mathbb{C}(C)$, we have $X(\mathfrak{f})((g)) = ((\mathfrak{f}^*)^{-1}(g))$. In particular, $X(\mathfrak{f})$ maps principal divisors to principal divisors.*

This follows from 4.3.17 and 4.4.24.   ◇

Proposition 4.4.26 *Assume that a linear series $\mathcal{L}$ is defined on C by a divisor D and a subspace F, namely, $\mathcal{L} = \{D + (g) \mid g \in F - 0\}$. Then:*

(a) $X(\mathfrak{f})(\mathcal{L}) = \{\mathfrak{f}(D) + (g') \mid g' \in (\mathfrak{f}^*)^{-1}(F) - 0\}$, *which therefore is a linear series of the same dimension and degree as $\mathcal{L}$.*

(b) *The restriction $X(\mathfrak{f})_{|\mathcal{L}} : \mathcal{L} \to X(\mathfrak{f})(\mathcal{L})$ is a projectivity.*

(c) *For any effective divisor T on C, $X(\mathfrak{f})(\mathcal{L} - T) = X(\mathfrak{f})(\mathcal{L}) - \mathfrak{f}(T)$.*

PROOF: For any non-zero $g \in F$, by 4.4.24 and 4.4.25,

$$X(\mathfrak{f})(D + (g)) = X(\mathfrak{f})(D) + ((\mathfrak{f}^*)^{-1}(g)).$$

This gives claim (a) and also claim (b), because then $X(\mathfrak{f})_{|\mathcal{L}}$ is the projectivity induced by $(\mathfrak{f}^*)_{|F}^{-1}$. The proof of claim (c) is straightforward from 4.4.24. ◇

Linear series on C may also be obtained from an external – non-intrinsic – projective construction:

Proposition 4.4.27 *Let Λ be an r-dimensional linear system of curves of $\mathbb{P}_2$. Assume that no curve in Λ contains C, and that T is a divisor on C such that $T \leq C' \cdot C$ for any $C' \in \Lambda$. Then:*

(a) *$\mathcal{L} = \{C' \cdot C - T \mid C' \in \Lambda\}$ is an r-dimensional linear series on C.*

(b) *The map*

$$\sigma : \Lambda \longrightarrow \mathcal{L}$$
$$C' \longmapsto C' \cdot C - T$$

is a projectivity.

(c) *Any linear r-dimensional series on C may be obtained in this way from suitable Λ and T as above, Λ with no fixed part.*

PROOF: The case $r = -1$ being obvious, we assume $r \geq 0$ in the sequel. After fixing homogeneous coordinates x_0, x_1, x_2 on $\mathbb{P}_2$, assume the linear system Λ to be

$$\Lambda = \{C_\lambda : \lambda_0 F_0 + \cdots + \lambda_r F_r = 0 \mid \lambda = (\lambda_0, \ldots, \lambda_r) \in \mathbb{C}^{r+1} - \{0\}\}$$

where $F_0, \ldots, F_r \in \mathbb{C}[x_0, x_1, x_2]$ are linearly independent homogeneous polynomials of the same degree. Since no curve in Λ contains C, in particular F_0 is not zero on C and therefore each F_i/F_0 restricts to a rational function $f_i \in \mathbb{C}(C)$ for $i = 1, \ldots, r$. Then, for any $\lambda \in \mathbb{C}^{r+1} - \{0\}$, the rational function $f_\lambda = \lambda_0 + \lambda_1 f_1 + \cdots + \lambda_r f_r$ is the restriction of $(\lambda_0 F_0 + \lambda_1 F_1 + \cdots + \lambda_r F_r)/F_0$.
Note first that $f_\lambda \neq 0$ for any $\lambda \in \mathbb{C}^{r+1} - \{0\}$, as otherwise the polynomial

$$\lambda_0 F_0 + \lambda_1 F_1 + \cdots + \lambda_r F_r$$

would be zero on C, against the hypothesis that $C \not\subset C_\lambda$. It follows that (f_λ) is defined for all $\lambda \in \mathbb{C}^{r+1} - \{0\}$, and also that $1, f_1, \ldots, f_r$ are linearly independent.

Now, if $\bar{C}$ denotes the curve $C_{1,0,\ldots,0} : F_0 = 0$, it follows from 4.4.6(b) that for any $\lambda \in \mathbb{C}^{r+1} - \{0\}$

$$C_\lambda \cdot C - T = (f_\lambda) + \bar{C} \cdot C - T,$$

and this divisor is effective due to the hypothesis on Λ and T. As a consequence, $\mathcal{L}$ is the linear series defined by $\bar{C} \cdot C - T$ and $\langle 1, f_1, \ldots, f_r \rangle$. The map σ is a projectivity just because mapping λ to f_λ is linear and bijective.

Conversely, if a non-empty linear series $\mathcal{L}$ is defined by a divisor D and a subspace F, then, by 4.4.12, we may assume without restriction that D is effective and F has a basis containing 1, say $1, f_1, \ldots, f_r$. Assume the f_i, $i = 1, \ldots, r$, to be the restrictions of rational functions F_i/F_0, $i = 1, \ldots, r$, where the homogeneous polynomials F_i, $i = 0, \ldots, r$, are coprime, have the same degree and F_0 is not zero on C.

If for some $\lambda = (\lambda_0, \ldots, \lambda_r) \in \mathbb{C}^{r+1} - \{0\}$ the polynomial

$$F_\lambda = \lambda_0 F_0 + \lambda_1 F_1 + \cdots + \lambda_r F_r$$

is zero or defines a curve containing C, the restriction of F_λ/F_0 to C is zero, namely

$$\lambda_0 + \lambda_1 f_1 + \cdots + \lambda_r f_r = 0,$$

against the linear independence of $1, f_1, \ldots, f_r$. Thus

$$\Lambda = \{C_\lambda : \lambda_0 F_0 + \lambda_1 F_1 + \cdots + \lambda_r F_r = 0 \mid \lambda = (\lambda_0, \ldots, \lambda_r) \in \mathbb{C}^{r+1} - \{0\}\}$$

is an r-dimensional linear system, with no fixed part, from which no curve contains C.

Using 4.4.6(b) again, for any non-zero $f_\lambda \in F$, $f_\lambda = \lambda_0 + \lambda_1 f_1 + \cdots + \lambda_r f_r$,

$$D + (f_\lambda) = D - \bar{C} \cdot C + C_\lambda \cdot C,$$

where $\bar{C}$ is still $\bar{C} : F_0 = 0$. Since the above divisor is by hypothesis effective, this proves that $\mathcal{L}$ arises as claimed, from the linear system Λ and the divisor $D - \bar{C} \cdot C$.   ◇

In the particular case when $T = 0$, 4.4.27 asserts that the effective divisors $C' \cdot C$, $C' \in \Lambda$, compose a linear series: it is called the linear series *traced* or *cut out* by Λ on C; we will denote it by $\Lambda \cdot C$. The more relevant part of the 4.4.27 is that any linear series on C arises as the residual linear series of a linear series $\Lambda \cdot C$, traced on C by a linear system, with respect to a group T of fixed points of $\Lambda \cdot C$. Taking all those residual linear series and not just the linear series traced by linear systems is essential: the latter are not all linear series on C because their degrees are constrained by Bézout's theorem. Next is an example:

Example 4.4.28 Let C be the cubic $C : x_0 x_1 x_2 + (x_1 + x_2)^3 = 0$. Its affine part in the chart $x_0 \neq 0$, with the usual coordinates $x = x_1/x_0$, $y = x_2/x_0$, has equation $xy + (x + y)^3 = 0$. An easy computation shows that C is irreducible; it clearly has a node at the origin $(0, 0) = [1, 0, 0]$, with tangents the coordinate axes $\ell_1 : x_2 = 0$ and $\ell_2 : x_1 = 0$. Denote by q_1 and q_2 the branches of C at the node tangent to ℓ_1 and ℓ_2, respectively. The lines through the origin describe a pencil and cut out divisors of degree three (4.4.4) on C, which are $\ell_1 \cdot C = 2q_1 + q_2$, $\ell_2 \cdot C = q_1 + 2q_2$ and $\ell \cdot C = q_1 + q_2 + p$, $p \neq q_1, q_2$, if $\ell \neq \ell_1, \ell_2$. Then, subtracting $q_1 + q_2$ from the groups cut out by the lines through the origin on C gives a linear series with both the degree and the dimension equal to one. After Bézout's theorem (or 4.4.4), it is clear that no divisor in this series – no effective divisor of degree one in fact – may be cut out by another curve on C.

Remark 4.4.29 The hypothesis and notations being as in 4.4.27, we have seen in its proof that $\mathcal{L}$ is defined by $\bar{C} \cdot C - T$ and $\langle 1, f_1, \ldots, f_r \rangle$, in such a way that

$$C_\lambda \cdot C - T = (\lambda_0 + \lambda_1 f_1 + \cdots + \lambda_r f_r) + \bar{C} \cdot C - T$$

for all $C_\lambda \cdot C - T \in \mathcal{L}$. Therefore $\lambda_0, \ldots, \lambda_r$ may be taken as homogeneous coordinates of $C_\lambda \cdot C - T$ in $\mathcal{L}$ for any $C_\lambda \cdot C - T \in \mathcal{L}$.

Dropping the hypothesis that no $C' \in \Lambda$ contains C in 4.4.27 may cause some of the divisors $C' \cdot C$ to be not defined. However, those which are defined still describe a linear series. Before giving a precise claim, note that if the linear system Λ still is

$$\Lambda = \{C_\lambda : \lambda_0 F_0 + \lambda_1 F_1 + \cdots + \lambda_r F_r = 0 \mid \lambda = (\lambda_0, \ldots, \lambda_r) \in \mathbb{C}^{r+1} - \{0\}\},$$

then the polynomials $\sum_{i=0}^r \lambda_i F_i$ that are zero on C describe a subspace S of $\langle F_0, \ldots, F_r \rangle$: as a consequence, the curves in Λ containing C describe a linear system Λ_C contained in Λ.

Corollary 4.4.30 *Assume that Λ is a linear system of curves of $\mathbb{P}_2$ and T a divisor on C such that $T \leq C' \cdot C$ for all $C' \in \Lambda$, $C' \not\supset C$. If Λ_C is the linear system of the curves in Λ that contain C, then*

$$\mathcal{L} = \{C' \cdot C - T \mid C' \in \Lambda - \Lambda_C\}$$

is a linear series on C of dimension $\dim \Lambda - \dim \Lambda_C - 1$.

PROOF: If $\Lambda_C = \Lambda$, we have $\mathcal{L} = \emptyset$ and the claim is satisfied. We thus assume $\Lambda_C \neq \Lambda$ in the sequel. Call E the vector space of polynomials defining Λ and, as above, S the subspace of E defining Λ_C. Take S' to be a supplementary of S in E. Then $\dim S' = \dim E - \dim S = \dim \Lambda - \dim \Lambda_C > 0$ and therefore S' defines a linear system $\Lambda' \subset \Lambda$ of dimension $\dim \Lambda - \dim \Lambda_C - 1$ that has no curve containing C. Any $F \in E$ which is not zero on C is $F \notin S$; therefore it may be written $F = G_1 + G_2$ where $G_1 \in S$ and $G_2 \in S' - \{0\}$. We claim that the curves $F = 0$ and $G_2 = 0$ have the same intersection multiplicity with any $\gamma \in X(C)$, and therefore cut out the same group on C: indeed, if $G_1 = 0$ this is obvious; otherwise the curve $G_1 = 0$ has C as an irreducible component, hence has γ as a branch and then it is enough to use 2.6.2. Our claim being proved, the set $\mathcal{L}$ may be rewritten

$$\mathcal{L} = \{C' \cdot C - T \mid C' \in \Lambda'\}$$

and so, by 4.4.27, it is a linear series of dimension $\dim \Lambda' = \dim \Lambda - \dim \Lambda_C - 1$, as claimed. $\diamond$

If $T = 0$ we will still refer to the linear series of 4.4.30 as the linear series *traced* or *cut out* by Λ on C, and denote it by $\Lambda \cdot C$.

4.5   Complete Linear Series, Riemann's Problem

As before, C denotes an arbitrary irreducible curve of $\mathbb{P}_2$. We have already seen in 4.4.7 that the principal divisors on C compose a subgroup of the group $\mathrm{Div}(C)$ of all divisors on C. Then the rule $D \equiv D'$ if and only if $D - D'$ is a principal divisor defines an equivalence relation on $\mathrm{Div}(C)$: it is called *linear equivalence*. The classes of divisors modulo linear equivalence thus describe the quotient group of $\mathrm{Div}(C)$ by the subgroup of the principal divisors, which is called the *Picard group* of C, usually denoted $\mathrm{Pic}(C)$. Next are some remarks about linear equivalence:

Remark 4.5.1 Since the principal divisors compose a subgroup, for any divisors D, T, D', T' on C, $D \equiv T$ and $D' \equiv T'$ imply $D + D' \equiv T + T'$ and $D \equiv T$ implies $-D \equiv -T$.

Remark 4.5.2 Two linearly equivalent divisors have the same degree, by 4.4.8; this allows us to define the *degree* of any element of $\mathrm{Pic}(C)$ as being the degree of any of its representatives. Clearly, the degree-zero elements of $\mathrm{Pic}(C)$ compose a subgroup, which is usually denoted $\mathrm{Pic}_0(C)$.

Remark 4.5.3 The divisors of zeros and poles of any $f \in \mathbb{C}(C) - \{0\}$ are linearly equivalent. Conversely, if D and D' are linearly equivalent effective divisors sharing no point, then, by the definition, $D - D' = (f)$ for some non-zero $f \in \mathbb{C}(C)$; by equating the positive and negative parts on both sides we have $D = (f)_0$ and $D' = (f)_\infty$.

Remark 4.5.4 By the definition of linear series, any two divisors in the same linear series are linearly equivalent. Conversely, if D and D' are linearly equivalent effective divisors, then $D - D' = (f)$ for some non-zero $f \in \mathbb{C}(C)$, by which both D and D' belong to the linear series defined by D' and $\langle 1, f \rangle$. Thus two effective divisors are linearly equivalent if and only if they both belong to the same linear series; in addition, this series may always be taken to be one-dimensional.

Remark 4.5.5 We have seen above (4.5.4) that two effective divisors D' and D'' are linearly equivalent if and only if they belong to the same linear series. Since two different curves of $\mathbb{P}_2$, of the same degree, both belong to the pencil they span, 4.4.27 yields that D' and D'' are linearly equivalent if and only if there exist curves C' and C'' of $\mathbb{P}_2$, both of the same degree and neither containing C, and an effective divisor T such that $D' + T = C' \cdot C$ and $D'' + T = C'' \cdot C$.

Example 4.5.6 If C is the cubic of Example 4.4.28, any two points of $X(C)$ – taken as degree one divisors – are linearly equivalent.

For any divisor D on C, we are interested in the set of all effective divisors linearly equivalent to D. Our first goal in this section is to prove that it is a linear series. To this end, we need:

Lemma 4.5.7 *If two linear series share an effective divisor, then they are both contained in a third linear series.*

PROOF: Assume the linear series to be $\mathcal{L}$ and $\mathcal{L}'$. If there is a $D \in \mathcal{L} \cap \mathcal{L}'$, then, by 4.4.14, D may be used to define both series, namely there are finite-dimensional vector subspaces $F, F' \subset \mathbb{C}(C)$ such that

$$\mathcal{L} = \{D + (f) \mid f \in F - \{0\}\}$$

and

$$\mathcal{L}' = \{D + (f) \mid f \in F' - \{0\}\}.$$

Then the linear series defined by D and $F + F'$ contains both $\mathcal{L}$ and $\mathcal{L}'$. ◇

Theorem 4.5.8 *The set $|D|$, of all effective divisors linearly equivalent to a given divisor D, is a linear series.*

Remark 4.5.9 The reader may note that for any divisor D' linearly equivalent to D, $|D'| = |D|$. The converse is true in the case when $|D| \neq \emptyset$, as then, for any $D'' \in |D| = |D'|$ we have $D \equiv D'' \equiv D'$. Two divisors D, D', with unequal degrees, both negative, have $|D| = |D'| = \emptyset$ and $D \not\equiv D'$, by 4.5.2.

PROOF OF 4.5.8: The case $|D| = \emptyset$ is clear. We thus assume $|D| \neq \emptyset$ in the sequel. Up to replacement of D with an element of $|D|$ we assume also that D is effective (4.4.14). We will construct embodied non-empty linear series with strictly increasing dimensions, all contained in $|D|$. Starting from the zero-dimensional linear series $\{D\}$ and proceeding inductively, assume that $\mathcal{L}$ is a linear series and $D \in \mathcal{L} \subset |D|$. If $\mathcal{L} = |D|$, then the claim is proved. Otherwise take $D' \in |D| - \mathcal{L}$: since $D, D' \in |D|$, they are linearly equivalent and so, by 4.5.4 both belong to a certain linear series $\mathcal{H}$. The series $\mathcal{L}$ and $\mathcal{H}$ share the divisor D and therefore, by 4.5.7, they are both contained in a linear series $\mathcal{L}'$. Since $D \in \mathcal{L}'$, by 4.5.4 any element of $\mathcal{L}'$ is linearly equivalent to D; this proves that $\mathcal{L}' \subset |D|$. In addition, since $D' \in \mathcal{L}' - \mathcal{L}$, $\dim \mathcal{L}' > \dim \mathcal{L}$ by 4.4.16. Now, linearly equivalent divisors having the same degree, any non-empty linear series contained in $|D|$ has its degree equal to $\deg D$ and therefore its dimension bounded by $\deg D$, by 4.4.21. This proves that the above construction needs to stop after finitely many steps by getting a linear series equal to $|D|$, and hence the claim. ◇

The linear series of the form $|D|$, for D a divisor on the curve C, are called *complete linear series*. We will refer to $|D|$ as the complete linear series *defined* or *determined* by D. The empty set appears as a complete linear series, as $\emptyset = |D|$ for any D with $\deg D < 0$, by 4.5.2. According to the definition, a non-empty linear series $\mathcal{L}$ is complete if and only if it contains all effective divisors linearly equivalent to an arbitrarily chosen $D \in \mathcal{L}$.

Remark 4.5.10 By 4.5.9, if a complete linear series $|D|$ is non-empty, then it determines D up to linear equivalence. This fails to be true if $|D| = \emptyset$.

Remark 4.5.11 If

$$\mathfrak{f} : C \longrightarrow C'$$

is a birational map, then it is clear from 4.4.24 and 4.4.25 that the isomorphism $X(\mathfrak{f})$ preserves linear equivalence and maps any complete linear series $|D|$ to the complete linear series $|\mathfrak{f}(D)|$.

The notion of a complete linear series is central to the intrinsic geometry on algebraic curves and calls for some further comments:

Leaving the case of an empty linear series aside, if $|D| \neq \emptyset$ we may replace D with any element of $|D|$ and assume without restriction that D is effective. Then $D' \in |D|$ if and only if $D' = D + (f)$ for some non-zero $f \in \mathbb{C}(C)$ for which $D + (f) \geq 0$. Since D is effective and $(f)_0$ and $(f)_\infty$ share no points, the latter condition is equivalent to having $(f)_\infty \leq D$, that is, f has as poles at most the points of D with their multiplicities: in such a case it is said that f has poles *allowed by* D. Determining the structure and dimension of the set of all rational functions on C with poles allowed by a given effective divisor is known as the *Riemann problem*. Theorem 4.5.8 above describes this set as the set of non-zero elements of a finite-dimensional subspace of $\mathbb{C}(C)$. Indeed, $|D|$ being a linear series and D an element of it, there is a finite-dimensional subspace $F \subset \mathbb{C}(C)$ such that

$$|D| = \{D + (f) \mid f \in F - \{0\}\}.$$

Then any $f \in F - \{0\}$ has $D + (f) \in |D|$ and therefore $D + (f) \geq 0$. Conversely, if $g \in \mathbb{C}(C)$ is $g \neq 0$ and $D + (g) \geq 0$, then, by the definition of $|D|$, $D + (g) \in |D|$ and there is an $f \in F - \{0\}$ for which $D + (g) = D + (f)$. It follows that $(g) = (f)$ and so, by 4.4.10, $g \in F$.

The subspace F above has been seen to be the set of all non-zero rational functions with poles allowed by D plus the zero function: it will be denoted by $\mathbf{L}(D)$ in the sequel, that is,

$$\mathbf{L}(D) = \{f \in \mathbb{C}(C) - \{0\} \mid D + (f) \geq 0\} \cup \{0\}. \tag{4.13}$$

It is easy to extend the above discussion to the case of a not necessarily effective divisor D, which includes the case $|D| = \emptyset$. If D is now an arbitrary divisor, still take $\mathbf{L}(D)$ defined by the equality (4.13) above. Then, the same arguments used in the case $D \geq 0$ prove that $\mathbf{L}(D)$ is the subspace that, together with D, defines $|D|$, and so, in particular, that $\mathbf{L}(D)$ is finite-dimensional. The only real difference in this case is the interpretation of the non-zero elements of $\mathbf{L}(D)$: if $D = \sum_p a_p p$, then $D + (f) \geq 0$ is equivalent to $o_p(f) \geq -a_p$ for any $p \in X(C)$; if $a_p > 0$, f is allowed to have a pole of order at most a_p at p, as before, while for $a_p < 0$, f is forced to have a zero of order at least $-a_p$ at p. The non-zero elements of $\mathbf{L}(D)$ are thus the functions with poles allowed by the positive part of D and, at least, the zeros prescribed by the negative part of D.

Using $\mathbf{L}(D)$ gives rise to the following more precise version of 4.5.8:

Corollary 4.5.12 *For any divisor D on an irreducible curve C, $\mathbf{L}(D)$, defined by (4.13) above, is a finite-dimensional vector space and*

$$|D| = \{D + (f) \mid f \in \mathbf{L}(D) - \{0\}\}.$$

A complete answer to Riemann's problem requires determining $\dim \mathbf{L}(D)$ or, equivalently, $\dim |D| = \dim \mathbf{L}(D) - 1$ (Brill and Noether's version). The Riemann–Roch theorem does this; it will be proved throughout the forthcoming sections in two steps: first we will give a lower bound for $\dim |D|$ which is its exact value in most cases (Riemann's inequality); then, after introducing the differentials and the canonical series on C, Riemann's inequality will be turned into an equality, which is Roch's contribution to the Riemann–Roch theorem. The rest of this section is devoted to proving a previous result which, in Brill and Noether's approach, is the key to Riemann's inequality: the Brill and Noether Restsatz.

Up to the end of this section we will assume that the curve C has only ordinary singularities. Then a curve A of $\mathbb{P}_2$ is said to be *adjoint* to C if and only if, for any $e > 1$, A has a point of multiplicity at least $e - 1$ at each – by hypothesis ordinary – singular point of multiplicity e of C. The curves adjoint to C are called *adjoint curves* – or just *adjoints* – to C. The conditions of having a point of multiplicity at least $e - 1$ at each e-fold point of C are called *adjunction conditions*. Of course the definition remains the same if the adjunction conditions are prescribed for all points of C and not only for the singular ones. A far more involved definition may be given for curves with arbitrary singularities, but we will not need it; the interested reader may see [3, Section 4.8].

Example 4.5.13 If C is smooth, then any curve is adjoint to C. For C not necessarily smooth, the curve C itself, as well as any curve containing C, are obvious – and rather useless – examples of adjoint curves. Far more interesting examples are given by 3.5.12, after which all polar curves of C are adjoint to C.

Remark 4.5.14 By 1.5.7 and 1.5.8, the adjoint curves to C of a given degree m describe a linear system $\mathfrak{A}_m$, for any positive integer m

$$\dim \mathfrak{A}_m \geq \frac{m^2 + 3m}{2} - \sum_{q \in C} \frac{e_q(C)(e_q(C) - 1)}{2} \tag{4.14}$$

and there exists an m_0 such that equality holds in (4.14) for any $m \geq m_0$.

The next lemma will provide a better handling of the adjoint curves:

Lemma 4.5.15 *A curve A is adjoint to C if and only if for each singular point q of C and each branch p of C with origin q, $[A.p] \geq e_q(C) - 1$.*

PROOF: Throughout the proof, dots $\cdots$ at the end of an expression will indicate terms of higher order. Fix a singular point q of C, take an affine chart containing q and, in it, affine coordinates x, y with origin q and the second axis non-tangent to C at q. Take $e = e_q(C)$. The singularity of C at q being ordinary, C has e different branches p_i, $i = 1, \ldots, e$, at q, which are smooth and have different tangent lines. Thus, each p_i has a Puiseux parameterization of the form $x = t$, $y = \alpha_i t + \cdots$, $i = 1, \ldots, e$ and $\alpha_i \neq \alpha_j$ if $i \neq j$ (2.4.6). Write $r = e_q(A)$; then an equation of the affine part of A has the form $h = h_r + \ldots$, where h_r is a non-zero homogeneous polynomial of degree r. If A is an adjoint, then $r \geq e-1$ and $[A \cdot p_i] = o_t(h_r(t, \alpha_i t) + \cdots) \geq e-1$ as claimed. Conversely, for any i it holds $o_t h(t, \alpha_i t + \cdots) \geq e - 1$. If $r < e - 1$, this yields $h_r(1, \alpha_i) = 0$ for all $i = 1, \ldots, e$. Since the α_i are all different, the polynomial $h_r(1, y)$ has e different roots, against the fact that it is non-zero and has degree $r < e - 1$. Hence $r \geq e - 1$, as claimed.   ◇

For any $p \in X(C)$, still denote by $\chi(p)$ the origin of p. The *adjunction divisor* $\Delta(C)$ of C is defined by the equality

$$\Delta(C) = \sum_{p \in X(C)} (e_{\chi(p)}(C) - 1)p.$$

It thus contains the branches of C with origin at a singular point, each with multiplicity one less than the multiplicity of its origin. The reader may easily check that

$$\deg \Delta(C) = \sum_{q \in C} e_q(C)(e_q(C) - 1).$$

Using the adjunction divisor, Lemma 4.5.15 may be rewritten as follows:

Lemma 4.5.16 *A curve A of $\mathbb{P}_2$ is adjoint to C if and only if it contains C or, else, $A \cdot C \geq \Delta(C)$.*

Brill and Noether's original formulation of the Restsatz contains in fact two results. One, birationally invariant and quite easy, is as follows:

Lemma 4.5.17 *For any two divisors D, T on an irreducible curve C of $\mathbb{P}_2$, T effective, $|D| - T = |D - T|$. In particular, the residual series of any complete linear series with respect to a given effective divisor T is complete, and depends only on the linear equivalence class of T.*

PROOF: By the definition of residual series, $D' \in |D| - T$ if and only if $D' + T \geq T$ and $D' + T \equiv D$. By 4.5.1 this is equivalent to $D' \geq 0$ and $D' \equiv D - T$, which in turn is satisfied if and only if $D' \in |D - T|$, by the definition of $|D - T|$.   ◇

This direct consequence of 4.5.17 will be useful later on:

Lemma 4.5.18 *Assume that $\mathcal{L}$ and $\mathcal{L}'$ are complete linear series on C. Then the following conditions are equivalent:*

(i)   *There is a group of $\mathcal{L}$ contained in a group of $\mathcal{L}'$.*

(ii)   *$\mathcal{L} \neq \emptyset$ and there is a divisor $T \geq 0$ such that $\mathcal{L} = \mathcal{L}' - T$.*

(iii) $\mathcal{L} \neq \emptyset$ *and each group of* $\mathcal{L}$ *is contained in a group of* $\mathcal{L}'$, *that is,* $\mathcal{L}$ *is non-empty and partially contained in* $\mathcal{L}'$.

PROOF: If there is a $D \in \mathcal{L}$ then, obviously $\mathcal{L} \neq \emptyset$. If there is a $D' \in \mathcal{L}'$ with $D \leq D'$, then $D' = D + T$ with $T \geq 0$ and, by 4.5.17,

$$\mathcal{L} = |D| = |D' - T| = |D'| - T = \mathcal{L}' - T.$$

This proves (i) $\Rightarrow$ (ii). Both (ii) $\Rightarrow$ (iii) and (iii) $\Rightarrow$ (i) are clear.   ◇

The other – harder – part of the Restsatz is not birationally invariant and follows from Noether's fundamental theorem 3.8.14. It ensures that the residual linear series, with respect to the adjunction divisor, of the linear series traced on C by the adjoints of any fixed degree, is complete:

Lemma 4.5.19 *For any integer* $m > 0$, *the set* $\mathcal{A}_m$ *of all divisors* $A \cdot C - \Delta(C)$, *for* A *an adjoint to* C *of degree* m *not containing* C, *is a complete linear series.*

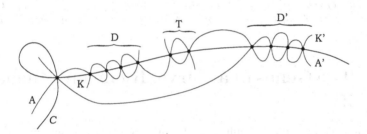

Figure 4.2: Curves and divisors in the proof of 4.5.19.

PROOF: Assume that C has homogeneous equation $F = 0$ and degree d. That $\mathcal{A}_m$ is a linear series follows from 4.5.14, 4.4.30 and 4.5.16. The case $\mathcal{A}_m = \emptyset$ being clear, assume that D' is an effective divisor linearly equivalent to $D \in \mathcal{A}_m$. Then, on one hand, there is an adjoint $A : G = 0$ to C, $A \not\supset C$, $\deg A = m$, such that $A.C = D + \Delta(C)$. On the other, by the linear equivalence of D and D' and 4.5.5, there are curves $K : H = 0$ and $K' : H' = 0$, of the same degree k, neither containing C, and an effective divisor T such that

$$D + T = K \cdot C \quad \text{and} \quad D' + T = K' \cdot C.$$

The divisor cut out on C by the composed curve $A + K'$ is then

$$(A + K') \cdot C = \Delta(C) + D + D' + T = \Delta(C) + D' + K \cdot C.$$

The last equality ensures that the curves C, K and $A + K'$ satisfy the hypothesis of Noether's fundamental theorem 3.8.14 (conditions (b)). There are thus homogeneous polynomials P and Q, of respective degrees $m + k - d$ and m, such that

$$GH' = PF + QH.$$

Call A' the curve $A' : Q = 0$; it has degree m. The difference $GH' - QH$ being a multiple of the equation F of C, by 4.4.5

$$A' \cdot C + K \cdot C = (A' + K) \cdot C = (A + K') \cdot C = \Delta(C) + D' + K \cdot C$$

and it follows that

$$A' \cdot C = \Delta(C) + D'.$$

By 4.5.16, the last equality proves that A' is an adjoint to C of degree m, and, this proved, the same equality shows that $D' \in \mathcal{A}_m$, as wanted.   ◇

By 4.5.17 any residual linear series of the complete linear series $\mathcal{A}_m$ of 4.5.19 with respect to an arbitrary effective divisor is complete too. Hence, we have proved:

Theorem 4.5.20 (Brill–Noether Restsatz, 1874) *For any integer $m > 0$ and any effective divisor T, the effective divisors $A \cdot C - \Delta(C) - T$, for A an adjoint to C of degree m not containing C, describe a complete linear series.*

4.6   The Genus of a Curve, Riemann's Inequality

Riemann's inequality will result from the Brill–Noether Restsatz by just computing dimensions of linear series residual of linear series traced by adjoints. Still assume that C is an irreducible curve of $\mathbb{P}_2$ with only ordinary singularities, and take $d = \deg C$ and $\delta = \sum_{q \in C} e_q(C)(e_q(C) - 1)/2 = \frac{1}{2} \deg \Delta(C)$. First of all we have:

Lemma 4.6.1 *Assume $m \geq d - 2$. Then the linear series $\mathcal{A}_m$, residual with respect to the adjunction divisor of the linear series traced on C by the adjoints to C of degree m, satisfies*

$$\dim \mathcal{A}_m \geq dm - \frac{d^2 - 3d + 2}{2} - \delta,$$

and there is an integer $\bar{m}$ such that equality holds for $m \geq \bar{m}$.

PROOF: As already noticed in 4.5.14, the dimension of the linear system $\mathfrak{A}_m$, of the curves of degree m adjoint to C, is at least $(m^2 + 3m)/2 - \delta$, and equality holds for m high enough. Since the adjunction conditions are obviously satisfied by the curve C, and hence for any curve containing C, the adjoints to C of degree m, $m \geq d$, containing C are the curves composed of C and an arbitrary curve of degree $m - d$. They thus describe a linear system of dimension $((m - d)^2 + 3(m - d))/2$. Since this expression takes value -1

for $m = d - 1, d - 2$, it still gives the correct dimension of the – empty – linear systems of the adjoint curves of degrees $d - 1$ and $d - 2$ containing C. Therefore (4.4.30) the linear series traced on C by the adjoints to C of degree m, $m \geq d - 2$, has dimension at least

$$\frac{m^2 + 3m}{2} - \delta - \frac{(m - d)^2 + 3(m - d)}{2} - 1 = dm - \frac{d^2 - 3d + 2}{2} - \delta,$$

and the equality holds for m high enough. By 4.4.19, the same holds for the linear series $\mathcal{A}_m$ residual of the above with respect to the adjunction divisor $\Delta(C)$, because all groups cut out on C by adjoint curves contain $\Delta(C)$ (4.5.16). ◇

We will use 4.6.1 to prove a corollary of the Restsatz. It provides a rather explicit construction of the complete linear series $|D|$ determined by an effective divisor D and is needed to prove Riemann's inequality.

Corollary 4.6.2 (Construction of complete linear series) *If C is an irreducible curve of $\mathbb{P}_2$ with only ordinary singularities, then:*

(a) *For any effective divisor D on C, there is a positive integer m_0 such that for any $m \geq m_0$, there is an adjoint curve A_m to C, of degree m, $A_m \not\supset C$, for which $A_m \cdot C \geq D + \Delta(C)$.*

(b) *Let D be any effective divisor on C. Assume that A_m is – as in (a) – an adjoint to C of degree m that does not contain C and has $A_m \cdot C \geq D + \Delta(C)$. Then $T = A_m \cdot C - D - \Delta(C)$ is effective and the residual series, with respect to $T + \Delta(C)$, of the linear series traced on C by the adjoints to C of degree m is $|D|$.*

PROOF: Lemma 4.6.1 shows that $\dim \mathcal{A}_m \geq \deg D$ provided

$$m \geq \frac{d^2 - 3d + 2}{2d} + \frac{\delta}{d} + \frac{\deg D}{d}.$$

If this inequality is satisfied, then, by 4.4.22, $\dim(\mathcal{A}_m - D) \geq 0$; hence $\mathcal{A}_m - D$ contains at least one effective divisor and so there is an adjoint A_m, of degree m, for which $A_m \cdot C$ is defined and satisfies $A_m \cdot C \geq D + \Delta(C)$. This proves claim (a).

Claim (b) is a direct consequence of the Restsatz: Taking $T = A_m \cdot C - D - \Delta(C)$, the residual linear series of $\mathcal{A}_m$ with respect to T contains D because $A_m \cdot C = D + \Delta(C) + T$, and is complete by 4.5.20. ◇

Assume that an effective divisor D on C has been given and, by 4.6.2(a), take an adjoint A_m to C, of degree $m \geq d - 2$ and not containing C, for which $T = A_m \cdot C - D - \Delta(C)$ is effective. Then, on one hand $\deg T = md - \deg D - \deg \Delta(C)$. On the other, by 4.6.2(b), $|D|$ is the residual series of $\mathcal{A}_m$ with respect to T and therefore, by 4.4.22 and 4.6.1, its dimension satisfies

$$\dim |D| \geq \dim \mathcal{A}_m - \deg T$$

$$\geq dm - \frac{d^2 - 3d + 2}{2} - \delta - md + \deg D + \deg \Delta(C)$$

$$= \deg D - \frac{d^2 - 3d + 2}{2} + \delta, \quad (4.15)$$

because $\deg \Delta(C) = 2\delta$. Furthermore the equality holds if, for m high enough, we take $D = \mathcal{A}_m \cdot C - \Delta(C)$ for an adjoint $\mathcal{A}_m$ of degree m, as then $T = 0$. This gives an inequality which is in fact Riemann's in a not yet birationally invariant form:

Proposition 4.6.3 *If C is an irreducible curve of $\mathbb{P}_2$ of degree d, with only ordinary singularities, and $|D|$ is a complete linear series on C, then*

$$\dim |D| \geq \deg D - g(C), \quad (4.16)$$

where

$$g(C) = \frac{(d-1)(d-2)}{2} - \sum_{q \in C} \frac{e_q(C)(e_q(C) - 1)}{2}.$$

Furthermore, (4.16) is an equality for at least one complete linear series $|D|$.

PROOF: If the complete linear series $|D|$ is non-empty, then D may be assumed to be effective and the claimed equality is the already proved (4.15), which we know to be an equality for a certain $|D|$. If $|D| = \emptyset$ we have to prove $-1 \geq \deg D - g(C)$. Assume otherwise, that is, to have a divisor D with $0 \leq \deg D - g(C)$ and decompose D into its positive and negative parts, $D = D_+ - D_-$. Then $\deg D_+ \geq \deg D_- + g(C)$ and, D_+ being effective, we may use the inequality of the claim to get

$$\dim |D_+| \geq \deg D_+ - g(C) \geq \deg D_-.$$

Using 4.5.17 and 4.4.22, this in turn gives

$$\dim |D| = \dim |D_+ - D_-| = \dim(|D_+| - D_-) \geq \dim |D_+| - \deg D_- \geq 0,$$

against the hypothesis $|D| = \emptyset$. ◇

Theorem 4.6.4 (Riemann's inequality; Riemann, 1857) *If C is any irreducible curve of $\mathbb{P}_2$, then there is a non-negative integer $g(C)$ for which*

(a)

$$\dim |D| \geq \deg D - g(C)$$

for any complete linear series $|D|$ on C, and

(b)

$$\dim |D| = \deg D - g(C)$$

for at least one complete linear series $|D|$ on C.

PROOF: By 4.4.24 and 4.5.11, if the claim holds for a curve C' birationally equivalent to C, then it holds for C too. By 4.2.15, C' may be taken with only ordinary singularities, and for such a C' the claim has been proved in 4.6.3: it suffices to take

$$g(C) = \frac{(d'-1)(d'-2)}{2} - \sum_{q \in C'} \frac{e_q(C')(e_q(C')-1)}{2},$$

where $d' = \deg C'$, and then notice that the inequality $g(C) \geq 0$ follows from 3.5.23. ◇

Remark 4.6.5 The integer $g(C)$ of 4.6.4 is called the *genus* of C, and also, sometimes, the *geometric genus* of C. In 4.6.4 the genus of C is characterized as being the maximal value of the differences $\deg D - \dim |D|$ for all complete linear series $|D|$ on C, which in particular proves that it is a birational invariant (see 4.4.24 and 4.5.11). Birationally equivalent curves thus have the same genus, a property often referred to as the *invariance of the genus*. The genus is the most important numerical birational invariant of an irreducible algebraic curve, but not the only one, see for instance Exercise 4.31.

While proving 4.6.4, we have obtained an explicit formula for $g(C)$; it is called the *genus formula*:

Corollary 4.6.6 (of the proof of 4.6.4) *If C is an irreducible curve of degree d of $\mathbb{P}_2$ and all singularities of C are ordinary, then*

$$g(C) = \frac{(d-1)(d-2)}{2} - \sum_{q \in C} \frac{e_q(C)(e_q(C)-1)}{2}. \tag{4.17}$$

The genus formula of 4.6.6 cannot be applied, in general, to curves having non-ordinary singularities, see Exercise 4.12. Actually, the right-hand side of (4.17) already appeared in (3.5.24) under the name deficiency, and the proof of 4.2.14 shows that it may be modified by the action of standard quadratic transformations, so it is not a birational invariant. A correct formula for curves with arbitrary singularities requires the summation in (4.17) to be extended not only to the (singular) points of C, but also to the singular points they give rise to when blown up in the procedure of reduction of singularities of Section 4.2 (*infinitely near singularities*). The reader may see [16] or [3, Section 3.11] and also Exercise 4.12(4).

4.7 Differentials on a Curve

Denote by C an irreducible curve of $\mathbb{P}_2$. In order to obtain the correction term that will turn Riemann's inequality into an equality, we need to introduce new objects intrinsically related to C: these are the *differentials on C* and a linear series they give rise to, called the *canonical series* of C.

The differentials on C will be introduced by an algebraic procedure that uses only the field $\mathbb{C}(C)$, as an extension of $\mathbb{C}$. We will call a *derivation* of $\mathbb{C}(C)$ any map from $\mathbb{C}(C)$ into a vector space N over $\mathbb{C}(C)$,

$$\rho : \mathbb{C}(C) \longrightarrow N,$$

which, for any $a \in \mathbb{C}$ and any $g, g' \in \mathbb{C}(C)$, satisfies

(a) $\rho(a) = 0$,

(b) $\rho(g + g') = \rho(g) + \rho(g')$ and

(c) $\rho(gg') = g'\rho(g) + g\rho(g')$.

From the above it easily follows that any derivation ρ also satisfies the usual rule for the derivative of a quotient, namely

(d) $\rho(g/g) = (g'\rho(g) - g\rho(g'))/(g')^2$ for any $g, g' \in \mathbb{C}(C)$, $g' \neq 0$.

The zero map from $\mathbb{C}(C)$ to an arbitrary $\mathbb{C}(C)$-vector space is an obvious example of a derivation. In the sequel we will of course deal with more interesting examples.

A homomorphism from a derivation $\rho : \mathbb{C}(C) \to N$ to a derivation $\rho' : \mathbb{C}(C) \to N'$ is any linear map $\varphi : N \to N'$ satisfying $\rho' = \varphi \circ \rho$. It is clear that the identical maps and the compositions of homomorphisms of derivations are homomorphisms of derivations. The homomorphism φ above is called an isomorphism if and only if it is an isomorphism of vector spaces. The reader may easily check that, according to the categorial meaning of the word isomorphism, φ is an isomorphism if and only if it has an inverse, namely a homomorphism of derivations from ρ' to ρ that composed with φ in both senses gives the corresponding identity maps.

The relevant fact is that there exists a universal derivation which, as such, is uniquely determined by $\mathbb{C}(C)$ up to isomorphism. The next proposition gives a precise claim; its proof follows the standard algebraic pattern used in many other similar cases.

Proposition 4.7.1 *Assume that C is an irreducible curve of $\mathbb{P}_2$ and $\mathbb{C}(C)$ its field of rational functions. Then:*

(a) *There is a derivation*

$$\boldsymbol{d} : \mathbb{C}(C) \longrightarrow \Omega(C)$$

such that for any derivation

$$\rho : \mathbb{C}(C) \longrightarrow N$$

there is a unique linear map

$$\bar{\rho} : \Omega(C) \longrightarrow N$$

such that $\rho = \bar{\rho} \circ \boldsymbol{d}$.

(b) *The property above determines $\boldsymbol{d}$ up to isomorphism, namely if a derivation*

$$\boldsymbol{d'} : \mathbb{C}(C) \longrightarrow \Omega'(C)$$

satisfies the same property, then there is an isomorphism of vector spaces $\phi : \Omega(C) \to \Omega'(C)$ that satisfies $\boldsymbol{d'} = \phi \circ \boldsymbol{d}$.

PROOF: **Existence:** In order to have a copy of $\mathbb{C}(C)$ with no algebraic structure, consider a symbol $[g]$ for each $g \in \mathbb{C}(C)$, different functions giving different symbols. Let $\mathbf{V}$ be the vector space freely generated over $\mathbb{C}(C)$ by all these symbols: the elements of $\mathbf{V}$ are thus formal linear combinations of symbols

$$\sum_{g \in \mathbb{C}(C)} c_g [g]$$

with $c_g \in \mathbb{C}(C)$, $c_g = 0$ for all but finitely many g, and $\sum_g c_g[g] = \sum_g c'_g[g]$ if and only if $c_g = c'_g$ for all $g \in \mathbb{C}(C)$.

Take the subspace H of $\mathbf{V}$ generated by all the elements of any of the forms

(1) $[a]$, for $a \in \mathbb{C}$,

(2) $[g + g'] - [g] - [g']$, for $g, g' \in \mathbb{C}(C)$,

(3) $[gg'] - g'[g] - g[g']$, for $g, g' \in \mathbb{C}(C)$,

and $\Omega(C) = \mathbf{V}/H$. Then, due to the definition of H, mapping each $f \in \mathbb{C}(C)$ to the class of $[g]$ modulo H defines a derivation $\boldsymbol{d} : \mathbb{C}(C) \to \Omega(C)$. Next we will check that $\boldsymbol{d}$ thus defined satisfies the claim. Assume that $\rho : \mathbb{C}(C) \to N$ is any derivation of $\mathbb{C}(C)$. The vector space $\mathbf{V}$ being freely generated by the symbols $[g]$, $g \in \mathbb{C}(C)$, ρ induces the linear map

$$\widetilde{\rho} : \mathbf{V} \longrightarrow N$$

$$\sum_g c_g[g] \longmapsto \sum_g c_g \rho(g).$$

Furthermore, ρ being a derivation, $\widetilde{\rho}$ maps each of the above generators of H to 0 and hence $\widetilde{\rho}(H) = \{0\}$. We may thus take $\bar{\rho}$ to be the linear map induced by $\widetilde{\rho}$, from the quotient $\Omega(C) = \mathbf{V}/H$ to N; from this, the equality $\rho = \bar{\rho} \circ \boldsymbol{d}$ is clear. Since the symbols $[g]$, $g \in \mathbb{C}(C)$, generate V, their classes modulo H, $\boldsymbol{d}(g)$, $g \in \mathbb{C}(C)$, generate $\Omega(C)$. Therefore, the equality $\rho = \bar{\rho} \circ \boldsymbol{d}$ uniquely determines $\bar{\rho}$, because it determines the images under $\bar{\rho}$ of all $\boldsymbol{d}(g)$, $g \in \mathbb{C}(C)$.

Uniqueness: Note first that if $\boldsymbol{d} : \mathbb{C}(C) \longrightarrow \Omega(C)$ is a derivation and satisfies the condition of claim (a), then $\boldsymbol{d}$ itself has the obvious factorization $\boldsymbol{d} = \mathrm{Id}_{\Omega(C)} \circ \boldsymbol{d}$. Therefore, by the claimed uniqueness of such a factorization in the particular case $\rho = \boldsymbol{d}$, the only linear map $\varphi : \Omega(C) \to \Omega(C)$ satisfying $\boldsymbol{d} = \varphi \circ \boldsymbol{d}$ is $\varphi = \mathrm{Id}_{\Omega(C)}$.

Assume now we have two derivations $\boldsymbol{d} : \mathbb{C}(C) \to \Omega(C)$ and $\boldsymbol{d'} : \mathbb{C}(C) \to \Omega'(C)$, both satisfying the property of claim (a). By this property applied to $\boldsymbol{d}$, there is a $\bar{\boldsymbol{d}}' : \Omega(C) \to \Omega'(C)$ such that $\boldsymbol{d}' = \bar{\boldsymbol{d}}' \circ \boldsymbol{d}$. Swapping over the roles of the derivations, there is a $\bar{\boldsymbol{d}} : \Omega'(C) \to \Omega(C)$ such that $\boldsymbol{d} = \bar{\boldsymbol{d}} \circ \boldsymbol{d}'$. It follows that

$$\boldsymbol{d}' = \bar{\boldsymbol{d}}' \circ \bar{\boldsymbol{d}} \circ \boldsymbol{d}' \quad \text{and} \quad \boldsymbol{d} = \bar{\boldsymbol{d}} \circ \bar{\boldsymbol{d}}' \circ \boldsymbol{d}$$

which, as noticed before, yields

$$\operatorname{Id}_{\Omega'(C)} = \bar{\boldsymbol{d}}' \circ \bar{\boldsymbol{d}} \quad \text{and} \quad \operatorname{Id}_{\Omega(C)} = \bar{\boldsymbol{d}} \circ \bar{\boldsymbol{d}}'.$$

As a consequence, $\bar{\boldsymbol{d}}$ and $\bar{\boldsymbol{d}}'$ are reciprocal isomorphisms of derivations, as wanted.   ◇

The elements of $\Omega(C)$ are called the *differentials* – or the *differential forms* – on C. In the sequel we will write $\boldsymbol{d}(g) = dg$, for any $g \in \mathbb{C}(C)$, and call it the *differential* of g. The differentials of the form dg for $g \in \mathbb{C}(C)$ are called, as usual, *exact differentials*.

For future reference we state a fact that has been obtained while proving 4.7.1:

Corollary 4.7.2 (of the proof of 4.7.1) *The exact differentials dg, $g \in \mathbb{C}(C)$, generate $\Omega(C)$.*

If $\mathfrak{f} : C \to C'$ is a non-constant rational map, take $\Omega(C)$ as a $\mathbb{C}(C')$-vector space using the product $g'\eta = \mathfrak{f}^*(g')\eta$ for $g' \in \mathbb{C}(C')$ and $\eta \in \Omega(C)$. Then the composition of the monomorphism $\mathfrak{f}^*$ with the universal derivation $\boldsymbol{d}$ of $\mathbb{C}(C)$ is a derivation of $\mathbb{C}(C')$ which factors according to 4.7.1, giving a commutative diagram

$$
\begin{array}{ccc}
\mathbb{C}(C) & \xleftarrow{\;\mathfrak{f}^*\;} & \mathbb{C}(C') \\
{\scriptstyle \boldsymbol{d}} \downarrow & & \downarrow {\scriptstyle \boldsymbol{d}'} \\
\Omega(C) & \xleftarrow{\;\overline{\boldsymbol{d} \circ \mathfrak{f}^*}\;} & \Omega(C').
\end{array}
$$

The induced $\mathbb{C}(C')$-linear map $\overline{\boldsymbol{d} \circ \mathfrak{f}^*}$ is called the *pull-back of differentials* (by $\mathfrak{f}$). It will be denoted using the same notation as for the pull-back of rational functions, namely $\mathfrak{f}^* = \overline{\boldsymbol{d} \circ \mathfrak{f}^*}$, this causing no confusion. For any form $\omega \in \Omega(C')$, $\mathfrak{f}^*(\omega)$ is called the *pull-back* or the *inverse image* of ω. Due to the commutativity of the diagram above, we have $\mathfrak{f}^*(dg) = d(\mathfrak{f}^*(g))$ for any $g \in \mathbb{C}(C')$. Since the exact differentials generate the spaces of differentials (4.7.2), this fact easily gives:

Proposition 4.7.3 *If $\mathfrak{f} : C \to C'$ and $\mathfrak{g} : C' \to C''$ are non-constant rational maps between irreducible curves of $\mathbb{P}_2$, then:*

(a) $(\operatorname{Id}_C)^* = \operatorname{Id}_{\Omega(C)}$.

(b) $(\mathfrak{f} \circ \mathfrak{g})^* = \mathfrak{g}^* \circ \mathfrak{f}^*$.

(c) *If $\mathfrak{f}$ is birational, then $\mathfrak{f}^*$ is bijective and has inverse $(\mathfrak{f}^{-1})^*$.*

In all cases, the star * denotes pull-back of differentials.

Our first goal in this section is to prove that $\Omega(C)$ is a one-dimensional vector space (over $\mathbb{C}(C)$), but this will take a while. For the moment we will prove two lemmas, namely:

Lemma 4.7.4 *If x, y are affine coordinates on an arbitrary affine chart of $\mathbb{P}_2$ containing at least one point of C, and $\bar{x}, \bar{y}$ are their restrictions to C, then $d\bar{x}, d\bar{y}$ generate $\Omega(C)$.*

PROOF By 4.7.2, it is enough to see that any exact differential dg, $g \in \mathbb{C}(C)$ is a linear combination of $d\bar{x}$ and $d\bar{y}$. According to 3.7.12, any $g \in \mathbb{C}(C)$ may be written $g = P(\bar{x}, \bar{y})/Q(\bar{x}, \bar{y})$ where P and Q are complex polynomials in two variables and $Q(\bar{x}, \bar{y}) \neq 0$. Then, $\boldsymbol{d}$ being a derivation, rule (d) at the beginning of this section shows that dg is a linear combination of $dP(\bar{x}, \bar{y})$ and $dQ(\bar{x}, \bar{y})$. In turn, rules (a), (b) and (c) easily give

$$dP(\bar{x}, \bar{y}) = (\partial_1 P)(\bar{x}, \bar{y})d\bar{x} + (\partial_2 P)(\bar{x}, \bar{y})d\bar{y},$$

where $\partial_1 P$ and $\partial_2 P$ are the usual derivatives of the polynomial P. A similar equality holding for $dQ(\bar{x}, \bar{y})$, the proof is complete.   ◇

Lemma 4.7.5 $\dim \Omega(C) \leq 1$.

PROOF: Take an affine chart, coordinates x, y and its restrictions $\bar{x}, \bar{y}$ as in 4.7.4, and let $f \in \mathbb{C}[x, y]$ be an equation of the affine part of C. Then in $\mathbb{C}(C)$ we have $f(\bar{x}, \bar{y}) = 0$. Computing as in the proof of 4.7.4,

$$0 = df(\bar{x}, \bar{y}) = (\partial_1 f)(\bar{x}, \bar{y})d\bar{x} + (\partial_2 f)(\bar{x}, \bar{y})d\bar{y}.$$

The above is a non-trivial linear dependence relation between the generators $d\bar{x}$ and $d\bar{y}$ of $\Omega(C)$. Indeed, f having positive degree, at least one of its derivatives $\partial_i f$ is not zero, after which it is not a multiple of f due to its degree. Therefore $(\partial_i f)(\bar{x}, \bar{y}) \neq 0$ (by 3.7.10).   ◇

Given $p \in X(C)$, a rational function $u \in \mathbb{C}(C)$ is called a *uniformizing parameter* of C at p if and only if $o_p(u) = 1$.

Remark 4.7.6 Due to 4.3.17, if $\mathfrak{f}: C \to C'$ is a birational map, then $\mathfrak{f}^*(u)$, $u \in \mathbb{C}(C')$, is a uniformizing parameter of C at $p \in X(C)$ if and only if u is a uniformizing parameter of C' at $\mathfrak{f}(p)$.

Lemma 4.7.7 *For any $p \in X(C)$ there is a uniformizing parameter t of C at p.*

PROOF: By 4.3.14 and 4.7.6, it is non-restrictive to assume that the origin q of p is a non-singular point of C. Then take t to be the restriction of L_0/L_1 where L_0 and L_1 are homogeneous equations of, respectively, a line ℓ_0 going through the origin of p and non-tangent to C there, and a line ℓ_1 missing the origin of p: t is a uniformizing parameter at p due to 4.4.6(b).   ◇

Remark 4.7.8 Given $p_1, \ldots, p_m \in X(C)$, all different from p, the uniformizing parameter t of 4.7.7 may be chosen to have $t(p_j) \neq 0, \infty$ for $j = 1, \ldots, m$. Indeed, proceeding as in the proof of 4.7.7, neither of the origins of $p_1, \ldots, p_m$ is the origin of p, because the origin of p is a smooth point of C. Then, just take both ℓ_0 and ℓ_1 missing the origins of $p_1, \ldots, p_m$.

Let as before C be an irreducible curve of $\mathbb{P}_2$ and $p \in X(C)$. Next we will present a construction – referred to in the sequel as *local representation at p* – that relates the universal derivation $\boldsymbol{d}$ to the standard derivation of series in a single variable. This construction appears as dependent on many choices besides those of C and p. Actually, modifying these choices produces a quite similar situation, related to the initial one by isomorphisms, but we do not need to care about that, as the local representation will only be used to prove three results – 4.7.10, 4.7.12 and 4.7.15 below – whose claims are clearly independent of the choices made.

Assume we have fixed $p \in X(C)$. As allowed by 4.2.15, choose a birational map $\mathfrak{f} : C \to C'$, where C' is an irreducible curve of $\mathbb{P}_2$ with no singularities other than ordinary singular points, and take $p' = \mathfrak{f}(p)$: the singularities being ordinary, p' is a smooth branch of C'. We choose an affine chart $\mathbb{A}_2$ of $\mathbb{P}_2$ containing the origin q' of p'. Take on $\mathbb{A}_2$ affine coordinates x, y with origin q' and the second axis $\ell_y : x = 0$ not tangent to p'. Let $f \in \mathbb{C}[x, y]$ be an equation of the affine part of C'. The branch p' being smooth and the axis ℓ_y non-tangent to p', $o_p x = [\ell \cdot p'] = 1$. The restriction $\bar{x}$ is thus a uniformizing parameter of C' at p' and therefore (4.7.6) $z = \mathfrak{f}^*(\bar{x})$ is a uniformizing parameter of C at p.

Since p', as a branch, is smooth, it has a single Puiseux series $s \in \mathbb{C}\{x\}$ relative to the coordinates x, y. If a non-zero polynomial $P \in \mathbb{C}[x, y]$ satisfies $P(x, s) = 0$, then C' and the projective closure T of $P = 0$ share infinitely many points $(x, s(x))$, by which C' is an irreducible component of T and P is a multiple of f. The converse is obviously true, because $f(x, s) = 0$. It follows that the kernel of the homomorphism of $\mathbb{C}$-algebras

$$\mathbb{C}[x, y] \longrightarrow \mathbb{C}((x))$$
$$Q(x, y) \longmapsto Q(x, s(x))$$

is $f\mathbb{C}[x, y]$. The homomorphism thus extends (3.6.2) to

$$\mathcal{R}_{C'} = \mathbb{C}[x, y]_{(f)} \longrightarrow \mathbb{C}((x))$$
$$\frac{Q(x, y)}{Q'(x, y)} \longmapsto \frac{Q(x, s(x))}{Q'(x, s(x))}$$

which in turn induces a monomorphism of fields,

$$\varphi' : \mathbb{C}(C') = \mathbb{C}[x, y]_{(f)} / f\mathbb{C}[x, y]_{(f)} \longrightarrow \mathbb{C}((x)),$$

through which the image of any rational function g' on C' is the result of replacing y with s in any of its representatives.

We will take $\varphi = \varphi' \circ (\mathfrak{f}^*)^{-1} : \mathbb{C}(C) \to \mathbb{C}((x))$: it is also a monomorphism of fields which, like φ' and $(\mathfrak{f}^*)^{-1}$, leaves all the elements of $\mathbb{C}$ invariant and therefore is a monomorphism of $\mathbb{C}$-algebras. Note also that it maps z to x. The monomorphism φ is a quite interesting one: it associates to each rational function g on C a sort of local – at p – representation of g as a Laurent series; further, it preserves orders, namely:

Lemma 4.7.9 *For any* $g \in \mathbb{C}(C)$,

$$o_p(g) = o_x \varphi(g).$$

PROOF: Take $g' = (\mathfrak{f}^*)^{-1}(g)$. Then, by 4.3.17, $o_p(g) = o_{p'}(g')$ and it suffices to see that $o_{p'}(g') = o_x \varphi'(g)$. This equality in turn is clear from the definition of $o_{p'}(g')$ because $(x, s(x))$ is a Puiseux parameterization of p' and therefore the series $\varphi'(g)$ defines the germ of analytic map g_p of 4.3.6. ◇

From now on, the field $\mathbb{C}((x))$ will be taken with the $\mathbb{C}(C)$-vector space structure induced by φ; the product of $g \in \mathbb{C}(C)$ and $S \in \mathbb{C}((x))$ is thus taken to be $gS = \varphi(g)S$.

Denote by $\partial_x : \mathbb{C}((x)) \to \mathbb{C}((x))$ the usual derivation of Laurent series. It is straightforward to check that the composition $\partial_x \circ \varphi$ is a derivation of $\mathbb{C}(C)$. By 4.7.1, $\partial_x \circ \varphi$ factors through $\boldsymbol{d}$ giving rise to a commutative diagram

$$\begin{array}{ccc} \mathbb{C}(C) & \xrightarrow{\varphi} & \mathbb{C}((x)) \\ \boldsymbol{d} \downarrow & & \downarrow \partial_x \\ \Omega(C) & \xrightarrow{\psi} & \mathbb{C}((x)) \end{array} \qquad (4.18)$$

where ψ is a $\mathbb{C}(C)$-linear map and, by the commutativity,

$$\psi(dz) = \psi(\boldsymbol{d}(z)) = \partial_x(\varphi(z)) = \partial_x(x) = 1.$$

This gives the first result we are looking for, namely:

Proposition 4.7.10 *For any irreducible curve* C *of* $\mathbb{P}_2$, $\dim \Omega(C) = 1$.

PROOF: We know from 4.7.5 that $\dim \Omega(C) \leq 1$, and the equality above shows that $dz \neq 0$. ◇

Once we know that $\dim \Omega(C) = 1$, we introduce a notation: if $g, u \in \mathbb{C}(C)$ and $du \neq 0$, then $dg = hdu$ for some $h \in \mathbb{C}(C)$. The coefficient h being determined by g and u, we will write $h = dg/du$, this giving rise to the familiar equality $dg = (dg/du)du$.

Back to our construction, we have seen in the proof of 4.7.10 above that $dz \neq 0$; hence, it makes sense to consider dg/dz for any $g \in \mathbb{C}(C)$. The following lemma relates the formal derivatives dg/dz and the usual ones $\partial_x(\varphi(g))$:

Lemma 4.7.11 *For any* $g \in \mathbb{C}(C)$, $\varphi(dg/dz) = \partial_x(\varphi(g))$.

PROOF: Computing the image of g both ways on the diagram (4.18) and using that ψ is $\mathbb{C}(C)$-linear,

$$\partial_x(\varphi(g)) = \psi(dg) = \psi\left(\frac{dg}{dz}dz\right) = \varphi\left(\frac{dg}{dz}\right)\psi(dz)$$

$$= \varphi\left(\frac{dg}{dz}\right)\partial_x(\varphi(z)) = \varphi\left(\frac{dg}{dz}\right)\partial_x(x) = \varphi\left(\frac{dg}{dz}\right).$$

$\diamond$

The second result we want is:

Proposition 4.7.12 *If u is any uniformizing parameter of a curve C of $\mathbb{P}_2$ at a point $p \in X(C)$, then $du \neq 0$. If u, v are uniformizing parameters of C at p, then $o_p(du/dv) = 0$.*

PROOF: If u is a uniformizing parameter, $1 = o_pu = o_x\varphi(u)$ by 4.7.9. Then, by 4.7.11,

$$o_p(du/dz) = o_x\partial_x\varphi(u) = o_x\varphi(u) - 1 = 0. \tag{4.19}$$

This proves that $du = (du/dz)dz \neq 0$, as we already know that $dz \neq 0$. For the second claim, we know form the first one that $dv \neq 0$. Then

$$\frac{du}{dz}dz = du = \frac{du}{dv}dv = \frac{du}{dv}\frac{dv}{dz}dz,$$

hence

$$\frac{du}{dz} = \frac{du}{dv}\frac{dv}{dz},$$

after which the second claim follows from equality (4.19) applied to both u and v. $\diamond$

Assume we have a non-zero differential form $\eta \in \Omega(C)$ and fix any $p \in X(C)$. By 4.7.7 there is a uniformizing parameter u of c at p and, by 4.7.12, $du \neq 0$. Since $\dim \Omega(C) = 1$ (4.7.10), there is an $\eta_u \in \mathbb{C}(C) - \{0\}$ such that $\eta = \eta_u du$. If another uniformizing parameter v at p is used, then we have $\eta = \eta_v dv = \eta_v(dv/du)du$; it follows that $\eta_u = \eta_v(dv/du)$ and, by 4.7.12 again, $o_p(\eta_u) = o_p(\eta_v)$. The integer $o_p(\eta_u)$ thus being independent of the choice of the uniformizing parameter u, the *order* of η at p is defined by the rule

$$o_p\eta = o_p\eta_u$$

if u is any uniformizing parameter at p and $\eta = \eta_u du$. The order of the zero differential form is defined as being ∞.

Lemma 4.7.13 *For any $\eta, \eta' \in \Omega(C)$, any $g \in \mathbb{C}(C)$ and any $p \in X(C)$,*

(a) $o_p(\eta + \eta') \geq \min(o_p\eta, o_p\eta')$, *with equality if $o_p\eta \neq o_p\eta'$.*

(b) $o_p(g\eta) = o_pg + o_p\eta$.

PROOF: Direct from the definition above and 4.3.10.   ◇

As for rational functions, $p \in X(C)$ is said to be a *zero* of a non-zero differential form η if and only if $o_p\eta > 0$. If this is the case, $o_p\eta$ is called the *order* or *multiplicity* of p as a zero of η. Similarly, $p \in X(C)$ is said to be a *pole* of η if and only if $o_p\eta < 0$. Then, $-o_p\eta$ is called the *order* or *multiplicity* of p as a pole of η.

Remark 4.7.14 Assume that $\mathfrak{f} : C \to C'$ is a birational map, $p \in X(C)$, $p' = \mathfrak{f}(p)$, $\eta' \in \Omega(C')$ and $\eta = \mathfrak{f}^*(\eta')$. Then, if $\eta' = \eta'_v dv$ with v a uniformizing parameter of C' at p',

$$\eta = \mathfrak{f}^*(\eta') = \mathfrak{f}^*(\eta'_v)\mathfrak{f}^*(dv) = \mathfrak{f}^*(\eta'_v)d(\mathfrak{f}^*(v))$$

and $\mathfrak{f}^*(v)$ is a uniformizing parameter of C at p by 4.7.6. It follows that $o_p\eta = o_{p'}\eta'$. In particular, for any $\eta' \in \Omega(C')$, $\mathfrak{f}^{-1}$ induces multiplicity-preserving bijections between the sets of zeros and the sets of poles of η' and $\eta = \mathfrak{f}^*(\eta')$.

We close this section with a determination of the multiplicities of the zeros and poles of an exact differential:

Proposition 4.7.15 *Let C be, as before, an irreducible curve of $\mathbb{P}_2$ and take any $g \in \mathbb{C}(C)$. Then $dg = 0$ if and only if g is constant. Otherwise, for any $p \in X(C)$,*

(a) $o_p dg = o_p(g - g(p)) - 1$ *if p is not a pole of g, and*

(b) $o_p dg = o_p g - 1$ *if p is a pole of g.*

PROOF: Take any $p \in X(C)$. We will use the local representation at p, as described above, with the same notations. If $g \in \mathbb{C}$, then $dg = 0$ because d is a derivation. Conversely, if $dg = 0$, then $dg/dz = 0$ and so, by 4.7.11, $\partial_x\varphi(g) = 0$; this implies $\varphi(g) \in \mathbb{C}$ and hence $g \in \mathbb{C}$, because φ is injective and leaves the elements of $\mathbb{C}$ invariant.

Assume now that p is not a pole of g. Then $g(p) \in \mathbb{C}$ and obviously $g - g(p)$ has a zero at p. Thus $0 < o_p(g - g(p)) = o_x\varphi(g - g(p))$. Since obviously $dg = d(g - g(p))$, 4.7.9, 4.7.11 and the standard series derivation rules give

$$o_p dg = o_p d(g - g(p)) = o_p(d(g - g(p))/dz) =$$
$$o_x\partial_x\varphi(g - g(p)) = o_x\varphi(g - g(p)) - 1 = o_p(g - g(p)) - 1,$$

as claimed. If p is a pole of g, then $o_x\varphi(g) < 0$ and a similar argument applies.   ◇

4.8   The Canonical Series

To begin with, we need to prove an important fact, namely that the set of zeros and poles of any non-zero differential is finite. To this end we will compute the zeros and poles of a particular differential in a particular projective situation. We will also keep a record of the multiplicities for future use.

Let C be an irreducible curve of $\mathbb{P}_2$ with no singularities other than ordinary singular points. As the reader may easily check using 3.5.14 and 3.5.16, we may take coordinates x_0, x_1, x_2 relative to a projective reference of $\mathbb{P}_2$ such that:

1.- The third vertex $T = [0, 0, 1]$ does not belong to C or to any of the tangents to C at its singular points.

2.- The line $L : x_0 = 0$ is not tangent to C and contains no singular point of C.

On the affine chart $x_0 \neq 0$ take affine coordinates $x = x_1/x_0$ and $y = x_2/x_0$. Let $q_1, \dots, q_r$ be the contact points of the tangents to C through T: there are finitely many such points by 3.5.16; all of them belong to $\mathbb{A}_2$ and are smooth points of C due to the choice of the reference. Write t_i for the tangent to C at q_i and p_i for the only branch of C with origin q_i. On the other hand, let $q_1', \dots, q_d'$ be the points of L on C. Also these are smooth points of C and we call p_i' the branch of C with origin q_i', $i = 1, \dots, d$.

Lemma 4.8.1 *Choices and notations being as above, if $\bar{x}$ is the restriction to C of the first affine coordinate x, then $d\bar{x}$ is not zero, has zeros $p_1, \dots, p_r$, each p_i with multiplicity $[t_i.C]_{q_i} - 1$, and poles $p_1', \dots, p_d'$, all with multiplicity two.*

PROOF: From the conditions imposed on the point T and the line L it is clear that $T \neq q_i'$ and $[L \cdot p_i'] = [L \cdot C]_{q_i'} = 1$ for $i = 1, \dots, d$. Then, by 4.4.6(b), the poles of $\bar{x} = \overline{x_1/x_0}$ are $p_1', \dots, p_d'$, all simple; by 4.7.15 all of them are poles of $d\bar{x}$ of multiplicity two, as claimed.

Let p be any branch of C other than $p_1', \dots, p_d'$. Then the origin q of p is a point of $\mathbb{A}_2$, say with affine coordinates (a, b). We have $\bar{x}(p) = a$ and, again by 4.4.6(b), $o_p(\bar{x} - a)$ equals the intersection multiplicity of p and the line $L_a : x_1 - ax_0 = 0$. If this intersection multiplicity is one, on one hand L_a is not tangent to p and on the other, by 4.7.15, p is neither a zero, nor a pole, of dx. Otherwise L_a is tangent to p – as a branch – and hence is tangent to C at the origin q of p. Since the lines L_a, $a \in \mathbb{C}$, are all lines through T other than L, this occurs if and only if, for some $i = 1, \dots, r$, $L = t_i$, $q = q_i$ and $p = p_i$. Using 4.7.15 once again, $d\bar{x}$ has a zero of multiplicity $[t_i \cdot p_i] - 1$ at p_i. To end the proof just note that since the origin q_i of p_i is non-singular, p_i is the only branch of C with origin q_i and therefore $[t_i \cdot p_i] - 1 = [t_i \cdot C]_{q_i} - 1$. ⋄

Now, as wanted:

Proposition 4.8.2 *If C is an irreducible curve of $\mathbb{P}_2$ and η any non-zero differential on C, then the set of zeros and poles of η is finite.*

PROOF: By 4.7.14 and 4.2.15, it is not restrictive to assume that C has ordinary singularities. Then, with the choices and notations of 4.8.1, $d\bar{x}$ satisfies the claim. By 4.7.10, any non-zero $\eta \in \Omega(C)$ is $\eta = g d\bar{x}$, with $g \in \mathbb{C}(C) - \{0\}$. It follows, by 4.7.13(b), that any zero or pole of η must either be a zero or pole of dx or belong to the divisor (g). $\diamond$

The divisor (η) of a non-zero $\eta \in \Omega(C)$ is defined by the rule

$$(\eta) = \sum_{p \in X(C)} (o_p \eta) p,$$

there being finitely many non-zero summands due to 4.8.2. The divisors (η) of the non-zero differential forms $\eta \in \Omega(C)$ are called *canonical divisors*.

Proposition 4.8.3 *If C is an irreducible curve of $\mathbb{P}_2$:*

(a) *for any non-zero $\eta \in \Omega(C)$ and any $g \in \mathbb{C}(C)$,*

$$(g\eta) = (g) + (\eta).$$

(b) *Any two canonical divisors on C are linearly equivalent. Any divisor linearly equivalent to a canonical divisor is a canonical divisor too.*

(c) *The effective canonical divisors compose a complete linear series.*

PROOF: Claim (a) is straightforward from 4.7.13(b). Regarding claim (b), if $\eta, \eta' \in \Omega(C) - \{0\}$ then, by 4.7.10, $\eta' = g\eta$ for some non-zero $g \in \mathbb{C}(C)$ and by claim (a), $(\eta') = (g) + (\eta)$ is linearly equivalent to (η). Conversely, if $D \equiv (\eta)$, then for some non-zero $g \in \mathbb{C}(C)$,

$$D = (g) + (\eta) = (g\eta).$$

To close, claim (c) is a direct consequence of claim (b). $\diamond$

The complete linear series described by all the effective canonical divisors on C is called the *canonical series* of C. It will be denoted $\mathcal{K}_C$ in the sequel. Effective canonical divisors are often called *canonical groups*.

Remark 4.8.4 If $\mathfrak{f} : C \to C'$ is a birational map between projective plane curves, then, by 4.7.14, for any $\eta \in \Omega(C)$, $\mathfrak{f}^{-1}((\eta)) = (\mathfrak{f}^*(\eta))$. Therefore, by 4.7.3, $\mathfrak{f}$ induces a bijection between the sets of canonical divisors of C and C' and in particular maps the canonical series of C onto the canonical series of C'.

Once canonical divisors have been introduced we may complete Lemma 4.8.1 by giving a different presentation of the canonical divisor described there.

Lemma 4.8.5 *Take the coordinates, the affine chart and the notations as for 4.8.1 and let $\mathcal{P}(C)$ be the polar of C relative to the point T. Then*

$$(d\bar{x}) = \mathcal{P}(C) \cdot C - \Delta(C) - 2L \cdot C.$$

PROOF: Just use the description of $\mathcal{P}(C) \cdot C$ given by 3.5.20 and 3.5.21, and compare with the description of (dx) given by 4.8.1.   ◇

A birationally invariant result follows:

Proposition 4.8.6 *If C is an irreducible curve of $\mathbb{P}_2$, then the degree of any canonical divisor on C is $2g(C) - 2$, $g(C)$ being the genus of C.*

PROOF: By 4.2.15, the invariance of the genus (4.6.5) and 4.8.4, it is enough to prove the equality in the case of C having ordinary singularities only. Any two canonical divisors being linearly equivalent, we will just compute the degree of $(d\bar{x})$ as given by 4.8.5. If $d = \deg C$ and $\delta = \sum_{q \in C} e_q(C)(e_q(C) - 1)/2$ then

$$\deg(d\bar{x}) = d(d-1) - 2\delta - 2d = (d-1)(d-2) - 2\delta - 2$$

and using the genus formula (4.6.6) ends the proof.   ◇

Next is a not birationally invariant description of the canonical series for curves with ordinary singularities:

Proposition 4.8.7 *If an irreducible curve C of $\mathbb{P}_2$, of degree d, has no singular points other than ordinary singularities, then the canonical series of C is the linear series $\mathcal{A}_{d-3}$, residual with respect to the adjunction divisor of the linear series cut out on C by the adjoints to C of degree $d - 3$.*

PROOF: Still take $\delta = \sum_{q \in C} e_q(C)(e_q(C)-1)/2$. By 4.5.14, the linear system of the adjoint curves of degree $d - 3$ has dimension at least

$$\frac{(d-3)^2 + 3(d-3)}{2} - \delta = g(C) - 1,$$

and since there are no adjoint curves of degree $d - 3$ containing C,

$$\dim \mathcal{A}_{d-3} \geq g(C) - 1.$$

If $\mathcal{A}_{d-3}$ is empty, then, by the inequality above, $g(C) = 0$. Using 4.8.6, it follows that all canonical divisors have negative degree, and therefore also the canonical series is empty.

Thus assume $\mathcal{A}_{d-3} \neq \emptyset$. Then there is an effective divisor of the form

$$D = \mathcal{A}_{d-3} \cdot C - \Delta(C)$$

where $\mathcal{A}_{d-3}$ is a curve – an adjoint in fact – of degree $d - 3$ that, obviously, does not contain C. Again take the coordinates and the affine chart as for lemmas 4.8.1 and 4.8.5 and use the same notations. Since curves of the same degree cut out on C linearly equivalent divisors,

$$\mathcal{A}_{d-3} \cdot C + 2L \cdot C = (\mathcal{A}_{d-3} + 2L) \cdot C \equiv \mathcal{P}(\mathcal{C}) \cdot C$$

and hence

$$D \equiv \mathcal{P}(C) \cdot C - \Delta(C) - 2L \cdot C = (d\bar{x})$$

by 4.8.5. Now, both linear series being complete, $\mathcal{A}_{d-3}$ is the set of all effective divisors linearly equivalent to D, while $\mathcal{K}_C$ is the set of all effective divisors linearly equivalent to $(d\bar{x})$: since we have seen $D \equiv (d\bar{x})$, both series agree, as claimed. ⬦

A lower bound for the dimension of the canonical series follows from 4.8.7. It will be turned into an equality in 4.9.14.

Corollary 4.8.8 *If C is an irreducible curve of $\mathbb{P}_2$, then $\dim \mathcal{K}_C \geq g(C) - 1$.*

PROOF: The claim being birationally invariant (by 4.6.5 and 4.8.4), it may be assumed without restriction that C has only ordinary singularities. Then Proposition 4.8.7 ensures that $\mathcal{K}_C = \mathcal{A}_{d-3}$, while in its proof we have seen that $\dim \mathcal{A}_{d-3} \geq g(C) - 1$. ⬦

Next we will show how canonical divisors arise related to the multiple points of the groups of one-dimensional linear series, which is useful in many situations (cf. for instance Exercise 4.8). Assume that C is an irreducible curve of $\mathbb{P}_2$ and $\mathcal{L}$ a one-dimensional series on C with no fixed point. Then, by 4.4.20, for each $p \in X(C)$ there is a unique group in $\mathcal{L}$ containing p and p has a well-determined multiplicity in it, which is called the *multiplicity* of p in $\mathcal{L}$. When this multiplicity is higher than one, p is called a *multiple point* of $\mathcal{L}$ (*double point* if the multiplicity is two).

Proposition 4.8.9 *On an irreducible curve C of $\mathbb{P}_2$, let $\mathcal{L}$ be a one-dimensional linear series with no fixed point. Then:*

(a) *$\mathcal{L}$ has finitely many multiple points.*

(b) *If $J(\mathcal{L})$ is the divisor composed of the multiple points of $\mathcal{L}$, each with multiplicity one less than its multiplicity in $\mathcal{L}$, and $D \in \mathcal{L}$, then $J(\mathcal{L}) - 2D$ is a canonical divisor.*

PROOF: Take any $D \in \mathcal{L}$ and write $\mathcal{L} = \{D + (f) \mid f \in F\}$, where F is a subspace of $\mathbb{C}(C)$ which, as $D \in \mathcal{L}$, is generated by 1 and a non-constant function g. Since $\mathcal{L}$ is assumed to have no fixed point, we have $(g)_\infty = D$. The other groups in $\mathcal{L}$ are $D + (a + g) = (a + g)_0$ for $a \in \mathbb{C}$. Then, by 4.7.15, $p \in X(C)$ has multiplicity $r > 1$ in $D + (a + g)$ if and only if it is a zero of multiplicity $r - 1$ of dg. Also by 4.7.15, p has multiplicity $r > 1$ in D if and only if it is a pole of multiplicity $r + 1$ of dg. All together, p has multiplicity $r > 1$ in a group of $\mathcal{L}$ if and only if it belongs with multiplicity $r - 1$ to $(dg) + 2D$. Therefore there are finitely many multiple points of $\mathcal{L}$ and, furthermore, $J(\mathcal{L}) = (dg) + 2D$, which proves the second claim. ⬦

The effective divisor $J(\mathcal{L})$ defined in 4.8.8 is called the *Jacobian group* of $\mathcal{L}$. Classical presentations of Brill and Noether's theory do not make use of differentials, but introduce the canonical series via 4.8.8, by first proving that all the divisors $J(\mathcal{L}) - 2D$, for $\mathcal{L}$ a one-dimensional linear series without fixed part and $D \in \mathcal{L}$, are linearly equivalent, and then using any of them to define the canonical series. The reader may see [19, Ch. 4, Sect. 33] or [10, Book V, Ch. I, Sect. 8].

Proposition 4.8.9 provides an easy proof of an important theorem concerning the singular points of the curves of a pencil. In old books it is often stated by saying that the curves of a pencil with no fixed part cannot have a variable singular point:

Theorem 4.8.10 (Bertini, 1880) *If $\mathcal{P}$ is a pencil of curves of $\mathbb{P}_2$, then all but finitely many curves of $\mathcal{P}$ have no singular points other than the base points of $\mathcal{P}$.*

PROOF: Assume the equations of the curves of $\mathcal{P}$ to be $\lambda F + \mu G = 0$, $(\lambda, \mu) \in \mathbb{C}^2 - \{(0,0)\}$, where F and G are linearly independent homogeneous polynomials of degree d in the homogeneous coordinates. We will first determine the locus $\mathrm{Sing}(\mathcal{P})$ of the singular points of the curves of $\mathcal{P}$. By 1.3.12, a point $q \in \mathbb{P}_2$ is a singular point of a curve of $\mathcal{P}$ if and only if there is a non-trivial solution for the system of equations in λ, μ

$$\lambda(\partial_i F)_q + \mu(\partial_i G)_q = 0, \quad i = 0, 1, 2,$$

where $(\partial_i F)_q$, $(\partial_i G)_q$ denote the results of evaluating both partial derivatives at already fixed coordinates of q. This occurs if and only if the homogeneous polynomials

$$\Delta_k = \begin{vmatrix} \partial_i F & \partial_i G \\ \partial_j F & \partial_j G \end{vmatrix}, \quad i, j = 0, 1, 2, \quad i < j,$$

are all zero at q. Hence $\mathrm{Sing}(\mathcal{P})$ is the algebraic set defined by the equations $\Delta_k = 0$ for $k = 0, 1, 2$. If all these equations are zero, then $\mathrm{Sing}(\mathcal{P}) = \mathbb{P}_2$. Otherwise, by 3.2.5,

$$\mathrm{Sing}(\mathcal{P}) = |D_1| \cup \cdots \cup |D_r| \cup \{q_1, \ldots, q_h\}, \tag{4.20}$$

where $D_1, \ldots, D_r$ are irreducible curves, $q_1, \ldots, q_h \in \mathbb{P}_2$, $r, h \geq 0$.

Next we will see that any irreducible curve all whose points belong to $\mathrm{Sing}(\mathcal{P})$ must be contained in one of the curves of $\mathcal{P}$. For, assume otherwise that an irreducible curve T is contained in no $C \in \mathcal{P}$ and has all its points in $\mathrm{Sing}(\mathcal{P})$. It is clear that T contains finitely many base points of $\mathcal{P}$, as otherwise it would be an irreducible component of the fixed part of $\mathcal{P}$, by 3.2.6. Since there is no $C \in \mathcal{P}$ containing T, $\mathcal{P}$ cuts out on T a one-dimensional linear series $\mathcal{L}$. Assume that $q \in T$ is not a base point of $\mathcal{P}$; then there is a unique $C_q \in \mathcal{P}$ containing q and, since $q \in \mathrm{Sing}(\mathcal{P})$, q is a singular point of C_q. This implies that any $p \in X(C)$ with origin q is a multiple point of the group $C_q \cdot T$ of $\mathcal{L}$. Since q is not a base point of $\mathcal{P}$, p is not a fixed point of $\mathcal{L}$ and so still p is multiple in the group of the variable part $\mathcal{L}'$ of $\mathcal{L}$ containing it. Since the preceding argument applies to infinitely many $p \in X(C)$, this contradicts 4.8.9.

The argument above discards the possibility $\mathrm{Sing}(\mathcal{P}) = \mathbb{P}_2$, and therefore $\mathrm{Sing}(\mathcal{P})$ is as in (4.20) above. Assume that $C \in \mathcal{P}$ has a singular point q which is not a base point of $\mathcal{P}$. Then either $q = q_i$ for some $i = 1, \ldots, h$, and

C is one of the finitely many curves of $\mathcal{P}$ that contain one of the non-base points in $\{q_1, \ldots, q_h\}$, or, for some $i = 1, \ldots, r$, $q \in D_i$ and, therefore, such a D_i is not contained in the fixed part of $\mathcal{P}$. Since D_i needs to be contained in a curve of $\mathcal{P}$, and C is the only curve of $\mathcal{P}$ through $q \in D_i$, C is the only curve of $\mathcal{P}$ containing D_i. Again, this leaves finitely many possibilities for C, and therefore the proof is complete.   ⋄

4.9 The Riemann–Roch Theorem

We will deal with the easier cases $g(C) = 0, 1$ first. On a curve C with $g(C) = 0$ all non-empty complete linear series have dimension equal to its degree, as shown next:

Proposition 4.9.1 *If C is an irreducible curve of $\mathbb{P}_2$, then the following conditions are equivalent*

(i)   *$g(C) = 0$.*

(ii)   *Any non-empty complete linear series on C has dimension equal to its degree.*

(iii)   *There is on C a linear series whose dimension is positive and equal to its degree.*

(iv)   *There is on C a linear series with dimension and degree both equal to one.*

PROOF: That (i) ⇒(ii) follows from Riemann's inequality 4.6.4 and 4.4.21. To prove (ii) ⇒(iii) it suffices to take $|D|$ for D any non-zero effective divisor. If $\mathcal{L}$ is a linear series with $\dim \mathcal{L} = \deg \mathcal{L} > 0$, then, by 4.4.22, the residual series of $\mathcal{L}$ with respect to a suitable effective divisor of degree $\deg \mathcal{L} - 1$ has dimension and degree equal to one, which proves that (iii) ⇒(iv). To close, if $\mathcal{L}$ has $\dim \mathcal{L} = \deg \mathcal{L} = 1$, then clearly its Jacobian group $J(\mathcal{L})$ has no points, and so, by 4.8.9(b) and 4.8.6, $-2 = 2g(C) - 2$, thus proving (i).   ⋄

Remark 4.9.2 If $g(C) = 0$, then, for each positive integer n, the set of all effective divisors of degree n on C is a complete linear series of degree n which, therefore, is the only complete linear series of degree n on C. For, if D is any effective divisor of degree n on C, $|D|$ is clearly complete and of degree n. By 4.9.1 it has dimension n and therefore, by 4.4.22, for any effective divisor D' of degree n, $\dim(|D| - D') \geq 0$; there is thus a $D'' \in |D|$ such that $D'' \geq D'$, and then the equality of degrees gives $D' = D'' \in |D|$.

Remark 4.9.3 If $g(C) > 0$, then, by 4.9.1, any effective divisor D of positive degree n has $\dim|D| < n$. Hence, by 4.4.22, there is an effective divisor D', also of degree n, such that $\dim(|D| - D') < 0$ and so $D' \notin |D|$: no complete linear series of positive degree n thus contains all the effective divisors of degree n.

For the case $g(C) = 1$:

Proposition 4.9.4 *If C is an irreducible curve of $\mathbb{P}_2$ and $g(C) = 1$, then any complete linear series $|D|$ on C with $D > 0$ has $\dim |D| = \deg D - 1$.*

PROOF: In this case, 4.4.21 and Riemann's inequality 4.6.4 give

$$\deg D \geq \dim |D| \geq \deg D - 1$$

and $\deg |D| = \dim |D|$ would give $g(C) = 0$ by 4.9.1, hence the claim.   ◇

Remark 4.9.5 We know from 4.8.6 and 4.8.8 that the degree of any canonical divisor is $2g(C) - 2$, while $\dim \mathcal{K}_C \geq g(C) - 1$. If $g(C) = 0$ all canonical divisors have negative degree and therefore none is effective, hence $\mathcal{K}_C = \emptyset$. If $g(C) = 1$ there is at least one effective canonical divisor K because $\dim \mathcal{K}_C \geq 0$, but since $\deg K = 0$, necessarily $K = 0$ and $\mathcal{K}_C = \{0\}$.

In order to deal with the curves C with $g(C) > 1$ we will need

Lemma 4.9.6 (Noether's reduction lemma) *Assume we have, on an irreducible curve C of $\mathbb{P}_2$, an effective divisor D with $\dim(\mathcal{K}_C - D) \geq 0$. If $p \in X(C)$ is not a fixed point of $\mathcal{K}_C - D$, then p is a fixed point of $|D + p|$.*

PROOF: The claim being obviously birationally invariant, by 4.3.14 we may assume without restriction that C has only ordinary singularities and also that the origin q of p is a non-singular point of C.

Take $d = \deg C$. By the hypothesis there is a canonical divisor K such that $K - D$ is effective and does not contain p. By 4.8.7, there is an adjoint A, of degree $d - 3$, such that $A \cdot C = K + \Delta(C)$. Since the origin q of p is a non-singular point of C, we may pick a line ℓ through q such that the intersection $\ell \cap C$ is composed of d different points, say $q_1, \ldots, q_{d-1}, q$. Then all these points are non-singular points of C and so we call p_i the only branch of C with origin q_i, $i = 1, \ldots, d-1$. If $B = p_1 + \cdots + p_{d-1}$, then $\ell \cdot C = B + p$ and so we may write

$$(\ell + A) \cdot C = D + p + B + K - D + \Delta(C).$$

Note that $\ell + A$ is an adjoint of degree $d - 2$. Assume that D' is any effective divisor linearly equivalent to $D + p$: then, by the Restsatz 4.5.20, there is an adjoint A', of degree $d - 2$ such that

$$A' \cdot C = D' + B + K - D + \Delta(C). \tag{4.21}$$

In particular, $A' \cdot C \geq B$; this ensures that A' goes through the points $q_1, \ldots, q_{d-1}$, which are $d - 1$ different points on ℓ. Since the degree of A' is $d - 2$, this gives $\ell \subset A'$, hence $q \in A'$ and therefore p belongs to $A' \cdot C$. Back to equality (4.21), the point p does not belong to B – by the choice of ℓ – or to $K - D$ – by the hypothesis – or to $\Delta(C)$ – because q is a non-singular point of C. Then p belongs to D' and therefore, D' being an arbitrary divisor in $|D + p|$, p is a fixed point of $|D + p|$, as claimed.   ◇

Now we are able to state and prove the Riemann–Roch theorem, considered the most important result of the intrinsic geometry of algebraic curves. We will give a geometric interpretation of the difference $\dim |D| - \deg D + g(C)$ between the two sides of the Riemann inequality, which is sometimes called *Roch's part* of the Riemann–Roch theorem.

Theorem 4.9.7 (Riemann, 1857; Roch, 1864) *If D is any divisor on an irreducible curve C of $\mathbb{P}_2$, then, for any canonical divisor K,*

$$\dim |D| = \deg D - g(C) + \dim |K - D| + 1. \tag{4.22}$$

Remark 4.9.8 If the claimed equality (4.22) holds for a divisor D, then it holds for any D' linearly equivalent to D, because $D' \equiv D$ yields $|D'| = |D|$ (4.5.9), $\deg D' = \deg D$ (4.5.2) and also $K - D' \equiv K - D$ (4.5.1), which in turn gives $|K - D'| = |K - D|$.

Remark 4.9.9 Also, if the claimed equality (4.22) holds for a particular canonical divisor K, then it holds for any canonical divisor. Indeed, if K' is another canonical divisor, then $K' \equiv K$ by 4.8.3; as above, it follows that $K' - D \equiv K - D$ and therefore $|K' - D| = |K - D|$.

PROOF OF 4.9.7: We will consider first the case in which $|D|$ is non-empty. Then, by 4.9.8, after replacing D with $D' \in |D|$, we assume with no restriction that D is effective, in which case $|K - D| = \mathcal{K}_C - D$ by 4.5.17.

In the cases $g(C) = 0, 1$ the claim has been already proved. Indeed, if $g(C) = 0$, then, by 4.9.5, $\deg K - D < 0$ and so $\dim |K - D| = -1$ for any effective D. Using this, (4.22) is $\dim |D| = \deg D$, which has been proved in 4.9.1. If $g(C) = 1$, then, according to 4.9.5, $\mathcal{K}_C - D = \{0\}$ if $D = 0$ and $\mathcal{K}_C - D = \emptyset$ otherwise. In the former case the claimed equality is the obvious $\dim\{0\} = \deg 0 - 1 + 1$, while in the latter it is $\dim |D| = \deg D - 1$, already proved in 4.9.4. We thus continue the proof under the supplementary hypothesis $g(C) > 1$, which ensures $\dim \mathcal{K}_C > 0$, by 4.8.8. First of all we will prove that if the Riemann equality is strict, that is, $\dim |D| > \deg D - g(C)$, then the linear series $|D|$ is partially contained in the canonical series. We will use induction on $\dim |D|$. If $\dim |D| = 0$, then by the hypothesis $g(C) - 1 \geq \deg D$. Since, by 4.8.8, $\dim \mathcal{K}_C \geq g(C) - 1 > 0$, 4.4.22 applies showing that there is a canonical divisor containing D. Assume now $\dim |D| > 0$: then fix any $p \in X(C)$ not a fixed point of $|D|$. By 4.4.20 $\dim |D - p| = \dim |D| - 1$. On one hand this yields $\dim |D - p| \geq 0$ and so there is a $D' \in |D - p|$. On the other hand

$$\dim |D - p| = \dim |D| - 1 > \deg D - g(C) - 1 = \deg(D - p) - g(C).$$

By the induction hypothesis, $D' \in |D - p|$ is contained in a canonical group K. Therefore $K - D' \in \mathcal{K}_C - D'$ and in particular $\dim(\mathcal{K}_C - D') \geq 0$. Then the reduction lemma 4.9.6 applies: since p is not a fixed point of $|D' + p| = |D|$, it is a fixed point of $\mathcal{K}_C - D'$ and hence $K \geq D' + p$. Using 4.5.18, this shows that $|D| = |D' + p|$ is partially contained in $\mathcal{K}_C$, as claimed.

We have thus seen that if D is contained in no canonical group, that is $\dim(\mathcal{K}_C - D) = -1$, then the claimed equality holds. Now, this case being proved, we will use induction on $\dim(\mathcal{K}_C - D)$. Thus assume $\dim \mathcal{K}_C - D \geq 0$. Take $p \in X(C)$ to be not a fixed point of $\mathcal{K}_C - D$: then

$$\dim(\mathcal{K}_C - (D + p)) = \dim((\mathcal{K}_C - D) - p) = \dim(\mathcal{K}_C - D) - 1$$

and the induction hypothesis applies to $D + p$. Using again the reduction lemma 4.9.6, p is a fixed point of $|D + p|$ and therefore

$$\begin{aligned} \dim|D| = \dim|D + p| &= \deg(D + p) - g(C) + \dim(\mathcal{K}_C - (D + p)) + 1 \\ &= \deg D + 1 - g(C) + \dim(\mathcal{K}_C - D) - 1 + 1 \\ &= \deg D - g(C) + \dim(\mathcal{K}_C - D) + 1, \end{aligned}$$

as wanted.

Now, there remains the case $\dim|D| = -1$ with no hypothesis on $g(C)$. Assume first that $|K - D|$ is non-empty. Then the claimed equality (4.22) has already been proved to hold for the divisor $K - D$, with K as canonical divisor: we have

$$\begin{aligned} \dim|K - D| &= \deg(K - D) - g(C) + \dim|D| + 1 \\ &= 2g - 2 - \deg D - g(C) + \dim|D| + 1 \\ &= -\deg D + g(C) + \dim|D| - 1, \end{aligned}$$

just the equality to be proved.

To close, if both $\dim|D| = -1$ and $\dim|K - D| = -1$, Riemann's inequality 4.6.4 applied to both linear series gives

$$g(C) - 1 \geq \deg D \quad \text{and} \quad g(C) - 1 \geq \deg(K - D) = 2g(C) - 2 - \deg D.$$

It follows that $\deg D = g(C) - 1$, from which the claimed equality

$$-1 = \deg D - g(C) - 1 + 1$$

is clear.   ◇

The rest of this section is devoted to presenting some direct consequences of the Riemann–Roch theorem; as before, C denotes an arbitrary irreducible curve of $\mathbb{P}_2$.

Divisors D for which $\dim|D| - \deg D + g(C)$ – the difference between the two sides of Riemann's inequality – is positive are called *special*. Clearly, the condition depends only on the linear equivalence class of D and not on D itself.

Remark 4.9.10 By 4.9.1, there are no effective special divisors on C if $g(C) = 0$. If $g(C) = 1$, by 4.9.4, the only effective special divisor is 0.

The corollary below is a direct consequence of 4.9.7 and the definition of special divisor:

Corollary 4.9.11 *A divisor D is special if and only if there is a canonical divisor K such that $\dim|K - D| \geq 0$; in such a case, $\dim|K - D| \geq 0$ for any canonical divisor K.*

As noticed in 4.9.8 and 4.9.9, the linear series $|K - D|$ remains the same if K and D are replaced with linearly equivalent divisors. In particular, if K is any canonical divisor, the non-negative integer $\dim(K - D) + 1$ depends only on the linear equivalence class of D: it is called the *index of speciality* of D, denoted $i(D)$. According to 4.9.11, a divisor D is special if and only if $i(D) > 0$.

Remark 4.9.12 Using the index of speciality, the Riemann–Roch equality (4.22) reads
$$\dim|D| = \deg D - g(C) + i(D).$$

Assume that a complete linear series $|D|$ is non-empty; then D is determined by $|D|$ up to linear equivalence (4.5.10): it thus makes sense to say that the linear series $|D|$ is *special* if and only if the divisor D is special, and also to define the *index of speciality* $i(|D|)$ of $|D|$ as being the index of speciality of D.

Remark 4.9.13 Assume that the divisor D is effective and so, in particular, $|D| \neq \emptyset$. If K is any canonical divisor, then, by 4.5.17, $|K-D| = \mathcal{K}_C - D$. Due to this, D is special if and only if it is contained in a canonical group, which in turn is equivalent to having $|D| = \mathcal{K}_C - T$ for some effective divisor T, by 4.5.18. In addition, the index of speciality of D is $i(D) = \dim(\mathcal{K}_C - D) + 1$; by 4.4.23, it equals the maximal number of linearly independent canonical groups containing D.

Next are two corollaries of the Riemann–Roch theorem: the first one characterizes the canonical series by its degree and dimension. The second one bounds both the degree and the dimension of any special linear series.

Corollary 4.9.14 *If $g(C) > 0$, then the canonical series $\mathcal{K}_C$ is non-empty, has degree $2g(C) - 2$ and dimension $g(C) - 1$. Furthermore, $\mathcal{K}_C$ is the only non-empty linear series on C which has degree $2g(C) - 2$ and dimension $g(C) - 1$.*

PROOF: That $\mathcal{K}_C$ is non-empty follows from 4.8.8, and then $\deg \mathcal{K}_C = 2g(C) - 2$ by 4.8.6. If K is any canonical divisor, then $i(K) = \dim|K - K| + 1 = 1$, and just taking $D = K$ in 4.9.7 gives $\dim \mathcal{K}_C = g(C) - 1$.

If $\mathcal{L}$ is non-empty and has $\deg \mathcal{L} = 2g(C) - 2$ and $\dim \mathcal{L} = g(C) - 1$, take any $D \in \mathcal{L}$. Then $\mathcal{L} \subset |D|$ and
$$\dim|D| \geq \dim \mathcal{L} = g(C) - 1 > 2g(C) - 2 - g = \deg D - g(C).$$

Therefore the linear series $|D|$ is special, and hence partially contained in $\mathcal{K}_C$ by 4.9.13. Also $\mathcal{L}$ is thus partially contained in $\mathcal{K}_C$ and, both series having the same degree, $\mathcal{L} \subset \mathcal{K}_C$. Since, in addition, they have the same dimension, $\mathcal{L} = \mathcal{K}_C$, by 4.4.16.   ◇

Corollary 4.9.15 *If a non-empty complete linear series $\mathcal{L}$ is special, then* $\deg \mathcal{L} \leq 2g(C) - 2$ *and* $\dim \mathcal{L} \leq g(C) - 1$.

PROOF: By 4.9.13 there is an effective divisor T such that $\mathcal{L} = \mathcal{K}_C - T$, from which $\deg \mathcal{L} \leq \deg \mathcal{K}_C$ and $\dim \mathcal{L} \leq \dim \mathcal{K}_C$. The claim then follows from 4.9.14.   ◇

Remark 4.9.16 If D is an effective special divisor, then, by 4.4.23, the canonical groups containing D describe a linear series whose codimension in $\mathcal{K}_C$ is, using 4.4.23, 4.9.14 and 4.9.7,

$$\dim \mathcal{K}_C - \dim(\mathcal{K}_C - D) = g(C) - 1 - i(D) = \deg D - \dim |D|.$$

Such a codimension is often presented in old books as the number of independent conditions imposed by the points of D on the canonical groups required to contain them.

Fixed points of complete linear series give rise to special series, namely:

Lemma 4.9.17 *If a complete linear series $|D|$ has p as a fixed point, then the divisor $D - p$ is special.*

PROOF: Using Riemann's inequality,

$$\dim |D - p| = \dim |D| \geq \deg D - g(C) > \deg(D - p) - g(C),$$

hence the claim.   ◇

Corollary 4.9.18 (Clifford's theorem, Clifford, 1882) *For any non-empty complete and special linear series $\mathcal{L}$, $2 \dim \mathcal{L} \leq \deg \mathcal{L}$.*

PROOF: Take $D \in \mathcal{L}$. The divisor D being special, it is contained in a canonical group K (4.9.13) and so there is an effective divisor T such that $K = D + T$. Then, by 4.5.17, $|D| = \mathcal{K}_C - T$. The canonical groups containing D describe a linear variety $\mathcal{L}_1$ of $\mathcal{K}_C$ of dimension $\dim(\mathcal{K}_C - D) = i(D) - 1$ (4.4.23). Similarly, the canonical groups containing T describe a linear variety $\mathcal{L}_2$ of $\mathcal{K}_C$ of dimension $\dim(\mathcal{K}_C - T) = \dim |D|$. Since $\mathcal{L}_1 \cap \mathcal{L}_2 = \{K\}$, $\mathcal{L}_1$ and $\mathcal{L}_2$ span a linear variety of $\mathcal{K}_C$ which has dimension $\dim \mathcal{L}_1 + \dim \mathcal{L}_2$. Then

$$g(C) - 1 = \dim \mathcal{K}_C \geq \dim \mathcal{L}_1 + \dim \mathcal{L}_2 = i(D) - 1 + \dim |D|$$

and it is enough to replace $i(D)$ with $\dim |D| - \deg D + g(C)$ (4.9.7).   ◇

Corollary 4.9.19 *The canonical series $\mathcal{K}_C$ has no fixed point.*

PROOF: If $g(C) = 0$, $\mathcal{K}_C = \emptyset$ (4.9.1). Thus assume $g(C) > 0$ and therefore $\dim \mathcal{K}_C \geq 0$. If p is a fixed point of $\mathcal{K}_C$, then the residual series $\mathcal{K}_C - p$ is special, has the same dimension as $\mathcal{K}_C$, namely $g(C) - 1$, and degree $\deg \mathcal{K}_C - 1 = 2g(C) - 3$, against 4.9.18.   ◇

Example 4.9.20 If $g(C) = 0$, then $\dim \mathcal{K}_C = -1$ by 4.9.14; the canonical series is thus empty and therefore there are no special divisors. By Riemann–Roch 4.9.7, $\dim |D| = \deg D$ for any effective divisor D, as already seen in 4.9.1.

Example 4.9.21 If $g(C) = 1$, then $\dim \mathcal{K}_C = 0$ and $\deg \mathcal{K}_C = 0$ by 4.9.14. The null divisor 0 is then the only effective canonical divisor and so $\mathcal{K}_C = \{0\}$. No effective divisor other than 0 is special and so the Riemann–Roch theorem (4.9.7) gives $\dim |D| = \deg D - 1$ for any divisor D with $\deg D > 0$, as already seen in 4.9.4.

Example 4.9.22 If $g(C) = 2$, then $\dim \mathcal{K}_C = 1$ and $\deg \mathcal{K}_C = 2$ by 4.9.14. The canonical series having no fixed point (4.9.19), we have $i(p) = 1$ for any single-point divisor p, $p \in X(C)$, and, by Riemann–Roch 4.9.7, $\dim |p| = 0$, which is also clear from 4.9.1 and 4.4.21. The only special divisors D with $\deg D > 1$ are the canonical divisors, and, obviously, for them $\dim |D| = \dim \mathcal{K}_C = 1$. All non-canonical divisors with D with $\deg D > 1$ thus have $\dim |D| = \deg D - 2$, by Riemann–Roch 4.9.7.

4.10   Rational Maps Between Curves

Rational maps between curves were already introduced in Section 4.1. The main interest there was birational maps; now we will deal with the general case. Assume that

$$\mathfrak{f} : C \longrightarrow C'$$

is a non-constant rational map between irreducible curves C and C', of projective planes $\mathbb{P}_2$ and $\mathbb{P}_2'$. We will pay more attention to the map

$$X(\mathfrak{f}) : X(C) \longrightarrow X(C')$$

(cf. Section 4.3) than to $\mathfrak{f}$ itself, as the latter, as a map, may be not defined at finitely many points of C. As before we will write $X(\mathfrak{f})(p) = \mathfrak{f}(p)$ for any $p \in X(C)$, unless some confusion may result.

First of all we need:

Lemma 4.10.1 *For any $p' \in X(C')$, $X(\mathfrak{f})^{-1}(p')$ is a finite set.*

PROOF: Let $q' \in C'$ be the origin of p'. If $X(\mathfrak{f})(p) = p'$ and $\mathfrak{f}$ is defined at the origin of p, then p belongs to the set of branches with origin at a point of $\mathfrak{f}^{-1}(q')$ and this set is finite by 4.1.19. On the other hand, since the set of points of C at which $\mathfrak{f}$ is not defined is finite, so is the set of branches at whose origin $\mathfrak{f}$ is not defined.   ◇

Let us recall that the ramification index of $\mathfrak{f}$ at $p \in X(C)$, introduced in Section 4.3, is a positive integer r_p for which $o_p\mathfrak{f}^*(g') = r_p o_{\mathfrak{f}(p)}g'$ for any $g' \in \mathbb{C}(C')$ (4.3.15).

If D' is any divisor on C', $D = \sum_{p'} a_{p'} p'$, then its *inverse image* (or *pull-back*) by $\mathfrak{f}$ is defined as being the divisor on C

$$\mathfrak{f}^*(D) = \sum_{p \in X(C)} a_{\mathfrak{f}(p)} r_p p,$$

where the summation is finite due to 4.10.1.

Remark 4.10.2 Since the ramification indices r_p are always positive, a point $p \in X(C)$ belongs to $\mathfrak{f}^*(D)$ if and only if $a_{\mathfrak{f}(p)} > 0$, that is, if and only if $\mathfrak{f}(p)$ belongs to D.

According to the definition, taking inverse images of divisors is a map, called *pull-back of divisors*,

$$\mathfrak{f}^* : \mathrm{Div}(C') \to \mathrm{Div}(C),$$

which is clearly a group-homomorphism mapping effective divisors to effective divisors. Denoting it by $\mathfrak{f}^*$ will cause no confusion with the pull-backs of rational functions and differentials.

Remark 4.10.3 It is straightforward to check that if $\mathfrak{f}$ is birational, then the pull-back of divisors $\mathfrak{f}^*$ is the inverse of the map $X(\mathfrak{f})$ of 4.4.24.

The proof of the next lemma directly follows from 4.3.15 and 4.3.16; it is left to the reader:

Lemma 4.10.4 *For any C, C' and $\mathfrak{f}$ as above,*

(a) *For any non-zero $g' \in \mathbb{C}(C')$,*

$$\mathfrak{f}^*((g')) = (\mathfrak{f}^*(g')).$$

(b) *$\mathfrak{f}^*$ maps linearly equivalent divisors to linearly equivalent divisors.*

(c) *The inverse images of the divisors in a linear series $\mathcal{L}'$, defined on C' by a divisor D' and a subspace $F' \subset \mathbb{C}(C')$, are the divisors in the linear series $\mathcal{L}$ defined on C by $\mathfrak{f}^*(D')$ and $\mathfrak{f}^*(F')$. Thus, $\dim \mathcal{L} = \dim \mathcal{L}'$.*

(d) *$(\mathrm{Id}_C)^* = \mathrm{Id}_{\mathrm{Div}(C)}$.*

(e) *If $\mathfrak{f} : C \to C'$ and $\mathfrak{g} : C' \to C''$ are non-constant rational maps, then*

$$(\mathfrak{g} \circ \mathfrak{f})^*(D'') = \mathfrak{f}^*(\mathfrak{g}^*(D''))$$

for any divisor D'' on C''.

The linear series $\mathcal{L}$ of 4.10.4(c) is called the *pull-back* – or the *inverse image* – of $\mathcal{L}'$ by $\mathfrak{f}$, denoted $\mathcal{L} = \mathfrak{f}^*(\mathcal{L}')$. If $\mathfrak{f}$ is birational, then it follows from 4.10.3 that $\mathfrak{f}^*(\mathcal{L}') = \mathfrak{f}^{-1}(\mathcal{L}')$.

The inverse image of any $p' \in X(C')$, p' taken as a single-point divisor, is called the *fibre* of p' (by $\mathfrak{f}$), denoted $\mathfrak{f}^*(p')$. In other words,

$$\mathfrak{f}^*(p') = \sum_{\mathfrak{f}(p)=p'} r_p p,$$

which is clearly an effective divisor whose points are the points of $X(C)$ with image p'. In a short while it will turn out that it is always non-zero (see 4.10.8 or 4.10.13 below).

The next claim describes the fibres of the non-constant rational maps to $\mathbb{P}_1$. In it, $X(\mathbb{P}_1)$ and $\mathbb{P}_1$ are identified through $\chi_{\mathbb{P}_1}$.

Proposition 4.10.5 *Let $\mathfrak{f} : C \to \mathbb{P}_1$ be a non-constant rational map. Assume we have fixed an absolute coordinate z on $\mathbb{P}_1$ and take $g = \mathfrak{f}^*(z)$. Then the fibres of $\mathfrak{f}$ are the divisors of the linear series associated to g. More precisely, the fibre of the point with absolute coordinate $z = \alpha$ is $(g - \alpha)_0$ if $\alpha \neq \infty$, and $(g)_\infty$ if $\alpha = \infty$.*

PROOF: Take any $p \in X(C)$ and assume $q = \mathfrak{f}(p)$ to have coordinate $z(q) = \alpha$. Replacing z with $z - \alpha$ if $\alpha \neq \infty$, or z with $1/z$ if $\alpha = \infty$, will replace α with 0 and g with either $g - \alpha$, in the first case, or with $1/g$, in the second. It is thus not restrictive to assume that $z(q) = \alpha = 0$. This done, the proof will be complete after proving that if r_p is the ramification index of $\mathfrak{f}$ at p, then $r_p = o_p(g)$. Clearly mapping any $\beta \in \mathbb{C}$ to the point of $\mathbb{P}_1$ with absolute coordinate β is a uniformizing map ι of q (as a branch). Let $\varphi : U \to C$ be a uniformizing map of p, with U small enough that $\mathfrak{f}(\varphi(t))$ and $g(\varphi(t))$ are defined for all $t \neq 0$. If $\mathfrak{f}_U$ is the extension of $\mathfrak{f} \circ \varphi_{|U-\{0\}}$, then, by the definition of $\mathfrak{f}(p)$, $\mathfrak{f}_U(0) = q$. On the other hand, up to a further reduction of U, for $t \neq 0$, $\mathfrak{f}_U(t) = \mathfrak{f}(\varphi(t))$ has coordinate

$$z(\mathfrak{f}(\varphi(t))) = (\mathfrak{f}^*(z))(\varphi(t)) = g(\varphi(t)),$$

by 4.1.27. Using the last two equalities, $\mathfrak{f}_U = \iota \circ \tilde{\psi}$ where $\tilde{\psi}(0) = 0$ and $\tilde{\psi}(t) = g(\varphi(t))$ for $t \neq 0$. Then, according to the definitions of r_p and $o_p(g)$,

$$r_p = o_0 \tilde{\psi} = o_0 g(\varphi(t)) = o_p(g)$$

as wanted. $\diamond$

Next is an important result:

Theorem 4.10.6 *If $\mathfrak{f} : C \to C'$ is a non-constant rational map between irreducible plane curves, then the degree of any fibre $\mathfrak{f}^*(p')$, $p' \in X(C')$, equals the degree $[\mathbb{C}(C) : \mathfrak{f}^*(\mathbb{C}(C'))]$, of the extension of fields $\mathfrak{f}^*(\mathbb{C}(C')) \subset \mathbb{C}(C)$.*

The proof of 4.10.6 will make use of the next lemma:

Lemma 4.10.7 *If* $\mathfrak{f} : C \to \mathbb{P}_1$ *is a non-constant rational map,* z *an absolute coordinate on* $\mathbb{P}_1$ *and* $g = \mathfrak{f}^*(z)$, *then the degree of the divisor* $(g)_0$ *equals the degree of the extension of fields* $\mathbb{C}(g) \subset \mathbb{C}(C)$.

PROOF OF 4.10.7: The field $\mathbb{C}$ being algebraically closed, any non-constant $g \in \mathbb{C}(C)$ is free over $\mathbb{C}$. For, if $P(g) = 0$ for some non-zero $P \in \mathbb{C}[Z]$, then g annihilates some of the linear factors in which P decomposes, and therefore it is constant.

Write $d = \deg(g)_0$ and $(g)_0 = \sum_{i=1}^{k} r_i P_i$ with all $r_i > 0$. Then $d = \sum_{i=1}^{k} r_i$. Using 4.7.8, it is easy to construct rational functions $h_{i,j} \in \mathbb{C}(C)$, $i = 1, \ldots, k$, $j = 1, \ldots, r_i$ satisfying:

$$o_{P_s} h_{i,j} = r_s \quad \text{if } s < i$$
$$o_{P_s} h_{i,j} = j - 1 \quad \text{if } s = i$$
$$o_{P_s} h_{i,j} = 0 \quad \text{if } s > i.$$

If the $h_{i,j}$ are linearly dependent over $\mathbb{C}(g)$, then there are $P_{i,j} \in \mathbb{C}(g)$, not all zero, such that

$$\sum_{i,j} P_{i,j} h_{i,j} = 0. \tag{4.23}$$

After clearing denominators and possible factors g, we may assume that each $P_{i,j}$ has the form

$$P_{i,j} = a_{i,j} + g P'_{i,j}$$

where $P'_{i,j} \in \mathbb{C}[g]$, $a_{i,j} \in \mathbb{C}$ and $a_{i,j} \neq 0$ for at least one pair i, j. Choose α to be the least i for which there is some j with $a_{i,j} \neq 0$, and then take β to be the least of those j. Taking into account that $o_{P_\alpha} g = r_\alpha > \beta - 1$, it is straightforward to check that

$$o_{P_\alpha} \sum_{i,j} P_{i,j} h_{i,j} = \beta - 1,$$

against (4.23). Therefore, the $h_{i,j}$ are linearly independent and, there being d of them, we have proved

$$[\mathbb{C}(C) : \mathbb{C}(g)] \geq d.$$

For the reverse inequality, assume that $f_1, \ldots, f_\ell \in \mathbb{C}(C)$ are linearly independent over $\mathbb{C}(g)$. Choose an effective divisor T such that $(f_i) + T \geq 0$ for all i and write $D = (g)_\infty$; by 4.4.8, $\deg D = d$. For any non-negative integer r, consider the rational functions

$$g^j f_i, \quad i = 1, \ldots \ell, \quad j = 0, \ldots, r.$$

Using that g is free over $\mathbb{C}$, it is straightforward to check that they are linearly independent over $\mathbb{C}$. On the other hand we have

$$(g^j f_i) + rD + T \geq 0, \quad i = 1, \ldots \ell, \quad j = 0, \ldots, r.$$

Due to this, for any r high enough, by Riemann–Roch 4.9.7 and 4.9.15,

$$(r+1)\ell \leq \dim \mathbf{L}(rD+T) = \dim |rD+T| + 1 = rd + \deg T - g(C) + 1,$$

this giving in turn $\ell \leq d$ and so

$$[\mathbb{C}(C) : \mathbb{C}(g)] \leq d,$$

as wanted.   ◇

PROOF OF 4.10.6: We will prove first that the claim is satisfied in the particular case in which the target C' of the rational map is $\mathbb{P}_1$. Indeed, take an absolute coordinate z on $\mathbb{P}_1$ and $g = \mathfrak{f}^*(z)$. Then $\mathfrak{f}^*(\mathbb{C}(\mathbb{P}_1)) = \mathfrak{f}^*(\mathbb{C}(z)) = \mathbb{C}(g)$. By 4.10.5, the fibres of $\mathfrak{f}^*$ are the groups of the linear series associated to g and therefore all have the same degree. Since one of the fibres is $(g)_0$, the common degree of the fibres is $[\mathbb{C}(C) : \mathfrak{f}^*(\mathbb{C}(\mathbb{P}_1))]$ by 4.10.7.

Now, for the general case, take any $p \in X(C')$. By 4.9.15, for r high enough the divisor $(r-1)p$ is not special, and therefore, by 4.9.17, the linear series $|rp|$ does not have p as a fixed point. There is thus an effective divisor linearly equivalent to rp that does not contain p, and hence an – obviously non-constant – rational function $g \in \mathbb{C}(C')$ with $(g)_0 = rp$.

Fix an absolute coordinate z on a projective line $\mathbb{P}_1$ and take $\mathfrak{g} : C' \to \mathbb{P}_1$ to be the rational map with affine representation g, which maps q to the point with absolute coordinate $g(q)$ for all but finitely many $q \in C'$; then $g = \mathfrak{g}^*(z)$. If $\bar{p} = \mathfrak{g}(p)$, then, by 4.10.5, $\mathfrak{g}^*(\bar{p}) = rp$, as rp is the group of the linear series associated to g that contains p. Therefore, the target of $\mathfrak{g}$ being $\mathbb{P}_1$,

$$r = [\mathbb{C}(C') : \mathfrak{g}^*(\mathbb{C}(\mathbb{P}_1))].$$

Considering now the composite rational map $\mathfrak{g} \circ \mathfrak{f}$ and using the above, we have

$$[\mathbb{C}(C) : (\mathfrak{g} \circ \mathfrak{f})^*(\mathbb{C}(\mathbb{P}_1))] = [\mathbb{C}(C) : \mathfrak{f}^*(\mathbb{C}(C'))][\mathfrak{f}^*(\mathbb{C}(C')) : \mathfrak{f}^*(\mathfrak{g}^*(\mathbb{C}(\mathbb{P}_1))]$$
$$= [\mathbb{C}(C) : \mathfrak{f}^*(\mathbb{C}(C'))][\mathbb{C}(C') : \mathfrak{g}^*(\mathbb{C}(\mathbb{P}_1))] = r[\mathbb{C}(C) : \mathfrak{f}^*(\mathbb{C}(C'))].$$

On the other hand,

$$\deg((\mathfrak{g} \circ \mathfrak{f})^*(\bar{p})) = \deg(\mathfrak{f}^*(\mathfrak{g}^*(\bar{p}))) = \deg(\mathfrak{f}^*(rp)) = r \deg(\mathfrak{f}^*(p)).$$

Since the target of $\mathfrak{g} \circ \mathfrak{f}$ is also $\mathbb{P}_1$,

$$\deg((\mathfrak{g} \circ \mathfrak{f})^*(\bar{p})) = [\mathbb{C}(C) : (\mathfrak{g} \circ \mathfrak{f})^*(\mathbb{C}(\mathbb{P}_1))]$$

and combining the last three displayed equalities gives the claim.   ◇

Theorem 4.10.6 directly gives:

Corollary 4.10.8 *If* $\mathfrak{p} : C \to C'$ *is a non-constant rational map between irreducible curves, then the degree of the fibre* $\mathfrak{f}^*(p')$ *is the same for all* $p' \in X(C')$.

The degree of the fibres of a non-constant rational map $\mathfrak{f} : C \to C'$, between irreducible curves, is called the *degree of* $\mathfrak{f}$, denoted $\deg \mathfrak{f}$ in the sequel. By 4.10.6 it equals $[\mathbb{C}(C) : \mathfrak{f}^*(\mathbb{C}(C'))]$. Using that $\mathfrak{f}^*$ is a homomorphism between the groups of divisors yields:

Corollary 4.10.9 *If* $\mathfrak{f} : C \to C'$ *is a non-constant rational map between curves, then* $\deg \mathfrak{f}^*(D) = \deg \mathfrak{f} \deg D$ *for any divisor D on C'.*

The proof of the next corollary directly follows from either 4.10.8, or 4.10.6, or 4.10.9.

Corollary 4.10.10 *If* $\mathfrak{f} : C \to C'$ *and* $\mathfrak{g} : C' \to C''$ *are non-constant rational maps between irreducible curves, then* $\deg(\mathfrak{g} \circ \mathfrak{f}) = \deg \mathfrak{f} \cdot \deg \mathfrak{g}$.

The degree of a non-constant rational map being obviously positive due to 4.10.6, we have:

Corollary 4.10.11 *For any non-constant rational map* $\mathfrak{f}$ *between irreducible curves,* $X(\mathfrak{f})$ *is exhaustive.*

Corollary 4.10.12 *A non-constant rational map between irreducible plane curves is birational if and only if it has degree one.*

PROOF: Direct from 4.10.6 and 4.1.33.   ◇

All but finitely many fibres of any non-constant rational map between irreducible curves are composed of distinct points:

Proposition 4.10.13 *If* $\mathfrak{f} : C \to C'$ *is a non-constant rational map between curves, then for all but finitely many* $p' \in X(C')$, *all points belonging to the fibre* $\mathfrak{f}^*(p')$ *have multiplicity one in it.*

PROOF: After composing $\mathfrak{f}$ with a non-constant rational map $\mathfrak{g} : C' \to \mathbb{P}_1$, by 4.10.4(e) it is enough to prove the claim for $\mathfrak{g} \circ \mathfrak{f}$, whose fibres describe a one-dimensional linear series by 4.10.5. Since it has been seen in 4.8.9 that the number of multiple points of a one-dimensional series is finite, the claim follows.   ◇

The multiplicity of a point $p \in X(C)$ in the fibre of $\mathfrak{f}(p)$ being the ramification index r_p of $\mathfrak{f}$ at p, 4.10.13 may be equivalently stated by saying that $r_p = 1$ for all but finitely many $p \in X(C)$. The finitely many points $p \in X(C)$ which have ramification index $r_p > 1$ are called *ramification points* of $\mathfrak{f}$. It is also said that $\mathfrak{f}$ is *ramified* at each of these points.

The set of all fibres of a non-constant rational map between curves $\mathfrak{f} : C \to C'$ may be seen as a one-dimensional family of effective divisors parameterized by C': seen this way, it is called a *one-dimensional algebraic series*. By 4.10.8 all groups of a one-dimensional algebraic series have the same degree, which is called the *degree* of the algebraic series. If the curve C' is rational, then the series is called *rational*. If this is the case, C' may be replaced with a projective line with no modification to the fibres (reparameterizing the series), and so, by 4.10.5, any one-dimensional rational series is a linear series.

We have seen in 4.2.15 that $X(\mathfrak{f})$ is bijective for any birational $\mathfrak{f}$. Here is a sort of converse.

Corollary 4.10.14 *If the fibre $\mathfrak{f}^*(p')$ of a non-constant rational map $\mathfrak{f}: C \to C'$ between irreducible plane curves has a single point (no matter its multiplicity) for all but finitely many $p' \in X(C')$, then $\mathfrak{f}$ is birational.*

PROOF: By 4.10.13 all but finitely many fibres consist of a single point with multiplicity one; hence $\deg \mathfrak{f} = 1$ and 4.10.12 applies.   ◇

There is also a condition for birationality in terms of the points of the curves themselves. It contains the converse of 4.1.26:

Corollary 4.10.15 *Let $\mathfrak{f}: C \to C'$ be a rational map between projective plane curves. If for all but finitely many $q' \in C'$, $\mathfrak{f}^{-1}(q')$ is a single point, then $\mathfrak{f}$ is birational.*

PROOF: The hypothesis ensures that $\mathfrak{f}$ is non-constant. Let A be the set of points $q' \in C'$ for which $\mathfrak{f}^{-1}(q')$ is a single point; by the hypothesis $C - A$ is a finite set. From the set of branches of C' with origin in A, $\chi_{C'}^{-1}(A)$, remove the images under $X(\mathfrak{f})$ of the branches of C whose origin is a singular point of C, to get a subset $B \subset X(C')$. Clearly, $X(C') - B$ is finite. Take any $p' \in B$. If $p \in X(C)$ has $X(\mathfrak{f})(p) = p'$, then its origin $q = \chi_C(p)$ is a smooth point and therefore $\mathfrak{f}$ is defined at q, by 4.1.16, and $\mathfrak{f}(q) = q'$, by 4.3.11. Using again that q is a smooth point, p is the only branch with origin the only point of C mapped to q by $\mathfrak{f}$. The fibres of the points (branches) $p' \in B$ thus have a single point and 4.10.14 applies.   ◇

The irreducible curves of $\mathbb{P}_2$ which are birationally equivalent to $\mathbb{P}_1$ are called *rational curves*. An irreducible curve C is thus rational if and only if there is a birational map

$$\mathfrak{f}: \mathbb{P}_1 \longrightarrow C.$$

The reader may note that in this case $\mathfrak{f}$ is a well-defined map on the whole of $\mathbb{P}_1$ by 4.1.16, or by just taking a maximal representative, see section 4.1. The map $\mathfrak{f}$ is usually understood as a rational parameterization of C: through it, the homogeneous coordinates of a variable point $q \in C$ appear as polynomial functions of the two homogeneous coordinates (*homogeneous parameters*), or the single absolute coordinate (*absolute parameter*), of a point varying freely on $\mathbb{P}_1$. (Affine coordinates of q would be rational functions of the former, hence the parameterization is called *rational*.) It is also worth noting that the parameterization is not redundant, in the sense that all but finitely many $q \in C$ are each the image of a single point of $\mathbb{P}_1$, or, equivalently, each arises from a single value – up to proportionality in the homogeneous case – of the parameter(s).

Here is the fundamental characterization of rational curves, sometimes named Clebsch's theorem. Once proved, the reader may get many other results from 4.9.1.

Proposition 4.10.16 (Clebsch 1863) *An irreducible curve C is rational if and only if it has genus zero.*

PROOF: The *only if* part follows from the invariance of the genus (4.6.5). If $g(C) = 0$, then fix any $p \in X(C)$. By 4.9.1, the linear series $|p|$ has degree and dimension equal to one. If $|p| = p + \langle 1, f \rangle$, f may be used to define a non-constant rational map

$$\mathfrak{f} : C \longrightarrow \mathbb{P}_1$$
$$q \longmapsto (f(q))$$

on a projective line $\mathbb{P}_1$ with an already chosen absolute coordinate. By 4.10.5, $\mathfrak{f}$ has degree one, and therefore it is birational by 4.10.12.   ◇

The above characterization and the genus formula 4.6.6 together show that all rational curves of degree three or higher have singular points. In addition, if the singular points are ordinary, their number and multiplicities need to be enough to make the genus into zero. For instance, a rational quartic with ordinary singularities needs to have either three double points or a single triple point.

Theorem 4.10.17 (Riemann–Hurwitz formula; Hurwitz, 1891)
Assume that $\mathfrak{f} : C \to C'$ is a non-constant rational map between irreducible curves. If r_p denotes the ramification index of $\mathfrak{f}$ at $p \in X(C)$, then we have:

$$2g(C) - 2 = (2g(C') - 2) \deg \mathfrak{f} + \sum_{p \in X(C)} (r_p - 1).$$

PROOF Fix any non-zero differential form $\omega \in \Omega(C')$: we will compute the divisor $(\mathfrak{f}^*(\omega))$. To this end take any $p \in X(C)$ and $p' = \mathfrak{f}(p)$. Let h be a uniformizing parameter at p' and $\omega = f dh$, so that $o_{p'}\omega = o_{p'}f$. On one hand,

$$\mathfrak{f}^*(\omega) = \mathfrak{f}^*(f)\mathfrak{f}^*(dh) = \mathfrak{f}^*(f)d(\mathfrak{f}^*(h)). \tag{4.24}$$

On the other, $o_p\mathfrak{f}^*(h) = r_p o_{p'} h = r_p$, by 4.4.22. Hence, if t is a uniformizing parameter at p, then $\mathfrak{f}^*(h) = u t^{r_p}$, with $u \in \mathbb{C}(C)$ regular and non-zero at p, and so

$$d(\mathfrak{f}^*(h)) = r_p u t^{r_p - 1} dt + t^{r_p} du,$$

which in turn gives

$$o_p d(\mathfrak{f}^*(h)) = r_p - 1.$$

Since $o_p \mathfrak{f}^*(f) = r_p o_{p'} f = r_p o_{p'} \omega$, (4.24) yields

$$o_p \mathfrak{f}^*(\omega) = r_p o_{p'} \omega + r_p - 1$$

and so

$$(\mathfrak{f}^*(\omega)) = \sum_{p \in X(C)} (r_p o_{p'} \omega) p + \sum_{p \in X(C)} (r_p - 1) p$$
$$= \mathfrak{f}^*((\omega)) + \sum_{p \in X(C)} (r_p - 1) p.$$

Then the claim follows by taking degrees and using 4.10.9. ◇

Remark 4.10.18 The hypothesis and notations being as in 4.10.17, while proving it we have in fact proved the stronger equality

$$(\mathfrak{f}^*(\omega)) = \mathfrak{f}^*((\omega)) + \sum_{p \in X(C)} (r_p - 1)p,$$

for any non-zero $\omega \in \Omega(C')$.

Remark 4.10.19 We have already seen in 4.10.13 that there are finitely many ramification points of $\mathfrak{f}$ (which also follows from 4.10.17). Thus $\sum_{p \in X(C)} (r_p - 1)p$ is a divisor: it is called the *ramification divisor* of $\mathfrak{f}$.

Corollary 4.10.20 (Lüroth's theorem, Lüroth, 1876) *Let C be an irreducible curve of $\mathbb{P}_2$. If there is a non-constant rational map $\mathfrak{f}$ from $\mathbb{P}_1$ to C, then C is a rational curve.*

Just use the Riemann–Hurwitz formula 4.10.17 and $g(\mathbb{P}_1) = 0$ to get $g(C') = 0$, and then apply 4.10.16. ◇

The rational map $\mathfrak{f}$ of 4.10.20 may be seen as a parameterization of the curve C by which all but finitely many points of C arise each from $d = \deg \mathfrak{f}$ different values of the parameter, by 4.10.8 and 4.10.13. Then Lüroth's theorem asserts that if C has such a parameterization with $d > 1$ – a redundant parameterization – then C also has a birational – that is, non-redundant – parameterization.

4.11 Rational Maps Associated to Linear Series

This section provides an alternative presentation of the rational maps from an irreducible curve C into projective spaces, showing their close relationship with the linear series on C.

In this section, C will be an irreducible curve of a projective plane $\mathbb{P}_2$ on which homogeneous coordinates x_0, x_1, x_2 have been taken. Let $\mathcal{L}$ be a linear series on C with no fixed point and $r = \dim \mathcal{L} > 0$. Recall that, $\mathcal{L}$ being an r-dimensional projective space, its dual $\mathcal{L}^\vee$ is a projective space, also r-dimensional: the elements of $\mathcal{L}^\vee$ are the hyperplanes of $\mathcal{L}$ and, after taking homogeneous coordinates on $\mathcal{L}$, the coefficients of the equations of the hyperplanes of $\mathcal{L}$ may be taken as their homogeneous coordinates in $\mathcal{L}^\vee$ (see, for instance, [4, Ch. 4]).

Fix any $p \in X(C)$. The subset of $\mathcal{L}$

$$\mathcal{L}_p = \{D \in \mathcal{L} \mid D \geq p\}$$

is the linear series resulting from adding the point p to all groups of the residual series $\mathcal{L} - p$. By 4.4.20, it is $(r - 1)$-dimensional, and therefore, by 4.4.15, it is a hyperplane of $\mathcal{L}$. We may thus consider the map

$$\tilde{f}_{\mathcal{L}} : X(C) \longrightarrow \mathcal{L}^{\vee}$$
$$p \longmapsto \mathcal{L}_p.$$

Next we will see that $\tilde{f}_{\mathcal{L}}$ is the result of lifting to $X(C)$ a rational map $f_{\mathcal{L}} : C \to \mathcal{L}^{\vee}$ which is well determined by $\mathcal{L}$.

We have seen in 4.4.27 that there is an r-dimensional linear system Λ, of curves of $\mathbb{P}_2$, with no curve containing C and $\mathcal{L} = \Lambda \cdot C - T$ for a certain effective divisor T. After taking coordinates x_0, x_1, x_2 on $\mathbb{P}_2$, assume it to be

$$\Lambda = \{C_\lambda : \lambda_0 F_0 + \cdots + \lambda_r F_r = 0 \mid \lambda = (\lambda_0, \ldots, \lambda_r) \in \mathbb{C}^{r+1} - \{0\}\},$$

with $F_0, \ldots, F_r \in \mathbb{C}[x_0, x_1, x_2]$, linearly independent and homogeneous of the same degree. By 4.4.29, $\lambda_0, \ldots, \lambda_r$ may be taken as homogeneous coordinates of $C_\lambda \cdot C - T$ in $\mathcal{L}$; in particular $r = \dim \mathcal{L}$. Fix any $p \in X(C)$; assume that the origin of p, $\chi_C(p) = [\alpha_0, \alpha_1, \alpha_2] \in C$, has $\alpha_0 \neq 0$, the other cases being dealt with likewise. Let

$$\varphi : U \longrightarrow C$$
$$t \longmapsto [1, u(t), v(t)],$$

with U an open neighbourhood of 0 in $\mathbb{C}$ and u, v power series convergent in U, be a uniformizing map of p. Since $\mathcal{L}$ has no fixed point, the multiplicity of p in T is

$$\nu_p = \min_{i=0,\ldots,r} o_t F_i(1, u(t), v(t))$$

and therefore the group $C_\lambda \cdot C - T$, of $\mathcal{L}$, contains p if and only if

$$o_t \left(\lambda_0 F_0(1, u(t), v(t)) + \cdots + \lambda_r F_r(1, u(t), v(t)) \right) > \nu_p,$$

or, equivalently,

$$\lambda_0 \left(\frac{F_0(1, u(t), v(t))}{t^{\nu_p}} \right)_{t=0} + \cdots + \lambda_r \left(\frac{F_r(1, u(t), v(t))}{t^{\nu_p}} \right)_{t=0} = 0.$$

Remark 4.11.1 The above is thus an equation (in the variables $\lambda_0, \ldots, \lambda_r$) of $\mathcal{L}_p$, and therefore its coefficients are coordinates of $\mathcal{L}_p = \tilde{f}_{\mathcal{L}}(p)$, namely

$$\tilde{f}_{\mathcal{L}}(p) = \left[\left(\frac{F_0(1, u(t), v(t))}{t^{\nu_p}} \right)_{t=0}, \ldots, \left(\frac{F_r(1, u(t), v(t))}{t^{\nu_p}} \right)_{t=0} \right]. \quad (4.25)$$

In particular, if p does not belong to T, then $\nu_p = 0$ and so

$$\tilde{f}_{\mathcal{L}}(p) = [F_0(\alpha_0, \alpha_1, \alpha_2), \ldots, F_r(\alpha_0, \alpha_1, \alpha_2)].$$

Take now the rational map

$$\mathfrak{f}_{\mathcal{L}} : C \longrightarrow \mathcal{L}^\vee$$

defined by $F_0, \dots, F_r$, neither of these polynomials being zero on C due to the hypothesis on Λ. By 4.3.5, 4.3.4 and the equality (4.25) above, if $\mathfrak{f}_{\mathcal{L}}$ is defined at the origin $\chi_C(p)$ of p, then

$$\mathfrak{f}_{\mathcal{L}}(\chi_C(p)) = (\mathfrak{f}_{\mathcal{L}})_U(0)$$
$$= \left[\left(\frac{F_0(1, u(t), v(t))}{t^{\nu_p}} \right)_{t=0}, \dots, \left(\frac{F_r(1, u(t), v(t))}{t^{\nu_p}} \right)_{t=0} \right] = \tilde{\mathfrak{f}}_{\mathcal{L}}(p),$$

which in particular determines $\mathfrak{f}_{\mathcal{L}}$ by 4.1.15.

We have thus proved:

Proposition 4.11.2 *Let C be an irreducible plane curve of $\mathbb{P}_2$ and $\mathcal{L}$ a positive-dimensional linear series on C with no fixed point.*

(a) *There is a rational map $\mathfrak{f}_{\mathcal{L}} : C \to \mathcal{L}^\vee$, uniquely determined by $\mathcal{L}$, such that, for all $p \in X(C)$, $\mathfrak{f}_{\mathcal{L}}(\chi_C(p)) = \tilde{\mathfrak{f}}_{\mathcal{L}}(p)$, provided $\mathfrak{f}_{\mathcal{L}}$ is defined at the origin $\chi_C(p)$ of p.*

(b) *If $\mathcal{L} = \Lambda \cdot C - T$, T an effective divisor and*

$$\Lambda = \{C_\lambda : \lambda_0 F_0 + \cdots + \lambda_r F_r = 0 \mid \lambda = (\lambda_0, \dots, \lambda_r) \in \mathbb{C}^r - \{0\}\},$$

a linear system of curves of $\mathbb{P}_2$, none containing C, then, using suitable coordinates on $\mathcal{L}^\vee$, $\mathfrak{f}_{\mathcal{L}}$ is defined by $F_0, \dots, F_r$.

The rational map $\mathfrak{f}_{\mathcal{L}}$ will be called the *rational map associated to* (or *defined by*) $\mathcal{L}$. In the sequel, when saying that a rational map is defined by a linear series $\mathcal{L}$, we will sometimes implicitly assume that the series $\mathcal{L}$ is positive-dimensional and has no fixed point.

The reader may note that $\tilde{\mathfrak{f}}_{\mathcal{L}}$ provides a well-defined image for any branch of C, even in the case in which $\mathfrak{f}_{\mathcal{L}}$ is not defined at its origin. Due to 4.11.2(a), $\mathfrak{f}_{\mathcal{L}}$ is certainly not defined at the points q which are the origin of two branches which have different images under $\tilde{\mathfrak{f}}_{\mathcal{L}}$; the converse is not true, see Exercise 4.18.

The images of the maps $\tilde{\mathfrak{f}}_{\mathcal{L}}$, for $\mathcal{L}$ a positive-dimensional linear series on C with no fixed point, are called rational images of C; more precisely, the image of $\tilde{\mathfrak{f}}_{\mathcal{L}}$ will be called the *rational image of C by $\mathfrak{f}_{\mathcal{L}}$*. Rational images are often taken up to a projectivity on the target. The reader may easily check that the rational image of C under $\mathfrak{f}_{\mathcal{L}}$ is the result of adding to $\mathrm{Im}(\mathfrak{f}_{\mathcal{L}})$ finitely many points adherent to it: the added points are the images under $\tilde{\mathfrak{f}}_{\mathcal{L}}$ of the branches of C with origins the points at which $\mathfrak{f}_{\mathcal{L}}$ is not defined.

Example 4.11.3 Take $C : x_0 x_1 x_2 + x_1^3 + x_2^3$, which is an irreducible cubic with a node at $O = [1, 0, 0]$, as the reader may easily check. If Λ is the linear system of the conics through O, the branches of C at O are p, p', and $\mathcal{L} = \Lambda \cdot C - (p + p')$, then $\mathcal{L}$ has no fixed point and $\mathfrak{f}_{\mathcal{L}}$ is not defined at O.

Remark 4.11.4 The image of $\mathfrak{f}_{\mathcal{L}}$ (and therefore also the image of $\tilde{\mathfrak{f}}_{\mathcal{L}}$, due to 4.11.2(a)) spans $\mathcal{L}^{\vee}$. Indeed, if all points of the image lie in a hyperplane $\sum_{i=0}^{r} a_i y_i = 0$, then, by 4.11.2(b), the polynomial $\sum_{i=0}^{r} a_i F_i$ is zero at all but finitely many points of C, and therefore at all points of C (by 3.6.14), against the hypothesis on Λ. In particular, in no case $\mathfrak{f}_{\mathcal{L}}$ is constant.

Up to a projectivity, any non-constant rational map defined on C arises as the rational map associated to a linear series:

Proposition 4.11.5 *If* $\mathfrak{f} : C \to \mathbb{P}_m$ *is a non-constant rational map, then there is a linear series* $\mathcal{L}$ *on* C, *with no fixed point, such that* $\mathfrak{f} = \tau \circ \mathfrak{f}_{\mathcal{L}}$, *where* $\tau : \mathcal{L}^{\vee} \to \mathbb{P}_m$ *is a projectivity onto a linear variety of* $\mathbb{P}_m$.

PROOF: After fixing coordinates x_0, x_1, x_2 on $\mathbb{P}_2$ and $z_0, \ldots, z_m$ on $\mathbb{P}_m$, assume that $\mathfrak{f}$ is defined by homogeneous polynomials $G_0, \ldots, G_m \in \mathbb{C}[x_0, x_1, x_2]$. In $\mathbb{C}^{m+1}$, let $((a_j^0, \ldots, a_j^m))_{j=r+1,\ldots,m}$ be a basis of the subspace defined by the condition $\sum_{i=0}^{m} a^i G_i(q) = 0$ for all $q \in C$, and $((a_j^0, \ldots, a_j^m))_{j=0,\ldots,r}$ a basis of a supplementary of it. Then taking in $\mathbb{P}_m$ new coordinates $y_j = \sum_{i=0}^{m} a_j^i z_i$, $j = 0, \ldots, m$, and $F_j = \sum_{i=0}^{m} a_j^i G_i$ for $j = 0, \ldots, r$, $\mathfrak{f}$ is defined by the polynomials $F_0, \ldots, F_r, 0, \ldots, 0$ and no curve in the linear system $\Lambda : \lambda_0 F_0 + \cdots + \lambda_r F_r = 0$ contains C. Furthermore, $r > 0$ because $\mathfrak{f}$ is not constant. Then, taking $\mathcal{L}$ to be the variable part of $\Lambda \cdot C$, by 4.11.2, $\mathfrak{f}_{\mathcal{L}}$ is defined by $F_0, \ldots, F_r$ and the claim follows.   ◇

Let $\mathbb{P}_r, \mathbb{P}'_m$ be projective spaces of dimensions r and m, $r \geq m$, and with coordinates $y_0, \ldots, y_r$ and $z_0, \ldots, z_m$. If M is an $m \times r$ matrix with rank m, mapping $[y_0, \ldots, y_r]$ to the point of $\mathbb{P}'_m$ with coordinates $M(y_0, \ldots, y_r)^t$ is a rational map π we will call a *projection* onto $\mathbb{P}'_m$. If $m = r$, it is just a projectivity; otherwise it is undefined on the linear variety defined by the equations $M(y_0, \ldots, y_r)^t = (0, \ldots, 0)^t$, which is called the *centre* of π. An obvious example is the projection defined by the equalities $z_0 = y_0, \ldots, z_m = y_m$, and the reader may easily see that any projection may be written in this form using suitable coordinates.

Proposition 4.11.6 *If* $\mathcal{L}$ *is a linear series with no fixed point on an irreducible curve* C, *and* $\mathcal{S}$ *is a positive-dimensional linear series contained in* $\mathcal{L}$, *then we have* $\mathfrak{f}_{\mathcal{S}'} = \pi \circ \mathfrak{f}_{\mathcal{L}}$, *where* $\mathcal{S}'$ *is the variable part of* $\mathcal{S}$ *and* π *a projection.*

PROOF: Using for $\mathcal{L}$ and the linear system Λ the same notations as above, up to a linear substitution it is not restrictive to assume that the linear system $\Lambda' : \lambda_0 F_0 + \cdots + \lambda_m F_m = 0$ cuts out on C the linear series $\mathcal{S} + T$, after which the claim follows from 4.11.2(b).   ◇

Still assume that $\mathcal{L}$ is a positive-dimensional linear series on C with no fixed point and use the above notations, in particular $\mathcal{L}_p = \{D \in \mathcal{L} \mid D \geq p\}$. A pair of two different points $p_1, p_2 \in X(C)$ is said to be a *neutral pair* for $\mathcal{L}$ if and only if $\mathcal{L}_{p_1} = \mathcal{L}_{p_2}$. Since both sides of this equality are linear series of the same dimension, this occurs if and only if all groups of $\mathcal{L}$ containing one of the points, also contain the other, no matter the choice of the points. The two branches at a node of an irreducible curve C provide an easy example of a neutral pair for the linear series cut out on C by the curves of any fixed degree.

Remark 4.11.7 Directly from the definitions, it is clear that, still for $p_1 \neq p_2$, $\tilde{f}_{\mathcal{L}}(p_1) = \tilde{f}_{\mathcal{L}}(p_2)$ if and only if p_1, p_2 is a neutral pair for $\mathcal{L}$. In particular, $\tilde{f}_{\mathcal{L}}$ is injective if and only if $\mathcal{L}$ has no neutral pair. By 4.11.2(a), the latter condition also implies the injectivity of $f_{\mathcal{L}}$, as a map defined in a subset of C, but the converse fails to be true, see Exercise 4.20.

The notion of neutral pair may be extended to the case of more and possibly repeated points. To this end, we will prove a proposition:

Proposition 4.11.8 *If $\mathcal{L}$ is a positive-dimensional linear series with no fixed point, on an irreducible curve C, and N is an effective divisor on C with $\deg N \geq 2$, then the following conditions are equivalent:*

(i) *There is a point p of N such that, for any $D \in \mathcal{L}$, $D > p$ forces $D > N$.*

(ii) *For any point p of N and any $D \in \mathcal{L}$, $D \geq p$ forces $D \geq N$.*

(iii) $\dim(\mathcal{L} - \mathcal{N}) = \dim \mathcal{L} - 1$

PROOF: That (ii) $\Rightarrow$ (i) is clear. Assume (i); then $(\mathcal{L} - N) + N \supset (\mathcal{L} - p) + p$ and so, the reverse inclusion being obvious, $(\mathcal{L} - N) + N = (\mathcal{L} - p) + p$. Thus (see 4.4.23) $\dim(\mathcal{L} - N) = \dim(\mathcal{L} - p) = \dim \mathcal{L} - 1$, the last equality due to 4.4.20. To close, assume $\dim(\mathcal{L} - N) = \dim \mathcal{L} - 1$. For any p belonging to N, $(\mathcal{L} - N) + N \subset (\mathcal{L} - p) + p$. Since $\dim(\mathcal{L} - p) = \dim \mathcal{L} - 1$, both series have the same dimension and therefore agree. This proves (ii). ◇

Let N be an effective divisor with $\deg N \geq 2$ and $\mathcal{L}$ a positive-dimensional linear series with no fixed point, both on an irreducible curve C. The divisor N is said to be a *neutral divisor* – or a *neutral group* – for $\mathcal{L}$ if and only if N and $\mathcal{L}$ satisfy the equivalent conditions of 4.11.8 above. Clearly, the neutral pairs are the neutral divisors of the form $p_1 + p_2$, $p_1 \neq p_2$.

Example 4.11.9 If a linear series $\mathcal{L}$, with no fixed points, is one-dimensional and has $\deg \mathcal{L} \geq 2$, then any group of $\mathcal{L}$ is a neutral group for $\mathcal{L}$.

Of course, if N is neutral for $\mathcal{L}$, so is any effective divisor $N' \leq N$ provided $\deg N' \geq 2$. For any $p \in X(C)$, either p is the fixed part of $(\mathcal{L} - p) + p$ and then p belongs to no neutral divisor for $\mathcal{L}$, or, otherwise, the fixed part of $(\mathcal{L} - p) + p$ is the maximal neutral divisor for $\mathcal{L}$ containing p.

Example 4.11.10 Assume that C is an irreducible curve of $\mathbb{P}_2$ and that $p \in X(C)$ has origin $\chi_C(p) = q$. Let $\mathcal{S}$ be the linear series cut out on C by the lines of $\mathbb{P}_2$. Then q is a non-singular point of C if and only if p belongs to no neutral group for $\mathcal{S}$. Otherwise the maximal neutral group for

$\mathcal{S}$ containing p is $N = \sum_{i=1}^{r} e(p_i)p_i$, where $p = p_1, p_2, \ldots, p_r$, $r \geq 1$, are the branches of C with origin q and $e(p_i)$ is the multiplicity of the branch p_i. In particular, $\deg N = e_q(C)$.

All terms in its definition being birationally invariant, clearly the notion of neutral group is birationally invariant too:

Proposition 4.11.11 *For any birational map between curves* $\mathfrak{f} : C \to C'$ *and any positive-dimensional linear series* $\mathcal{L}$ *on* C, *with no fixed point,* N *is a neutral group for* $\mathcal{L}$ *if and only if* $\mathfrak{f}(N)$ *is a neutral group for* $\mathfrak{f}(\mathcal{L})$.

We will pay some attention to the neutral groups for the canonical series $\mathcal{K}_C$. As seen in 4.9.19, $\mathcal{K}_C$ has no fixed points. We will assume $g(C) > 1$ in order to have $\dim \mathcal{K}_C > 0$ (see 4.9.19). We have:

Proposition 4.11.12 *Assume that* C *is an irreducible curve and* $g(C) > 1$. *If* N *is a neutral divisor for the canonical series* $\mathcal{K}_C$, *then* $\deg N = 2$ *and* $\dim |N| = 1$. *Conversely, if there is on* C *a linear series* $\mathcal{G}$ *of dimension one and degree two, then* $\mathcal{G}$ *is partially contained in* $\mathcal{K}_C$ *and any divisor in* $\mathcal{G}$ *is a neutral divisor for* $\mathcal{K}_C$.

PROOF: By the definition of neutral group, $\deg N \geq 2$ and

$$i(N) = \dim(\mathcal{K}_C - N) + 1 = \dim \mathcal{K}_C = g(C) - 1.$$

Therefore, N is special and so, by Clifford's 4.9.18, $2 \dim |N| \leq \deg N$. Riemann–Roch 4.9.7 gives in turn

$$\dim |N| = \deg N - g(C) + g(C) - 1 = \deg N - 1.$$

It follows that $\deg N \leq 2$, and hence $\deg N = 2$. This proved, the last displayed equality yields $\dim |N| = 1$, as claimed.

For the converse, note first that $\mathcal{G}$ is complete due to 4.9.1. Then, by Riemann–Roch 4.9.7,

$$i(\mathcal{G}) = \dim \mathcal{G} - \deg \mathcal{G} + g(C) = g(C) - 1.$$

Since $g(C) > 1$, $\mathcal{G}$ is special and therefore partially contained in $\mathcal{K}_C$. Furthermore, for any $N' \in \mathcal{G}$,

$$\dim(\mathcal{K}_C - N') = i(\mathcal{G}) - 1 = g(C) - 2 = \dim \mathcal{K}_C - 1,$$

which shows that N' is neutral for $\mathcal{K}_C$. $\diamond$

To have a linear series $\mathcal{G}$ of degree two and dimension one is an interesting property of an irreducible curve C. In the case when $g(C) = 0$, the set of all effective divisors of degree two on C is a complete two-dimensional linear series $\mathcal{H}$, by 4.9.2. The linear series with dimension one and degree two are thus the one-dimensional projective subspaces of $\mathcal{H}$, by 4.4.15, and none of them is complete.

On an irreducible curve C with $g(C) \geq 1$, a linear series G, of dimension one and degree two, has no fixed point; otherwise the variable part of G would contradict 4.9.1. If $g(C) = 1$, then for any effective divisor G of degree two, $|G|$ is a (complete) linear series which has dimension one and degree two, due to 4.9.4 (or to Riemann–Roch 4.9.7); obviously these are all the linear series on C which have degree two and dimension one. The irreducible curves of genus one are called *elliptic curves*, because the elliptic functions appear as rational functions on them.

Irreducible curves C with $g(C) > 1$, on which there is a linear series $\mathcal{G}$ of dimension one and degree two, are called *hyperelliptic*. If C is hyperelliptic, then the above series $\mathcal{G}$ is complete (by 4.9.1) and clearly special; the lemma below shows that it is unique:

Lemma 4.11.13 *On an irreducible curve C with $g(C) > 1$, there is at most one linear series with degree two and dimension one.*

PROOF: Assume that $\mathcal{G}$ and $\mathcal{G}'$ are different linear series on C, both with degree two and dimension one. The series $\mathcal{G}$ and $\mathcal{G}'$ need to be disjoint, as otherwise, by 4.5.7, they would be both included in a strictly larger linear series, necessarily of degree two and dimension two, against 4.9.1. Fix any $p \in X(C)$ and assume that the groups of $\mathcal{G}$ and $\mathcal{G}'$ containing p are, respectively, $p+p_1$ and $p+p_2$; the series being disjoint, $p_1 \neq p_2$. By 4.11.12, both $p+p_1$ and $p+p_2$ are neutral groups for $\mathcal{K}_C$, after which, since $p_1 \neq p_2$, so is $p+p_1+p_2$, against 4.11.12. ◇

Remark 4.11.14 By 4.11.12, an irreducible curve C with $g(C) > 1$ is hyperelliptic if and only if there is a neutral group N for its canonical series $\mathcal{K}_C$. If this is the case, $|N|$ is the only linear series on C with degree two and dimension one, and any group of $|N|$ is neutral for $\mathcal{K}_C$.

For more on the canonical groups of a hyperelliptic curve, see Exercise 4.24. Any irreducible curve of genus two is hyperelliptic, because its canonical series has dimension one and degree two, by 4.9.14. By contrast, any smooth quartic C has genus three (by 4.6.6) and its canonical series $\mathcal{K}_C$ is cut out on C by the lines of its plane (by 4.8.7): this makes clear that there are no neutral groups for $\mathcal{K}_C$ and therefore, by 4.11.14, no smooth quartic is hyperelliptic; see also Exercise 4.21. For more examples of hyperelliptic curves, covering all values of the genus, see Exercise 4.22.

Section 4.12 below is devoted to the plane rational images of irreducible curves: there it will be proved that these images are irreducible curves. Next is a short note regarding the relationship between non-plane rational images of curves and space curves.

4.11.15 A note on algebraic space curves. By 3.2.8, the irreducible algebraic curves of $\mathbb{P}_2$ may be viewed as the algebraic sets defined by a single irreducible homogeneous polynomial. The irreducible algebraic curves of $\mathbb{P}_r$, $r > 2$, named in the sequel *space curves*, are not so easy to define: irreducible algebraic sets of $\mathbb{P}_r$ are those which may be defined by a set of homogeneous polynomials generating a prime ideal; however, selecting the irreducible algebraic sets which in some sense are one-dimensional requires introducing the notion of dimension, because in general there is not a direct relationship between dimension and minimal number of equations defining an algebraic set. This done, space curves are defined as the one-dimensional irreducible algebraic sets of $\mathbb{P}_r$, $r > 2$. The interested reader may see for instance [20, I.6] for the definition of dimension.

Classical geometers used to introduce space curves as the subsets of $\mathbb{P}_r$, $r > 2$, rationally parameterized by the points of an irreducible plane algebraic curve ([19, 0.V], [10, Book V, 43]). They were indeed fully aware that a naive definition using a projective line instead of an irreducible plane curve would be too restrictive, because if $r = 2$ this would give rise to rational curves only, by Lüroth's Theorem 4.10.20. Being more precise, in our terms, space curves may be equivalently defined as being the images of the maps $\tilde{\mathfrak{f}}_{\mathcal{L}}$, for $\mathcal{L}$ a linear series on an irreducible plane curve C, with no fixed point and dimension at least three, and also the images of these images by projectivities, as in 4.11.5.

We close this section by proving a result which will be useful in the next one.

Proposition 4.11.16 *If $\mathfrak{f} : C \to \mathbb{P}_r$ is a rational map from an irreducible curve C in a projective space $\mathbb{P}_r$, $r > 1$, then there is an irreducible homogeneous polynomial $P \in \mathbb{C}[y_0, \ldots, y_r]$ which is zero at all points of the image of $\mathfrak{f}$.*

PROOF: If $\mathfrak{f}$ is constant, then the claim is obviously true. Thus assume otherwise. Then $\mathrm{Im}(\mathfrak{f})$ is an infinite set by 4.1.19. Let us check first that if a homogeneous polynomial $P \in \mathbb{C}[y_0, \ldots, y_r]$ is zero at infinitely many points of $\mathrm{Im}(\mathfrak{f})$, then it is zero at all points of $\mathrm{Im}(\mathfrak{f})$. Indeed, take any $q' \in \mathrm{Im}(\mathfrak{f})$, $q' = \mathfrak{f}(q)$. There is thus a representative of $\mathfrak{f}$ defined at q, and therefore given by homogeneous polynomials $F_0, \ldots, F_r \in \mathbb{C}[x_0, x_1, x_2]$ for which $q' = [F_0(q), \ldots, F_r(q)]$. By the hypothesis on P, the polynomial $P(F_0, \ldots, F_r)$ is zero at infinitely many points of C, hence it is zero at all points of C, by 3.6.14. In particular it is zero at q, after which P is zero at q'.

Now, if a non-zero homogeneous polynomial $P \in \mathbb{C}[y_0, \ldots, y_r]$ is zero at all points of $\mathrm{Im}(\mathfrak{f})$, one of its irreducible factors is zero at infinitely many points of $\mathrm{Im}(\mathfrak{f})$, and therefore is zero at all points of $\mathrm{Im}(\mathfrak{f})$, by the above. It is thus enough to show that there is a non-zero homogeneous polynomial which is zero at all points of $\mathrm{Im}(\mathfrak{f})$. Assume otherwise: then mapping each rational function $Q(y_0, \ldots, y_r)/P(y_0, \ldots, y_r)$ to the restriction of $Q(F_0, \ldots, F_r)/P(F_0, \ldots, F_r)$ to C defines a $\mathbb{C}$-homomorphism of fields between $\mathbb{C}(\mathbb{P}_r)$ and $\mathbb{C}(C)$, which of course is a monomorphism. The field $\mathbb{P}_r$ being a field of rational functions in r variables (see Section 1.4), it has transcendence degree r over $\mathbb{C}$, while $\mathbb{C}(C)$ has transcendence degree 1 over $\mathbb{C}$, by 3.7.9. Therefore such a monomorphism cannot exist if $r > 1$. ⬦

Corollary 4.11.17 *Any rational map* $\mathfrak{f} : C \to \mathbb{P}'_2$, C *an irreducible curve of a projective plane* $\mathbb{P}_2$, *has its image contained in an irreducible curve* C' *of* $\mathbb{P}'_2$, *and may therefore we viewed as a map* $\mathfrak{f} : C \to C'$ *between irreducible curves. The irreducible curve* C' *is uniquely determined by* $\mathfrak{f}$ *provided* $\mathfrak{f}$ *is not constant.*

PROOF: For the existence, just take C' defined by the polynomial P of 4.11.16. If $\mathfrak{f}$ is not constant, then its image is infinite (4.11.4) and 3.2.6 applies.   ◇

4.12   Plane Rational Images of Curves

Still let C be an irreducible curve of a projective plane $\mathbb{P}_2$, with coordinates x_0, x_1, x_2, and $\mathcal{L}$ a positive-dimensional linear series on C with no fixed point. From now on, we will fix our attention on the cases in which $r = \dim \mathcal{L} = 1, 2$. As in the preceding section, assume that $\mathcal{L} = \Lambda \cdot C - T$ for an effective divisor T and an r-dimensional linear system Λ, of curves of $\mathbb{P}_2$, with no curve containing C,

$$\Lambda = \{C_\lambda : \lambda_0 F_0 + \cdots + \lambda_r F_r = 0 \mid \lambda = (\lambda_0, \ldots, \lambda_r) \in \mathbb{C}^{r+1} - \{0\}\},$$

with $F_0, \ldots, F_r \in \mathbb{C}[x_0, x_1, x_2]$, linearly independent and homogeneous of the same degree. As seen in the preceding section, $F_0, \ldots, F_r$ define the rational map $\mathfrak{f}_{\mathcal{L}}$.

If $r = 1$, $\mathcal{L}^\vee$ is a projective line; after identifying it to the line $y_2 = 0$ of a projective plane $\mathbb{P}'_2$ and adding $F_2 = 0$ to the polynomials F_0, F_1 defining $\mathfrak{f}_{\mathcal{L}}$, the rational map $\mathfrak{f}_{\mathcal{L}}$, associated to $\mathcal{L}$, will be seen as a rational map into the line $y_2 = 0$ of $\mathbb{P}'_2$.

If $r = 2$, write $\mathbb{P}'_2 = \mathcal{L}^\vee$ in order to uniformize the notations. By 4.11.17, there is a uniquely determined irreducible curve C' of $\mathbb{P}'_2$ that contains the image of $\mathfrak{f}_{\mathcal{L}}$, and therefore $\mathfrak{f}_{\mathcal{L}}$ is a rational map between curves.

Thus, in both the cases $r = 1$ and $r = 2$, $\mathfrak{f}_{\mathcal{L}}$ will be taken as a rational map $\mathfrak{f}_{\mathcal{L}} : C \to C'$, where C' is a well-determined irreducible curve of a projective plane $\mathbb{P}'_2$.

Taken as such, the rational map $\mathfrak{f}_{\mathcal{L}}$ has associated its lifting $X(\mathfrak{f}_{\mathcal{L}})$: $X(C) \to X(C')$ between Riemann surfaces. As one may expect, it is closely related to the map $\tilde{\mathfrak{f}}_{\mathcal{L}}$, defined at the beginning of Section 4.11:

Proposition 4.12.1 *If* $\dim \mathcal{L} = 1, 2$, *then, for any* $p \in X(C)$, *the origin of* $X(\mathfrak{f}_{\mathcal{L}})(p)$ *is* $\tilde{\mathfrak{f}}_{\mathcal{L}}(p)$, *that is,*

$$\tilde{\mathfrak{f}}_{\mathcal{L}}(p) = \chi_{C'}(X(\mathfrak{f}_{\mathcal{L}})(p)).$$

PROOF: Let $p \in X(C)$ and assume that the coordinates have been taken such that the origin of p is $[1, 0, 0]$. Let $t \mapsto [1, u(t), v(t)]$, for t in an open neighbourhood U of 0 in $\mathbb{C}$, be a uniformizing map of p. Then, by 4.3.4, after a suitable reduction of U, the germ of analytic map $(\mathfrak{f}_\mathcal{L})_p$ is represented by the map

$$t \longmapsto \left[\frac{F_0(1, u(t), v(t))}{t^{\nu_p}}, \frac{F_1(1, u(t), v(t))}{t^{\nu_p}}, \frac{F_2(1, u(t), v(t))}{t^{\nu_p}} \right],$$

$\nu_p = \min_i o_t F_i(1, u(t), v(t))$, which factors through a uniformizing map of $p' = X(\mathfrak{f}_\mathcal{L})(p)$, by the definition of $X(\mathfrak{f}_\mathcal{L})$ (see Section 4.3). Therefore, the origin of p' is

$$\left[\left(\frac{F_0(1, u(t), v(t))}{t^{\nu_p}} \right)_{t=0}, \left(\frac{F_1(1, u(t), v(t))}{t^{\nu_p}} \right)_{t=0}, \left(\frac{F_2(1, u(t), v(t))}{t^{\nu_p}} \right)_{t=0} \right],$$

which is $\tilde{\mathfrak{f}}_\mathcal{L}(p)$, by 4.11.1.   ◇

Remark 4.12.2 By 4.12.1, the commutative diagram of 4.3.11 splits into two commutative triangles

$$
\begin{array}{ccc}
X(C) & \xrightarrow{X(\mathfrak{f}_\mathcal{L})} & X(C') \\
\downarrow{\scriptstyle \chi_C} & \searrow{\scriptstyle \tilde{\mathfrak{f}}_\mathcal{L}} & \downarrow{\scriptstyle \chi_{C'}} \\
C & \xrightarrow[\mathfrak{f}_\mathcal{L}]{} & C',
\end{array}
$$

the same convention as in 4.3.11 being taken for the lower one. In particular, C' is the rational image of C under $\mathfrak{f}_\mathcal{L}$, because $X(\mathfrak{f}_\mathcal{L})$ is exhaustive by 4.10.11. The curve C' is often called just the *image of C under $\mathfrak{f}_\mathcal{L}$.*

The next proposition shows how to recover the linear series $\mathcal{L}$ from the rational map $\mathfrak{f}_\mathcal{L}$. It has many applications.

Proposition 4.12.3 *Let $\mathcal{L}$ be a linear series on an irreducible curve C, with no fixed point and dimension $r = 1, 2$. Let $\mathfrak{f}_\mathcal{L} : C \to C'$, where C' is an irreducible curve of a projective plane $\mathbb{P}_2'$, be the rational map associated to $\mathcal{L}$. If $\mathcal{S}$ is the linear series cut out on C' by the lines of $\mathbb{P}_2'$, then $\mathcal{L} = \mathfrak{f}_\mathcal{L}^*(\mathcal{S})$.*

PROOF: Assume $r = 2$. We will prove that for any $\lambda = (\lambda_0, \lambda_1, \lambda_2) \in \mathbb{C}^3 - \{0\}$, the inverse image of the divisor D', cut out on C' by the line $\ell_\lambda : \sum_{i=0}^3 \lambda_i y_i = 0$, is the divisor $D = C_\lambda \cdot C - T$, where C_λ has equation $\sum_{i=0}^3 \lambda_i F_i = 0$. Fix such a λ and select a line $\ell_\mu : \sum_{i=0}^3 \mu_i y_i = 0$ of $\mathbb{P}_2'$ such that $\ell_\lambda \cap \ell_\mu \cap C = \emptyset$. If g is the restriction to C' of the rational function $(\sum_{i=0}^3 \lambda_i y_i)/(\sum_{i=0}^3 \mu_i y_i)$ and $L = \ell_\mu \cdot C'$, then $D' = (g) + L$ and so, by 4.10.4,

$$\mathfrak{f}^*(D') = (\mathfrak{f}^*(g)) + \mathfrak{f}^*(L). \tag{4.26}$$

On one hand, the hypothesis about ℓ_μ ensures that D' and L share no points (by 4.4.2) and therefore neither do $\mathfrak{f}^*(D')$ and $\mathfrak{f}^*(L)$. On the other hand, if p

belongs to D, taking a uniformizing map of p and the other notations as in the proof of 4.12.1, the multiplicity of p in D is

$$o_t \left(\sum_{i=0}^{2} \lambda_i F_i(1, u(t), v(t)) \right) - \nu_p > 0.$$

As a consequence, the origin q' of $\mathfrak{f}(p)$, which has been seen to be

$$\left[\left(\frac{F_0(1, u(t), v(t))}{t^{\nu_p}} \right)_{t=0}, \left(\frac{F_1(1, u(t), v(t))}{t^{\nu_p}} \right)_{t=0}, \left(\frac{F_2(1, u(t), v(t))}{t^{\nu_p}} \right)_{t=0} \right]$$

in the proof of 4.12.1, belongs to ℓ_λ and therefore does not belong to ℓ_μ. Two consequences follow: first, the coordinates of q' do not satisfy the equation of ℓ_μ and therefore

$$o_t \left(\sum_{i=0}^{2} \mu_i F_i(1, u(t), v(t)) \right) = \nu_p; \tag{4.27}$$

second, p does not belong to $\mathfrak{f}^*(L)$.

We have thus seen that no point belonging to either D or $\mathfrak{f}^*(D')$, belongs to $\mathfrak{f}^*(L)$. Using this, in order to check the equality $D = \mathfrak{f}^*(D')$, it is enough to check that any $p \in X(C)$ not belonging to $\mathfrak{f}^*(L)$ has the same multiplicity in both divisors. By the equality (4.26) above, the multiplicity of such a p in $\mathfrak{f}^*(D')$ is $o_p(\mathfrak{f}^*(g))$. The function $\mathfrak{f}^*(g)$ being the restriction of $(\sum_{i=0}^{2} \lambda_i F_i)/(\sum_{i=0}^{2} \mu_i F_i)$, using (4.27) it turns out to be

$$o_p(\mathfrak{f}^*(g)) = o_t \left(\sum_{i=0}^{2} \lambda_i F_i(1, u(t), v(t)) \right) - \nu_p,$$

which we have already noticed to be the multiplicity of p in D.

This completes the proof for $r = 2$. The case $r = 1$ follows from a similar argument using a pencil instead of the whole net of lines of $\mathbb{P}'_2$; it is left to the reader.   ⬦

Corollary 4.12.4 *If $\mathfrak{f}_\mathcal{L} : C \to C'$ is a rational map between irreducible curves defined by a linear series $\mathcal{L}$ on C, then*

$$\deg \mathcal{L} = \deg \mathfrak{f}_\mathcal{L} \deg C'.$$

In particular, $\mathfrak{f}_\mathcal{L}$ is birational if and only if $\deg \mathcal{L} = \deg C'$.

PROOF: Just use 4.10.9 and the fact that the degree of C' is the degree of the linear series cut out on C' by the lines of $\mathbb{P}'_2$.   ⬦

Here is a characterization of the linear series defining birational maps:

Proposition 4.12.5 *A rational map between curves, $\mathfrak{f}_\mathcal{L} : C \to C'$, defined by a linear series $\mathcal{L}$ on C, is birational if and only if $\mathcal{L}$ has at most finitely many neutral pairs.*

PROOF: By 4.11.7, the condition of the claim is equivalent to the existence of at most finitely many pairs of distinct points in $X(C)$ which have the same image by $\tilde{\mathfrak{f}}_{\mathcal{L}}$. If $\mathfrak{f}_{\mathcal{L}}$ is birational, then $X(\mathfrak{f}_{\mathcal{L}})$ is injective (4.2.15) and therefore the latter condition holds since $\tilde{\mathfrak{f}}_{\mathcal{L}} = \chi_{C'} \circ X(\mathfrak{f}_{\mathcal{L}})$ (4.12.2). The same equality shows that if, conversely, there are at most finitely many pairs of distinct points in $X(C)$ with coincident images by $\tilde{\mathfrak{f}}_{\mathcal{L}}$, then the same holds for $X(\mathfrak{f}_{\mathcal{L}})$, and 4.10.14 applies.   ◇

A linear series defining a non-birational map has all the fibres of that map as neutral groups:

Proposition 4.12.6 *If $\mathfrak{f}_{\mathcal{L}} : C \to C'$ is a rational map between irreducible curves defined by a linear series $\mathcal{L}$ on C and $\deg \mathfrak{f}_{\mathcal{L}} > 1$, then all the fibres of $\mathfrak{f}_{\mathcal{L}}$ are neutral divisors for $\mathcal{L}$.*

PROOF: Write $\mathfrak{f} = \mathfrak{f}_{\mathcal{L}}$. Take any $p \in X(C)$ and $p' = \mathfrak{f}(p)$. By the hypothesis $\deg \mathfrak{f}^*(p) = \deg \mathfrak{f} > 1$. Assume that $D \geq p$ for some $D \in \mathcal{L}$. By 4.12.3, $D = \mathfrak{f}^*(S)$ where S is the divisor cut out on C' by a line; then, by 4.10.2, $S \geq p'$ and therefore $D = \mathfrak{f}^*(S) \geq \mathfrak{f}^*(p)$, as wanted.   ◇

The groups of $\mathcal{L}$ are sums of fibres of $\mathfrak{f}_{\mathcal{L}}$ due to 4.11.1. When $\mathfrak{f}_{\mathcal{L}}$ is not birational, it is said that $\mathcal{L}$ is *composed with* the algebraic series (of degree higher than one) described by the fibres of $\mathfrak{f}$. Otherwise, the linear series $\mathcal{L}$ is called *simple*.

In the case in which $\mathfrak{f}_{\mathcal{L}} : C \to C'$ is birational, the neutral groups for $\mathcal{L}$ are closely related to the singularities of C':

Proposition 4.12.7 *If the rational map between curves $\mathfrak{f}_{\mathcal{L}} : C \to C'$, defined by a two-dimensional linear series without fixed points $\mathcal{L}$ on C, is birational, then $p \in X(C)$ belongs to a neutral divisor for $\mathcal{L}$ if and only if $q = \tilde{\mathfrak{f}}_{\mathcal{L}}(p)$, the origin of $X(\mathfrak{f}_{\mathcal{L}})(p)$, is a singular point of C'. If this is the case and N_p is the maximal neutral divisor for $\mathcal{L}$ containing p, then $e_q(C') = \deg N_p$ and the branches of C' with origin q are the images of the branches of C that belong to N_p.*

PROOF: Let $\mathcal{S}$ be the linear series cut out on C' by the lines of its plane. The map $\mathfrak{f}$ being birational, by 4.12.3 and 4.11.11, N is a neutral divisor for $\mathcal{L}$ containing p if and only if $\mathfrak{f}(N)$ is a neutral divisor for $\mathcal{S}$ containing $\mathfrak{f}(p)$, and the latter situation has already been considered in 4.11.10.   ◇

Next are presented simple projective models for curves of genus one, two and three, and also for hyperelliptic curves. The reader may note how the projective properties of the models result from birationally invariant properties of the linear series defining the birational maps.

4.12.8 Curves of genus one: If C is an irreducible curve and $g(C) = 1$, fix any effective divisor of degree three, D, on C, and take $\mathcal{L} = |D|$. Then $\mathcal{L}$ has $\dim \mathcal{L} = 2$ (4.9.4) and no fixed point, as a fixed point of $\mathcal{L}$ would give rise to a residual series $\mathcal{L}'$ with $\dim \mathcal{L}' = \deg \mathcal{L}' = 2$, against 4.9.1. Similarly, a neutral group T for $\mathcal{L}$ would give $\dim(\mathcal{L} - T) = 1$ and $\deg(\mathcal{L} - T) \geq 1$, again contradicting 4.9.1. We may consider the rational map

$$\mathfrak{f}_{\mathcal{L}} : C \longrightarrow C',$$

associated to $\mathcal{L}$, which is birational by 4.12.5. The image C' is then a smooth cubic, by 4.12.7 and 4.12.4.

Needless to say, any smooth cubic of $\mathbb{P}_2$ is irreducible (by 3.3.8) and therefore has genus one, so we have seen:

Proposition 4.12.9 *Any irreducible curve of genus one is birationally equivalent to a smooth cubic of a projective plane, and any smooth cubic of a projective plane is irreducible and has genus one.*

4.12.10 Curves of genus two: Since any curve of genus two is hyperelliptic, this case is covered by 4.12.14 below; it is included here because it can be dealt with using easier arguments. The reader may compare both.

If C is irreducible and $g(C) = 2$, then its canonical series $\mathcal{K}_C$ has degree two, dimension one (4.9.14) and no fixed point (4.9.19). Choose two different points $p_1, p_2 \in X(C)$, with p_2 not belonging to the only group of $\mathcal{K}_C$ containing p_1; then $T = p_1 + p_2 \notin \mathcal{K}_C$. Therefore, T is not special and hence $\dim |T| = 0$. Take any $K \in \mathcal{K}_C$ and $\mathcal{L} = |K + T|$. Neither $\mathcal{L}$ nor $\mathcal{L} - p$, for any $p \in X(C)$, are special series due to their degrees. As a consequence, $\mathcal{L}$ has degree four, dimension two and no fixed point, by 4.9.17. Let us examine the neutral groups for $\mathcal{L}$. On one hand T is obviously a maximal neutral group because $\mathcal{L} - T = \mathcal{K}_C$ (by 4.5.17), which has no fixed point. On the other, if T' is any neutral group of degree two for $\mathcal{L}$, then $\mathcal{L} - T'$ has degree two and dimension one. Thus $\mathcal{L} - T' = \mathcal{K}_C = \mathcal{L} - T$ (4.9.14), which yields $T' \equiv T$ and hence $T' = T$ because $\dim |T| = 0$; T being maximal, there is no other neutral group for $\mathcal{L}$.

Since $\mathcal{L}$ has no fixed point, the rational map $\mathfrak{f}_{\mathcal{L}}$ is defined. It is birational because there is a single neutral group for $\mathcal{L}$ (4.12.5). The corresponding rational image C' of C is thus a quartic of $\mathbb{P}'_2$. As usual, we identify $X(C)$ and $X(C')$ through $X(\mathfrak{f}_{\mathcal{L}})$, and therefore the groups of $\mathcal{L}$ with their corresponding line-sections of C' (4.12.3). By 4.12.7, C' has a single singular point, which is a double point, namely the common origin $q \in C'$ of the two different branches p_1, p_2 composing T. To complete the description, we will check that q is a node. Indeed, for $i = 1, 2$, the tangent to the branch p_i at q is the only line L_i of $\mathbb{P}'_2$ for which $L_i \cdot C' \geq T + p_i$. As $L_1 = L_2$ we have $L_1 \cdot C' = L_2 \cdot C'$, and this in turn implies $L_1 \cdot C' = L_2 \cdot C' = 2p_1 + 2p_2$ because $p_1 \neq p_2$; thus $2p_1 + 2p_2 \in \mathcal{L}$, which gives $p_1 + p_2 \in \mathcal{L} - T = \mathcal{K}_C$, against the choice of T.

Conversely, any quartic C' of $\mathbb{P}'_2$ with a node q and no other singular point is irreducible, because if $C' = C_1 + C_2$, then C_1 and C_2 would have q as their only intersection point and then they could not be transverse at q due to Bézout's theorem 3.3.5. Furthermore, such a C' has genus two, by 4.6.6. All together, we have:

Proposition 4.12.11 *Any irreducible curve of genus two is birationally equiv-alent to a quartic with a node and no other singular point. Any quartic with a node and no other singular point is irreducible and has genus two.*

4.12.12 Non-hyperelliptic curves of genus three: If C is an irreducible curve, non-hyperelliptic and with $g(C) = 3$, then its canonical series $\mathcal{K}_C$ has degree four, dimension two (4.9.14) and no fixed point (4.9.19) or neutral group (4.11.14). Arguing as in 4.12.8, we may take the associated rational map $\mathfrak{f}_{\mathcal{K}_C}$, which is birational; the corresponding rational image of C is then a smooth quartic of $\mathbb{P}_2$. Conversely, any smooth quartic C' of $\mathbb{P}_2$ is irreducible, has genus three and is not hyperelliptic, because $\mathcal{K}_{C'}$ is cut out on C' by the lines of $\mathbb{P}_2$ (4.8.7), and therefore it clearly has no neutral group. We thus have:

Proposition 4.12.13 *Any irreducible non-hyperelliptic curve of genus three is birationally equivalent to a smooth quartic of $\mathbb{P}_2$. Any smooth quartic of $\mathbb{P}_2$ is in turn irreducible, non-hyperelliptic and has genus three.*

4.12.14 Hyperelliptic curves: Let C be a hyperelliptic curve. We will write $g = g(C)$ for the genus of C; thus $g > 1$. Take $\mathcal{G}$ to be the only linear series on C with degree two and dimension one. As said before, by 4.9.1, $\mathcal{G}$ is complete and has no fixed point. In particular, each $p \in X(C)$ belongs to a unique group of $\mathcal{G}$; we will denote by p' the only point for which $p + p' \in \mathcal{G}$. Then, obviously, $(p')' = p$ and $p \mapsto p'$ is a bijection of $X(C)$.

We choose g different points $p_i \in X(C)$, $i = 1, \ldots, g$, such that for no $i, j = 1, \ldots, g$, $p_i + p_j \in \mathcal{G}$, which is always possible as the reader may easily check. Then the group $T = p_1 + \cdots + p_g$ is not special. Indeed, otherwise $T \leq K$ for a canonical group K (4.9.13) and then, by 4.11.12, all points $p_1', \ldots, p_g'$ would belong to K too. Since the points $p_1, \ldots, p_g, p_1', \ldots, p_g'$ are all different due to the choice of $p_1, \ldots, p_g$, we would have $\deg K \geq 2g$, against 4.9.14. Now, once T is known to be non-special, Riemann–Roch 4.9.7 gives $\dim |T| = 0$, that is, the only effective divisor linearly equivalent to T is T itself.

Take any $G \in \mathcal{G}$ and $\mathcal{L} = |G+T|$. The divisor T not being special, neither is $G + T$ (4.9.13) and, again by Riemann–Roch 4.9.7, $\dim \mathcal{L} = 2$. For any $p \in X(C)$, we have $p + p' \equiv G$ and therefore (4.5.17)

$$\mathcal{L} - (p + p') = \mathcal{L} - G = |T| = \{T\}. \tag{4.28}$$

Due to this, $\mathcal{L}$ has no fixed point, as a fixed point p would give $\dim(\mathcal{L}-p) = 2$ and therefore $\dim(\mathcal{L} - (p + p')) \geq 1$.

By 4.5.17, $\mathcal{L} - T = \mathcal{G}$; since $\mathcal{G}$ has dimension one and no fixed point, T is a maximal neutral group for $\mathcal{L}$. Furthermore, T is the only maximal neutral group for $\mathcal{L}$. For, assume that $\bar{p}$ belongs to a neutral group for $\mathcal{L}$. Then, for a certain $p \in X(C)$, $p + \bar{p}$ is neutral for $\mathcal{L}$. Since $p + p'$ is not neutral for $\mathcal{L}$ (due to (4.28) above), we have $\bar{p} \neq p'$. On the other hand, since $(p + p') + T \in \mathcal{L}$, we have $p + \bar{p} \leq (p + p') + T$, because $p + \bar{p}$ is neutral for $\mathcal{L}$; hence $\bar{p} \leq p' + T$.

Since $\bar{p} \neq p'$, this shows that $\bar{p}$ belongs to T, proving that any point of a neutral group for $\mathcal{L}$ belongs to T.

We have seen that the linear series $\mathcal{L}$ has no fixed point, dimension two, degree $g + 2$ and a single maximal neutral group. We may thus, as in the preceding cases, consider the associated rational map $\mathfrak{f}_{\mathcal{L}}$, which is birational, and the rational image C' of C under $\mathfrak{f}_{\mathcal{L}}$: C' is an algebraic curve of the projective plane $\mathbb{P}'_2 = \mathcal{L}^\vee$, of degree $g + 2$ and in the sequel we identify $X(C)$ and $X(C')$ through $X(\mathfrak{f}_{\mathcal{L}})$. The only singular point of C' is the common origin q of the branches $p_1, \ldots, p_g$ composing T, which is a g-fold point (4.12.7). The groups of $\mathcal{L}$ being cut out on C' by the lines of $\mathbb{P}'_2$, by 4.12.3, take $L_i \subset \mathbb{P}'_2$ to be the tangent line to p_i, p_i taken as a branch of C'. If $L_i = L_j$ for $i \neq j$, $T + p_i + p_j = L_i \cdot C' \in \mathcal{L}$, which yields $p_i + p_j \in \mathcal{G}$, against the choice of the points p_i, $i = 1, \ldots, g$. The point q is thus an ordinary g-fold point of C'.

Conversely, assume that $g > 1$ and C' is a curve of order $g + 2$, of a projective plane $\mathbb{P}_2$, with an ordinary g-fold point q and no other singular point. Then C' is irreducible. Indeed, assume otherwise, that $C' = C_1 + C_2$. Then necessarily $C_1 \cap C_2 \subset \{q\}$, by 1.3.3, and C_1 and C_2 share no tangent at q because the singularity of C' at q is ordinary. Using this, Bézout's theorem 3.3.5 and 2.6.8 yield $\deg C_1 \deg C_2 = e_q(C_1)e_q(C_2)$. An elementary computation left to the reader shows that this is not possible. Once checked that C' is irreducible, it obviously has genus g, by the genus formula 4.6.6, and is hyperelliptic because the variable part of the linear series cut out on C' by the lines through q has degree two and dimension one.

All together, we have proved:

Proposition 4.12.15 *Any hyperelliptic curve of genus g is birationally equivalent to an irreducible curve of degree $g + 2$ with an ordinary singular point of multiplicity g and no other singular point. Any curve of degree $g + 2$, $g \geq 2$, with an ordinary singular point of multiplicity g and no other singular point, is irreducible, hyperelliptic and has genus g.*

4.13   Exercises

4.1

(1) Prove that if C is an irreducible curve which has a single branch γ at a point O, then any rational function f on C which is regular and takes value zero at O has $o_\gamma f \geq e_O(C)$.

(2) Take $C : x_0 x_1^2 - x_2^3 = 0$. Prove that it is irreducible and that the restriction of x_1/x_2 to C is not regular at $[1, 0, 0]$.

4.2 Assume that C is an irreducible curve and $\gamma \in X(C)$. Prove that if γ is a zero (resp. a pole) of a rational function $f \in \mathbb{C}(C)$, then the origin of γ is either a zero (resp. a pole) or an indetermination point of f.

4.3 Use 4.4.12 to prove that if C is an irreducible plane curve, then two non-constant $g, g' \in \mathbb{C}(C)$ have the same associated linear series if and only if $g' = (ag + b)/(cg + d)$ for some $a, b, c, d \in \mathbb{C}$ with $ad - bc \neq 0$.

4.4 Prove that the genus formula of 4.6.6 still holds if the curve C, besides ordinary singularities, is allowed to have ordinary cusps. *Hint:* use induction on the number of cusps and a suitable standard quadratic transformation cancelling one of the cusps. An argument like the one in the proof of 4.2.13 will be needed to prove that each of the other ordinary cusps is transformed into an ordinary cusp.

4.5 Assume that D and D' are linearly equivalent divisors on an irreducible curve C of $\mathbb{P}_2$. Write D_+ and D'_+ for the positive parts of D and D', and D_- and D'_- for their negative parts. Prove that there is an effective divisor T and a non-zero $f \in \mathbb{C}(C)$ such that $D_+ + D'_- = (f)_0 + T$ and $D'_+ + D_- = (f)_\infty + T$. Compare with 4.5.3.

4.6 Let D be an effective divisor on an irreducible curve C. Prove that there is an $f \in \mathbb{C}(C)$ with $(f)_\infty = D$ if and only if the linear series $|D|$ has no fixed point.

4.7 Prove that for any $p \in \mathbb{P}_1 = X(\mathbb{P}_1)$, $-2p$ is a canonical divisor.

4.8 Plücker's first formula, an alternative view. Reprove Plücker's formula 3.5.19 by – with the notations therein – computing the Jacobian group of the linear series cut out on C by the lines through q, and using 4.8.9 and the extended genus formula of Exercise 4.4. Explore the cases in which q either is a non-singular point of C or belongs to a tangent at a singular point.

4.9 If C and L are, respectively, an irreducible non-singular plane quintic and a line of $\mathbb{P}_2$, and $D = C \cdot L$, then for any choice of $p \in X(C)$, p is a fixed point of $|D+p|$. Give three proofs of this fact, one using the Restsatz (4.6.2 in fact), another using Noether's reduction lemma 4.9.6, and a third one by computing $\dim |D + p|$ and $\dim |D|$ by means of the Riemann–Roch theorem 4.9.7.

4.10 Prove the following partial converse of 4.9.17: if a complete linear series $|D|$ is not special, while $|D - p|$ is, then the point p is a fixed point of $|D|$.

4.11 After fixing coordinates on $\mathbb{P}_2$, let C be the curve $C : Ax_2^2 + B = 0$, where $A, B \in \mathbb{C}[x_0, x_1]$ are homogeneous polynomials of degrees $d - 2$ and d, respectively, and AB has no multiple factor. Then:

(1) Prove that C is irreducible.

(2) Determine the singularities of C, prove that all are ordinary and that $g(C) = d - 2$.

(3) Take $E : x_2 = 0$ and the rational map

$$\mathfrak{f} : C \to E$$
$$[x_0, x_1, x_2] \mapsto [x_0, x_1, 0],$$

induced on C by projecting from $[0, 0, 1]$ on E. Determine the ramification points of $\mathfrak{f}$ and make a direct check of the Riemann–Hurwitz formula in this case.

4.12 After fixing coordinates on $\mathbb{P}_2$, let C be the curve $C : x_0^{d-2}x_2^2 + B = 0$, where $B \in \mathbb{C}[x_0, x_1]$ is a homogeneous polynomial of degree $d \geq 2$ which has neither factor x_0 nor multiple factor. Then:

(1) Prove that C is irreducible.

(2) Prove that C has no singular points other than $[0, 0, 1]$, which is a $(d-2)$-fold point; if $d > 2$, it is the origin of either one or two branches of C depending on the parity of d.

(3) As above, take $E : x_2 = 0$ and the rational map

$$\mathfrak{f} : C \to E$$
$$[x_0, x_1, x_2] \mapsto [x_0, x_1, 0].$$

Determine the ramification points of $\mathfrak{f}$ and, using the Riemann–Hurwitz formula 4.10.17, prove that either $g(C) = (d-1)/2$ or $g(C) = (d-2)/2$, depending on the parity of d.

(4) Check that the genus formula 4.6.6 would not give a correct result for $d > 3$ if applied to C.

4.13 Use exercise 4.11 or exercise 4.12 to prove that there are in $\mathbb{P}_2$ irreducible curves of genus g for any non-negative $g \in \mathbb{Z}$. Show that there are positive integers g for which any curve of genus g is singular.

4.14 Definitions and notations being as in Exercise 1.11, prove that if $\mathcal{N}$ is a net of curves of $\mathbb{P}_2$ containing an irreducible curve C, then its Jacobian curve $J(\mathcal{N})$ is defined (that is, $\mathfrak{J}(\mathcal{N}) \neq 0$). Following the steps below is advised:

(1) Prove that there are finitely many points belonging to all curves of $\mathcal{N}$ (*base points* of $\mathcal{N}$).

(2) Assume that $\mathbf{J}(\mathcal{N}) = \mathbb{P}_2$ and let q be a smooth point of C, not a base point of $\mathcal{N}$. Use Exercise 1.11(1) to prove that all $C' \in \mathcal{N}$ going through q satisfy $[C' \cdot C] \geq 2$.

(3) Still assuming that $\mathbf{J}(\mathcal{N}) = \mathbb{P}_2$, take $C', C'' \in \mathcal{N}$ such that C, C', C'' span $\mathcal{N}$, and use the variable part of the linear series cut out on C by the pencil spanned by C' and C'' to contradict 4.8.9.

4.15 Prove that if two cubics of $\mathbb{P}_2$ intersect at nine different points $p_1, \ldots, p_9$, then any cubic through $q_1, \ldots, q_8$ also goes through q_9. *Hint:* first use 3.4.8 and 4.8.10 to prove that there is a smooth – hence irreducible, by 3.3.8 – cubic C through $q_1, \ldots, q_9$. Then, identifying $C = X(C)$, use that the complete linear series $|q_9|$ on C has dimension zero.

4.16 Prove that if two different cubics of $\mathbb{P}_2$, one at least irreducible, both have a node at a point q and share different points $q_1, \ldots, q_5$, all different from q, then there is a cubic which has a double point at q, goes through $q_1, \ldots, q_4$ and misses q_5. Compare to Exercise 4.15.

4.17 Proceed as for Exercise 4.15 to prove the following more general statement (Lamé, 1818):

Let C and C' be curves of degree $d \geq 3$ of $\mathbb{P}_2$ that intersect in exactly d^2 distinct points. Any curve of degree d going through $\frac{d^2+3d}{2} - 1$ of the intersection points of

C and C' also goes through the remaining ones, provided the latter do not lie on a curve of degree $d - 3$. Give an example with $d = 4$ showing that the hypothesis about the remaining points (Bacharach, 1886) is necessary.

For an even more general statement, named the Cayley–Bacharach theorem, see [18, V.1.1]. Finite sets of points of $\mathbb{P}_2$ such that all curves of a given degree through part of them also pass through the remaining ones are called *sets of tied points*. The study of tied points has a long history, with contributions by Maclaurin (1720), Euler (1748), Cramer (1750), Lamé (1818), Gergonne (1827), Jacobi (1836), Cayley (1843) and Bacharach (1886).

4.18 Take $C : x_0 x_1^2 - x_2^3 = 0$, and check that it is an irreducible cubic with a single branch at its only singular point $q = [1,0,0]$. Let $\mathcal{L}$ be the variable part of the linear series cut out on C by the lines through q. Prove that $\tilde{\mathfrak{f}}_{\mathcal{L}}$ is a bijection while $\mathfrak{f}_{\mathcal{L}}$ is not defined at q. *Hint:* use 4.1.28 and Exercise 4.1.

4.19 Let C be an irreducible curve of $\mathbb{P}_2$ and $\mathcal{L}$ a one-dimensional linear series on C with no fixed points. Prove that the Jacobian group $J(\mathcal{L})$ equals the ramification divisor of $\mathfrak{f}_{\mathcal{L}}$. Reprove 4.8.9(b) using 4.10.18.

4.20 Assume that C is an irreducible curve of $\mathbb{P}_2$ whose only singular point is a triple point q at which C has two branches p_1, p_2 with the same tangent and a third branch, p_3, with a different tangent. Take Λ to be the linear system of all conics of $\mathbb{P}_2$ and $\mathcal{L} = \Lambda \cdot C - \sum_{i=1}^{3} p_i$. Prove that p_1, p_2 is a neutral pair for $\mathcal{L}$, and that $\mathfrak{f}_{\mathcal{L}}$ is not defined at q, while it is defined and injective on $C - \{q\}$.

4.21 Use 4.8.7 and 4.11.14 to prove that no smooth irreducible curve of $\mathbb{P}_2$ is hyperelliptic.

4.22 Let C be an hyperelliptic curve of genus g and C' its rational image as constructed in 4.12.14. Describe $\mathcal{K}_{C'}$ and directly check that each canonical group on C is a sum of $g - 1$ pairs of points, all belonging to $\mathcal{G}$ (see also Exercise 4.24).

4.23 Prove that all the curves of Exercise 4.11 are hyperelliptic, thus showing the existence of hyperelliptic curves of genus g for all $g \geq 2$.

4.24 Let C be a hyperelliptic curve of genus g. Call $\mathcal{G}$ the only (4.11.13) linear series on C which has degree two and dimension one.

(1) Use 4.12.3 (or 4.11.2 and 4.10.5) to prove that the fibres of the rational map $\mathfrak{f}_{\mathcal{G}} : C \to \mathbb{P}_1$, $(\mathbb{P}_1 = \mathcal{G}^\vee)$ are the groups of $\mathcal{G}$.

(2) If $\mathcal{S}$ is the only complete linear series of degree $g - 1$ on $\mathbb{P}_1$ (4.9.2), prove that $\mathfrak{f}^*(\mathcal{S}) = \mathcal{K}_C$.

(3) Use the above to prove that $K \in \mathcal{K}_C$ if and only if K is a sum of $g - 1$ groups of $\mathcal{G}$.

4.25 Prove that any irreducible curve of genus two is birationally equivalent to a quartic with an ordinary cusp and no other singular point. Prove also that any quartic with an ordinary cusp and no other singular point is irreducible and has genus two.

4.26 Additions on an elliptic curve. Let C be an elliptic curve and assume we have fixed a point $\bar{p} \in X(C)$.

(1) Prove that mapping each $p \in X(C)$ to the class of $p - \bar{p}$ in $\text{Pic}_0(C)$ (see 4.5.2) is a bijection $j_{\bar{p}}$ between $X(C)$ and $\text{Pic}_0(C)$.

(2) Prove that translating the addition of $\text{Pic}_0(C)$ to $X(C)$ through $j_{\bar{p}}$ turns $X(C)$ into an abelian group whose addition $\oplus$ is given by the rule:

For any $p_1, p_2 \in X(C)$, $p_1 \oplus p_2$ is the only point of $X(C)$ linearly equivalent to $p_1 + p_2 - \bar{p}$.

(3) Prove that the neutral element of $X(C)$ is $\bar{p}$, thus proving in particular that the group structure on $X(C)$ depends on the choice of $\bar{p}$. Give a rule determining the opposite of an arbitrary $p \in X(C)$.

(4) Fix $\hat{p} \in X(C)$ and denote by $\hat{\oplus}$ the corresponding addition on $X(C)$. Take $G = |\bar{p} + \hat{p}|$ and for each $p \in X(C)$ let $\varphi(p)$ be the only point of $X(C)$ satisfying $p + \varphi(p) \in G$. Prove that φ is an isomorphism between the group structures that $\oplus$ and $\hat{\oplus}$ define on $X(C)$. (In addition, $\varphi = X(\mathfrak{g})$ for a birational map $\mathfrak{g} : C \to C$, see Exercise 4.29.)

(5) Assuming that C is a smooth cubic, after identifying C and $X(C)$ through χ_C, give a projective construction of $\oplus$ using lines (taken as degree-one adjoints). For a purely projective definition of the group structures on a smooth cubic the reader may see [12, V.6].

4.27 Prove that if a curve C of genus g is elliptic or hyperelliptic, then it is birationally equivalent to a curve of $\mathbb{P}^2$ defined by an equation of the form

$$Ax_2^2 + Bx_2 + C = 0$$

where $A, B, C \in \mathbb{C}[x_0, x_1]$ are homogeneous of degrees g, $g+1$ and $g+2$, respectively.

4.28 Weierstrass's gap theorem (Weierstrass, 1856). Let C be an irreducible curve of $\mathbb{P}_2$, call g the genus of C and fix $p \in X(C)$.

(1) Prove that for any positive integer r,

$$\dim |rp| - \dim |(r-1)p| \leq 1$$

and that equality holds if $r \geq 2g$.

(2) Prove that for each integer $r \geq 2g$ there is an $f \in \mathbb{C}(C)$ which has $(f)_\infty = rp$.

(3) Prove that there are exactly g positive integers r for which there is no $f \in \mathbb{C}(C)$ with $(f)_\infty = rp$ (or, equivalently, $\dim |rp| = \dim |(r-1)p|$). Such positive integers are called the *Weierstrass gaps* of C at p, all of them are strictly less than $2g$ due to (1) or (2) above.

(4) Prove that if $r, r' \in \mathbb{N}$ are not Weierstrass gaps of C at p, then neither is $r+r'$.

(5) Prove that if C is not rational, then 1 is a Weierstrass gap of C at any $p \in X(C)$.

(6) Prove that 2 is not a Weierstrass gap of C at p if and only if $g \leq 1$ or else C is hyperelliptic and the only one-dimensional linear series of degree two on C is $|2p|$. Check that if this is the case, then the sequence of gaps of C at p is composed of the odd numbers $1, 3, \ldots, 2g - 1$.

For more on Weierstrass gaps the reader may see [8, 4.6] or [14, 2.4].

4.29 Prove that if an irreducible curve C has a one-dimensional linear series $\mathcal{G}$ of degree two, then there is a uniquely determined birational map $\mathfrak{g} : C \to C$ such that $p + \mathfrak{g}(p)$ belongs to $\mathcal{G}$ for all $p \in X(C)$. The uniqueness being direct, for the existence the following steps are suggested:

(1) If $g(C) = 0$, prove that it is not restrictive to replace C with $\mathbb{P}_1$ and proceed with $\mathbb{P}_1$ by direct computation, which will be easier if the coordinates on $\mathbb{P}_1$ are taken such that the double points of $\mathcal{G}$ are $[1, 0]$ and $[0, 1]$.

(2) If $g(C) > 0$, prove that it is not restrictive to replace C with the curve of Exercise 4.27 and, with the notations therein, use the rational map $\mathfrak{G} : \mathbb{P}_2 \to \mathbb{P}_2$ defined by the equalities

$$\bar{x}_0 = Ax_0, \quad \bar{x}_1 = Ax_1, \quad \bar{x}_2 = -Ax_2 - B,$$

after proving that $\mathfrak{G}^2 = \mathrm{Id}_{\mathbb{P}_2}$ and hence that $\mathfrak{G}$ is birational.

4.30 Use Exercise 4.29 to prove that if C is an elliptic curve, then the birational maps $\mathfrak{f} : C \to C$ act transitively on $X(C)$.

4.31 Birational classification of elliptic curves. Let C be an elliptic curve and denote by $\mathcal{G}$ a linear series on C with degree two and dimension one.

(1) Prove that $\mathcal{G}$ has four distinct double points, say $p_1, p_2, p_3, p_4 \in X(C)$. Take ρ to be the cross ratio in $\mathcal{G}$

$$\rho = (2p_1, 2p_2, 2p_3, 2p_4).$$

(2) Take $\bar{\rho}(\mathcal{G}) = \bar{\rho} = \{\rho, 1/\rho, 1 - \rho, 1/(1 - \rho), 1 - 1/\rho, \rho/(\rho - 1)\}$ and prove that it does not depend on the ordering of the points p_i. (See for instance [4, 2.10.1] for the action of the symmetric group on the cross-ratio.) The reader preferring a single-valued invariant may replace $\bar{\rho}(\mathcal{G})$ with $j = (\rho^2 - \rho + 1)^3/\rho^2(\rho - 1)^2$, but it is probably not worth the effort.

(3) Use that any $\mathcal{G}$ may be written $\mathcal{G} = |2p|$ for a $p \in X(C)$, as well as Exercise 4.30 and 4.4.26(b), to show that $\bar{\rho}(\mathcal{G})$ is the same for all $\mathcal{G}$ on C, and therefore depends only on C. Thus write $\bar{\rho}(C) = \bar{\rho}(\mathcal{G})$ and call it the *invariant* of C (see (4) below).

(4) Prove the birational invariance of $\bar{\rho}(C)$: if C and C' are birationally equivalent elliptic curves, then $\bar{\rho}(C) = \bar{\rho}(C')$.

(5) Prove that the cubics of Exercise 3.19 have $\bar{\rho}(C) = \bar{\alpha}$.

(6) Prove that there are infinitely many different birational classes of elliptic curves.

(7) Use 4.12.9, Exercise 3.19 and (5) above to show that two elliptic curves with the same invariant are birationally equivalent, thus showing that two elliptic curves are birationally equivalent if and only if they have equal invariants.

4.32 Projective classification of smooth cubics. Let C be a smooth cubic; it is an elliptic curve by the genus formula 4.6.6.

(1) Fix a point $q \in C$, not a flex, and prove that there are exactly four tangent lines to C through q other than the tangent at q.

(2) Prove that the set of the cross ratios of the above tangents, taken in all possible orders, equals the invariant of C as an elliptic curve (cf. Exercise 4.31), and so, in particular, it is independent of the choice of q.

(3) Use Exercise 3.19 to prove that there is a projectivity mapping two smooth
cubics to one another if and only if they have the same invariant.

4.33 The dual curve (Poncelet, Gergonne, Plücker, 1825–35). Let C be
an irreducible curve of $\mathbb{P}_2$ of degree $d > 1$.

(1) Prove that the variable part $\mathcal{L}$ of the linear series cut out on C by its polars
has dimension two.

(2) Identify $\mathcal{L}$ and $\mathbb{P}_2$ through the composition of the projectivities of 3.5.8 and
4.4.27, after which the rational map associated to $\mathcal{L}$ will be taken as a rational
map $\eth : C \to C^\vee$, where $C^\vee$ is an irreducible curve of degree higher than one
of the dual plane $\mathbb{P}_2^\vee$; $C^\vee$ is called the *dual curve* of C. Take coordinates on
$\mathbb{P}_2$, write down equations of $\eth$ and prove that for each smooth point q of C, $\eth$
is defined at q and $\eth(q)$ is the tangent to C at q.

(3) Assume that $\varphi : U \to C$ is a non-constant analytic map defined in an open
neighbourhood U of 0 in $\mathbb{C}$, given by

$$t \longmapsto [u_0(t), u_1(t), u_2(t)], \quad u_i \in \mathbb{C}\{t\}, \quad i = 0, 1, 2.$$

Prove that, up to a suitable reduction of U, $\eth \circ \varphi_{|U-\{0\}}$ is given by

$$t \longmapsto \left[\left| \begin{matrix} u_1(t) & u_2(t) \\ u_1'(t) & u_2'(t) \end{matrix} \right|, \left| \begin{matrix} u_2(t) & u_0(t) \\ u_2'(t) & u_0'(t) \end{matrix} \right|, \left| \begin{matrix} u_0(t) & u_1(t) \\ u_0'(t) & u_1'(t) \end{matrix} \right| \right],$$

the $'$ indicating derivative with respect to t.

(4) Let

$$\eth^\vee : C^\vee \longrightarrow (C^\vee)^\vee = C^{\vee\vee}$$

be the above rational map from $C^\vee$ into its dual curve, the latter taken as a
curve of $\mathbb{P}_2$ through the identification of Exercise 1.12. Prove that, still up to
a reduction of U, $\eth^\vee \circ \eth \circ \varphi_{|U-\{0\}} = \varphi_{|U-\{0\}}$. Use this fact to prove that $\eth$ is
birational and has inverse $\eth^\vee$.

(5) Use 4.12.4 to prove that if C has no singularities other than nodes and ordinary
cusps, then the degree of $C^\vee$ (called the *class* of C) is the integer $d(d-1) -
2\delta - 3\kappa$ appearing in Plücker's first formula 3.5.19.

4.34 Tangential singularities. (Continued from Exercise 4.33.)

(6) Prove that if φ above is a uniformizing map of a branch γ of C, then, up to
reducing U, $\eth \circ \varphi_{|U-\{0\}}$ extends to a uniformizing map $\varphi^\vee$ of $\gamma^\vee = \eth(\gamma)$.

(7) Prove that, using suitable coordinates, the uniformizing map φ of an arbitrary
branch γ of C may be assumed to have the form

$$t \longmapsto [1, t^e, t^{e+e'} + \cdots], \quad e > 1, \quad e' > 0$$

and then $\varphi^\vee$ has the form

$$t \longmapsto [e' t^{e+e'} + \cdots, -(e + e') t^{e'} + \cdots, e].$$

Deduce that:

- the origin of $\gamma^\vee$ is the tangent line to γ,
- the tangent line to $\gamma^\vee$ is the (pencil of lines through the) origin of γ,
- the multiplicity of $\gamma^\vee$ is the class of γ and
- the class of $\gamma^\vee$ is the multiplicity of γ.

(8) Prove that the points of $C^\vee$ are the tangent lines to C. Use Exercise 3.15 to prove that the class of C is the maximum reached by the number of distinct tangent lines to C through a point q, while q varies in $\mathbb{P}_2$.

(9) The singularities of $C^\vee$ are called *singular tangents*, and also *tangential singularities*, of C. Prove that, among them:

- The nodes of $C^\vee$ are the lines tangent to C at exactly two distinct points, both smooth and neither a flex. These lines are called *ordinary double tangents* of C.
- The ordinary cusps of $C^\vee$ are the tangent lines at the ordinary flexes of C.

4.35 The Plücker formulas (Plücker, 1835). Use the first and second Plücker formulas (3.5.19 and Exercise 3.18) to prove the third and fourth ones. The whole set of Plücker's formulas is

$$m = d(d-1) - 2\delta - 3\kappa,$$
$$\iota = 3d(d-2) - 6\delta - 8\kappa,$$
$$d = m(m-1) - 2\tau - 3\iota,$$
$$\kappa = 3m(m-2) - 6\tau - 8\iota,$$

and they hold for any irreducible curve of $\mathbb{P}_2$ of degree $d > 1$ and class m having:

- δ nodes and κ ordinary cusps as the only singular points,
- ι flexes, all ordinary, and
- τ ordinary double tangents and the tangents at the flexes as the only singular tangents.

4.36 Prove that any smooth curve of $\mathbb{P}_2$ of degree $d > 2$ has singular tangents. At the beginning of the nineteenth century, instances of the first and third Plücker formulas, together with the wrong assumption that a "general enough" curve should have neither singular points nor singular tangents led to a contradiction, called the *Poncelet paradox*.

Bibliography

[1] E. Bertini. *Introduzione alla geometria proiettiva degli iperspazi.* Giuseppe Principato, Messina, 1923.

[2] W. Brill and M. Noether. Ueber die algebraischen Functionen und ihre Anwendung in der Geometrie. *Math. Ann.*, **7**:269–310, 1874.

[3] E. Casas-Alvero. *Singularities of plane curves*, volume 276 of *London Math. Soc. Lecture Note Series*. Cambridge University Press, 2000.

[4] E. Casas-Alvero. *Analytic projective geometry*. EMS Textbooks in Mathematics. EMS Publishing House, 2014.

[5] C. Chevalley. *Introduction to the theory of algebraic functions in one variable.* Amer. Math. Soc., 1951.

[6] C. Ciliberto. Geometric aspects of polynomial interpolation in more variables and of Waring's problem. In *European Congress of Mathematics (Barcelona 2000)*, volume 201 of *Progr. Math.*, pages 289–316. Birkhäuser, 2001.

[7] C. Ciliberto, B. Harbourne, R. Miranda, and J. Roé. Variations on Nagata's conjecture. *Clay Mathematical Proceedings*, 18:185–203, 2013.

[8] P.N. Cohn. *Algebraic numbers and algebraic functions.* Chapman and Hall/CRC, 2017.

[9] J.L. Coolidge. *A history of geometrical methods.* Oxford University Press, London, 1940.

[10] F. Enriques and O. Chisini. *Lezioni sulla teoria geometrica delle equazioni e delle funzioni algebriche.* N. Zanichelli, Bologna, 1915-34.

[11] G. Fischer. *Plane algebraic curves.* American mathematical Society, 2001.

[12] W. Fulton. *Algebraic curves.* Benjamin Inc., New York, 1968.

[13] J. Gray. *Worlds out of nothing.* UMS. Springer Verlag, New York, Heidelberg, Berlin, 2007.

© Springer Nature Switzerland AG 2019
E. Casas-Alvero, *Algebraic Curves, the Brill and Noether Way*, Universitext,
https://doi.org/10.1007/978-3-030-29016-0

[14] P. Griffiths and J. Harris. *Principles of algebraic geometry*. John Wiley and Sons, 1978.

[15] S. Lang. *Algebra*. Addison Wesley, Reading, Massachusetts, 1965.

[16] D.G. Northcott. A note on the genus formula for plane curves. *J. London Math. Soc.*, **30**:376–382, 1955.

[17] J.G. Semple and G.T. Kneebone. *Algebraic curves*. Oxford University Press, London, 1959.

[18] J.G. Semple and L. Roth. *Introduction to algebraic geometry*. Oxford University Press, London, 1949.

[19] F. Severi. *Tratatto di Geometria algebrica*. N. Zanichelli, Bologna, 1926.

[20] I.R. Shafarewich. *Basic algebraic geometry*. Springer Verlag, Berlin, Heidelberg, New York, 1977.

[21] R. Walker. *Algebraic curves*. Dover, New York, 1962.

[22] C.T.C. Wall. *Singular points of plane curves*, volume 63 of *London Math. Soc. student texts*. Cambridge University Press, 2004.

[23] O. Zariski and P. Samuel. *Commutative algebra*. Van Nostrand, Princeton, New Jersey, 1960.

Index

© Springer Nature Switzerland AG 2019

E. Casas-Alvero, *Algebraic Curves, the Brill and Noether Way*, Universitext,

https://doi.org/10.1007/978-3-030-29016-0

Printed in the United States
By Bookmasters